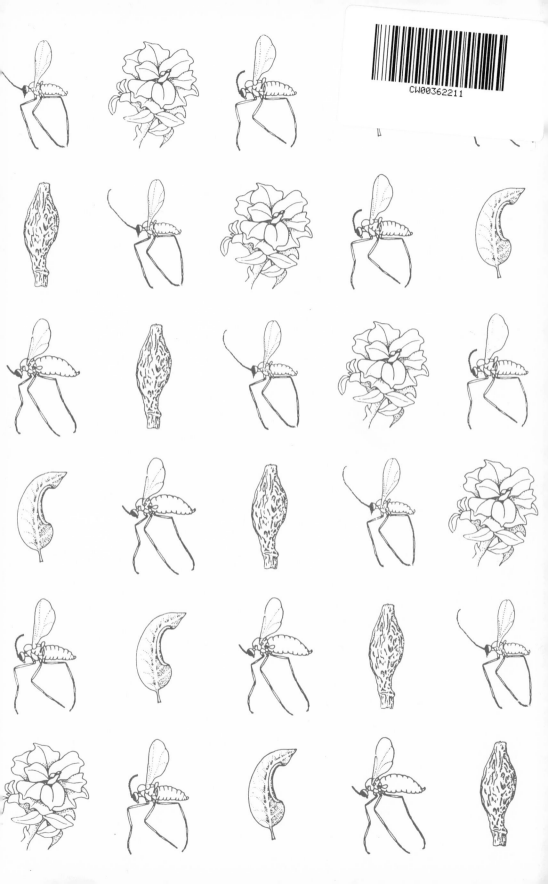

The Gall Midges of the
Neotropical Region

Also by Raymond J. Gagné

The Plant-Feeding Gall Midges of North America

THE GALL MIDGES
OF THE
NEOTROPICAL REGION

Raymond J. Gagné

Systematic Entomology Laboratory
Plant Sciences Institute
Agricultural Research Service
U.S. Department of Agriculture

Comstock Publishing Associates *a division of*

Cornell University Press   *Ithaca and London*

First published 1994 by Cornell University Press.

Printed in the United States of America

∞ The paper in this book meets the minimum requirements of the American National Standard for Information Sciences— Permanence of Paper for Printed Library Materials, ANSI Z39.48-1984.

Library of Congress Cataloging-in-Publication Data

Gagné, Raymond J.
    The gall midges of the neotropical region / Raymond J. Gagné.
        p.   cm.
    Includes bibliograpical references (p.   ) and index.
    ISBN 0-8014-2786-X
    1. Gall midges—Latin America.   2. Gall midges—Latin America—
Host plants.   3. Galls (Botany)—Latin America.   I. Title.
QL537.C33G33   1994
595.77′1—dc20   93-35824

To Pierre and Cybèle

Contents

Acknowledgments

The drawings in this book were executed by Susan Grupp of Chantilly, Virginia, who prepared figures 2–10, inked my line drawings of body parts, and made Tables 1–3; Elizabeth P. Roberts of Baltimore, Maryland, who illustrated the plant parts in figures 280–341; and Deborah Leather Roney of Vienna, Virginia, whose drawings of whole adults and immature stages are reproduced here from my earlier book (1989).

I am grateful to friends and colleagues who kindly and generously contributed to this book in various ways. They helped by offering their technical expertise, collecting and donating specimens, loaning specimens from collections in their care, inviting me to travel and providing hospitality and aid during my visits, and reviewing early drafts of the manuscript: Reinaldo Abas, Museo Argentina de Ciencias Naturales "Bernardino Rivadavia"; Dalton de Souza Amorim, Ribeirão Preto, Brazil; Axel Bachmann, Museo Argentina de Ciencias Naturales "Bernardino Rivadavia"; Jeffrey K. Barnes, New York State Biological Survey, Albany; Michel Baylac, Muséum National d'Histoire Naturelle, Paris, France; Fred Bennett, University of Florida, Gainesville; Paul E. Boldt, Agricultural Research Service, Temple, Texas; Hugo Cordo, Biological Control of Weeds Laboratory, Hurlingham, Argentina; Márcia Souto Couri, Museu Nacional, Rio de Janeiro, Brazil; Françoise Dreger-Jauffret, Institut de Botanique, Strasbourg, France; William G. Eberhard, Universidad de Costa Rica, San José; Hernan Espinoza, Fundación Hondureña de Investigatión Agricola, San Pedro Sula, Honduras; Neal L. Evenhuis, Bernice P. Bishop Museum, Honolulu, Hawaii; Jan Peter Feil, Aarhus Universitet, Denmark; Candy Feller, Georgetown University, Washington, D.C.; G. Wilson A. Fernandes, Universidade Federal de Minas Gerais, Belo Horizonte, Brazil; Douglas Futuyma, State University of New York, Stony Brook; Sarah E. Gagné, Silver Spring, Maryland; Medardo Galindo, Federación de Asociaciones de Productores y Exportadores Agropecuarios y Agroindustriales de Honduras, San Pedro Sula; Shirley Graham, Kent State University, Kent State, Ohio; David Grimaldi, American Museum of Natural History, New York, New York; Dale H. Habeck, University of Florida, Gainesville; Paul E. Hanson, Universidad de Costa Rica, San José; Keith M. Harris, Commonwealth Institute of Entomology, London; Richard Hoebeke, Cornell University, Ithaca, New York; Angela Christina Lara, Universidade Federal de Minas Gerais, Belo Horizonte, Brazil; Guillermo Logarzo, Biological Control of Weeds Laboratory, Hurlingham, Argentina; Rachel Cruttwell McFadyen, Queensland Department of Lands, Brisbane, Australia; Jean-Michel Maes, Museo Entomológico, Léon, Nicaragua; Valéria Cid Maia, Museu Nacional, Rio de Janeiro, Brazil; Phuongthang "Nit" Malikul, Agricultural Research Service, Washington, D.C.; Rogério P. Martins, Universidade Federal de Minas Gerais, Belo Horizonte, Brazil; Loïc Matile, Muséum National d'Histoire Naturelle, Paris, France; Frank W. Mead, Florida Department of Agriculture, Division of Plant Industry, Gainesville; Carlos Victor Mendonça Fo., Universidade Federal de Minas Gerais, Belo Horizonte, Brazil; Edwin Möhn, Staatliches Museum für Naturkunde, Stuttgart, Germany; Eugênio T. Neto, Universidade Federal de Minas Gerais, Belo Horizonte, Brazil; W. C. Nijveldt, Wagenigen, The Netherlands; Allen L. Norrbom, Systematic Entomology Laboratory, Agricultural Research Service, Washington, D.C.; Celuta Paganelli, São Paulo, Brazil; William A. Palmer, Queensland Department of Lands, Temple, Texas; Denise Pamplona, Museu Nacional, Rio de Janeiro, Brazil; Jorge E. Peña, University of Florida Tropical Research and Education Center, Homestead; Rolando Pretto, Federación de Asociaciones de Productores y Exportadores Agropecuarios y Agroindustriales de Honduras, San Pedro Sula; Cristobal J. H. Pruett H., Santa Cruz de la Sierra, Bolivia; Susanne Renner, Johannes Gutenberg Universität, Mainz, Germany; María Luisa A. Ríos de Saluso, Estación Experimental Agropecuaria Paraná, Entre Ríos, Argentina; Odette Rohfritsch, Institut de Botanique, Strasbourg, France; Curtis W. Sabrosky, Medford Leas, New Jersey; Michael E. Schauff, Systematic Entomology Laboratory, Agricultural Research Service, Washington, D.C.; Stephen F. Smith, Smithsonian Institution, Washington, D.C.; Gary J. Steck, Florida Department of Agriculture, Division of Plant Industry, Gainesville; Carl E. Stegmaier, Jr., Tallahassee, Florida; João R. Stehmann, Universidade Federal de Minas Gerais, Belo Horizonte, Brazil; Russell D. Stewart, Animal and Plant Health Inspection Service, Hyattsville, Maryland; George C. Steyskal, Gainesville, Florida; Manya B. Stoetzel, Systematic Entomology Laboratory, Agricultural Research Service, Beltsville, Maryland; Edgar Suárez, Universidad Nacional, Heredia, Costa Rica; Adolpho Vetrano, Estación Experimental Agropecuaria, Castellar, Argentina; Abraham Willink, Instituto Miguel Lillo, San Miguel de Tucumán, Argentina; Norman E. Woodley, Systematic Entomology Laboratory, Agricultural Research Service, Washington, D.C.; Allen M. Young, Milwaukee Public Museum, Milwaukee, Wisconsin; and Ingeborg Zenner-Polania, Instituto Colombiano Agropecuario, Bogota, Colombia.

RAYMOND J. GAGNÉ

Washington, D.C.

Abbreviations

Abbreviations for collections, institutions, and museums are as follows:

ASM Severtsov Institute of Animal Morphology, Russian Academy of Sciences, Moscow

BBM Bernice P. Bishop Museum, Honolulu, Hawaii

BMNH The Natural History Museum, London, England, formerly British Museum (Natural History)

CUI Cornell University, Ithaca, New York

Felt Coll. Felt Collection. This collection is the property of the New York State Museum, Albany, New York; it is presently on extended loan to the Systematic Entomology Laboratory, U.S. Department of Agriculture, in the USNM, Washington, D.C.

INTAC Laboratorio Zoologia Agricola (INTA), Castellar, Buenos Aires, Argentina

ITIC Instituto Tropical de Investigaciones Científicas of the University of El Salvador (not a collection)

ITZA Instituut voor Taxonomische Zoologie, Amsterdam, The Netherlands

IZAM Instituto de Zoologia Agricola, Maracay, Venezuela

MACN Museo Argentino de Ciencias Naturales "Bernardino Rivadavia," Buenos Aires, Argentina

MNHNP Muséum National d'Histoire Naturelle, Paris, France

MNHNS Museo Nacional de Historia Natural, Santiago, Chile

MZSP Museu de Zoologia, Universidade de São Paulo, São Paulo, Brazil

SMNS Staatliches Museum für Naturkunde, Stuttgart, Germany

Tavares Coll. Tavares Collection. The specimens that remain of this collection are the property of Instituto Nun'Alvres, Caldas da Saúde, Portugal. They are presently on loan to the SMNS.

UNLP Universidad Nacional de La Plata, Argentina

USNM National Museum of Natural History, Washington, D.C.

ZMHU Zoologisches Museum der Humboldt Universität, Berlin, Germany

Compass directions are given in capital letters without punctuation (e.g., E for east, SE for southeast). Other abbreviations used are as follows:

aut.   automatic
Co.   County
depos.   depository
descr.   description
distr.   distribution
elev.   elevation
ex   from
I., Is.   Island, Islands
km   kilometer
m   meter
mon.   monotypic
Mt.   Mountain
orig. des.   original designation
preocc.   preoccupied
Prov.   Province
R.   River
ref.   reference
subseq. des.   subsequent designation
vic.   vicinity

The Gall Midges of the
Neotropical Region

Introduction

This book reviews the work done on Neotropical Cecidomyiidae and provides a basic reference for further research. It is a practical identification guide to the genera and the plant damage of gall midges of the Neotropical Region. The tribes and genera are revised, many new taxonomic changes are proposed, and the fauna is related to that of the rest of the world. It is in many ways a companion to *The Plant-Feeding Gall Midges of North America* (Gagné 1989), but the present book is more comprehensive in its treatment of gall midge classification. All known cecidomyiids from the Neotropics are dealt with here, not only those associated with plants.

The Cecidomyiidae, or gall midges, are a family of flies in the order Diptera. The family belongs to the suborder Nematocera and, within that, to the superfamily Sciaroidea, the large and very old assemblage of fungus-feeding midges. Cecidomyiids were already diverse in the Cretaceous period, but have greatly flourished since then along with the flowering plants. Cecidomyiid larvae are primitively fungus feeding, and many have that habit still, but several lines have developed into vascular plant feeders and now make up the bulk of the family. Other lines have developed as predators of mites, aphids, and various other arthropods, but cecidomyiids are best known for the plant galls they cause, accounting for the family's common name, gall midges.

Most adult cecidomyiids are tiny and fragile, with a wing length usually less than 3 mm. Adults feed little if at all, and many are short-lived. Their main role is to mate and lay eggs near, on, or in the proper host. Females emerge from the pupal stage with their eggs fully developed. All or most feeding is done by the larvae. These are legless, elongate-spindleform maggots, and various groups have developed adaptations, including pseudopods, for diverse modes of life. Larvae have a

1

greatly reduced head and mouthparts that are modified for ingesting liquids.

Cecidomyiidae are possibly the largest family of flies but are poorly known, especially in the Neotropical Region. Almost 5,000 described species are recorded worldwide, but only 453 of them are known from the Neotropics. The actual number that must occur in this region is inestimable: a large number of unidentified species appears in the keys to plant damage, many additional undescribed species are represented in the National Museum of Natural History, and great numbers of individuals and species are commonly collected in insect flight traps.

Although numerous, ubiquitous, and economically important, gall midges have not had the attention they deserve. They are perceived as a difficult group, largely because of their small size and fragility. The necessity for microscopic slide mounting adds considerably to the investment of time required for study. Another factor that may have made the family forbidding is that few comprehensive revisions have been done of this group until recently. Most taxa of Neotropical gall midges were described before the 1920s, and many were known only from sketchy descriptions. The most recent keys to genera are by Kieffer (1913a) and Felt (1925).

Nevertheless, interest in Neotropical cecidomyiids is growing as additional species are found to be pests, pollinators, or biological control agents and as it becomes obvious how numerous and common a part of the fauna they are. The family includes many economically important plant pests. Chief of these may be *Prodiplosis longifila* Gagné, which kills buds of tomato, potato, green beans, cotton, and other crops. Introduced pests include *Stenodiplosis sorghicola* (Coquillett) on sorghum and *Erosomyia mangiferae* Felt on mango. Native species are being selected as biocontrol agents for Neotropical weedy plants that have become established in Africa, Australia, and southern North America and as predators of coccoids, thrips, and other arthropods. Cecidomyiids are becoming more prominent in ecological studies, mainly through their roles in pollination and gall making. Their host-specific galls are even being used as indicators of plants that are difficult to identify when not flowering.

The Neotropical Region is arbitrarily defined here to encompass the New World south of the Tropic of Cancer in Mexico and south of Lake Okeechobee in Florida and all of the West Indies. I include the southern tip of Florida because it shares much of its flora with the West Indies.

1 A History of the Study
of Neotropical Gall Midges

Attention came late to the cecidomyiids of the Neotropical Region. The first notice of a cecidomyiid was in 1852 by E. Blanchard, who described two species caught in flight in Chile. Until close to the turn of the century, fewer than a score of species had been described, most of them known only from specimens caught in flight, but a few from known plants. From 1899 until 1932 a great amount of work on Neotropical gall midges was done by three Europeans, J.-J. Kieffer, E. H. Rübsaamen, and J. da S. Tavares, and a North American, E. P. Felt. Each of them had already had some experience with gall midges before beginning work on the Neotropical fauna, with the result that their descriptions were generally of good quality for the period. Their most valuable legacy was in providing host identifications and descriptions of galls for most of the species they described. This information ensured that the species could be found again and identified with some certainty. After that early flush of discovery, little more was published on Neotropical gall midges until 1959, when the large and exemplary body of work by E. Möhn on the fauna of El Salvador began to appear. His student A. Wünsch later published a fine study of the representatives of two tribes that he had collected in northern Colombia (Wünsch 1979). Between 1968 and 1989, I described some species from the Neotropical Region, mainly in support of other researchers who needed the names to report results of their work.

In short, our knowledge of Neotropical gall midges results mainly from the work of a few individuals. Presented here are brief sketches of the lives of the 11 most prominent in the study of Neotropical cecidomyiids. Eight were or are taxonomists, all of whom, besides myself, described nine or more species from the Neotropics: J.-J. Kieffer, E. Rübsaamen, J. da S. Tavares, E. P. Felt, M. T. Cook, J. Brèthes, E. Möhn, A. Wünsch.

The other three individuals made a significant impact in different ways: P. Jörgensen for his collecting, C. A. V. Houard for his comprehensive compilation of data on Neotropical galls, and K. M. Harris, who is also a taxonomist but is included here for an influential revision rather than for having described any Neotropical species. Common to all these sketches is information on the acquisition, content, and eventual disposition of the various collections. The 11 sketches are in chronological order by the subjects' birth dates.

Jean-Jacques Kieffer

J.-J. Kieffer (1857–1925) was French, from the Province of Lorraine. He was a Catholic priest and for 45 years a teacher in a diocesan secondary school in the small town of Bitche. For a person with full-time teaching and regular priestly duties, his scientific activity was phenomenal. He wrote 470 scientific papers, including comprehensive treatises on plant galls of France and Algeria, on the dipterous families Cecidomyiidae, Ceratopogonidae, and Chironomidae, and on various groups of parasitic Hymenoptera.

Kieffer had no formal training in science and never left Lorraine except for two short trips to Algeria in the 1920s; yet he had a broad and comprehensive knowledge of cecidomyiids, their hosts, and the other insect groups he worked with. He published in both French and German, but mainly in French. Many of his taxonomic papers on gall midges include biological observations and interesting vignettes about his collecting. His writing was direct and forceful, especially in his responses to unjust attacks or, as apparently happened, when an in-press manuscript was partially copied without his permission (see Kieffer 1897, 1902).

Kieffer described 19 genera and 53 species of gall midges from Argentina, Chile, and Paraguay, all but one of the species in papers coauthored with the collectors: Herbst in Chile (Kieffer and Herbst 1905, 1906, 1909, 1911) and Jörgensen in Argentina (Kieffer and Jörgensen 1910). He further influenced work on the Neotropical fauna with his comprehensive works on family classification, summarized in his monograph on Cecidomyiidae (Kieffer 1913a). Kieffer had a talent for picking out and describing what was most distinctive about his taxa, but he was spare with illustrations. It is fortunate that he recorded the hosts of the majority of the Neotropical species he described, making it possible for future collectors to find fresh material based on the descriptions of the galls.

Kieffer is notorious for having left no collection. Most of his types, including those of his Neotropical species, are presumed lost. Exceptions include those from particular expeditions that he described and subsequently returned to their owners or lending institutions. Other extant specimens of over 50 European species, not necessarily types, were given to Felt during the latter's visits to Bitche in 1910 (Felt 1911k). Unfortunately, the Neotropical specimens he studied do not survive. Eleven years after Kieffer's death, F. W. Edwards of the British Museum visited Kieffer's school and determined that Kieffer left no collection there (Edwards 1938). Nevertheless, I have heard it said, and some stories never seem to die, that some Kieffer specimens still exist in Bitche. During my visit in 1987 to Kieffer's school, the Collège Saint Augustin, now in a building completed shortly after his death, I saw the collection to which this hearsay probably refers: two glass-topped insect drawers mounted on the walls of the natural history classroom. These drawers contain unlabeled, showy insects, mostly beetles, purportedly prepared by Kieffer, but there are no specimens of families on which Kieffer specialized and certainly no types. The attic storeroom of the school contains two shelves of Kieffer's books and copies of his reprints, but no insects. Kieffer's brass monocular microscope is stored in the school director's office. The director of the school at the time of my visit was surprised but gratified to learn that Kieffer's name is still well known.

The gift of specimens to Felt in 1909 suggests that Kieffer did maintain a collection during his lifetime, even though its extent is unknown. That a collection no longer exists may indicate only that there was none of any apparent value to those who took care of Kieffer's affairs after his sudden death from a stroke. Maintaining a collection takes money, space, and time, elements that Kieffer had in short supply. He apparently never had much money at his disposal. Some remuneration came from his publications (some scientific journals paid their authors), and, in his later years, he received modest amounts of money from prizes and grants. Kieffer's biographers allude to disputes between him and his bishop about his relatively low salary. He could have earned more as pastor of a parish church but, upon reaching an age when such a position should have been forthcoming to a diocesan priest, his bishop needed him and kept him in the teaching post. Kieffer roomed at times at the school and at times in town, which must have contributed to the difficulty of maintaining a collection. In addition he was for an extended period the sole support of his father and two spinster sisters who lived with him in Bitche. Lack of funds and space aside, it is also possible that when Kieffer died, whether he roomed in town or at the school, that a collection, perhaps consisting of random-size vials, unprepossessing microscopic slides, and dusty boxes of plant material, was thrown out in the name of housekeeping.

Further details about Kieffer's life and work can be found in Fleur 1929; Nominé 1926, 1929; and Tavares 1926.

Ewald Heinrich Rübsaamen

A German from Westphalia, E. H. Rübsaamen (1857–1919), trained as an illustrator and teacher. He developed a great interest in galls and gall makers, which helped him earn a position in the Berlin Museum around 1892. He continued his interest in galls there, working especially with gall midges from Germany, but also on the exotic material then being sent to Berlin by collectors all over the world. His drawings were of exceptionally high quality, and his papers are among the best illustrated on Cecidomyiidae. From 1902 until his death, Rübsaamen had a leadership position in the Rhineland grape phylloxera control program, during which great expanses of vineyards were destroyed in an effort to control the pest. He nevertheless found time to continue his gall midge studies. Between 1899 and 1916, he described 17 genera and 22 species of South American gall midges. Most of these species were collected from the upper Amazon regions of Brazil and Peru and the remainder from the provinces of Santa Catarina and Rio de Janeiro in southern Brazil.

Rübsaamen developed an animosity toward Kieffer that extended into print (see, for example, Rübsaamen 1896). The attacks were intemperate and censorious and were answered very forcefully by Kieffer. I believe Rübsaamen's dislike was partly due to envy of Kieffer's great productivity, which was impossible for a man with Rübsaamen's commitment to carefully illustrated papers. Some of his original illustrations graced papers by Tavares (e.g., Tavares 1908) and even an early paper by Kieffer (1895d). The finest example of Rübsaamen's work is the posthumously published, exceptionally beautifully illustrated study of the gall-making cecidomyiids of Germany (Rübsaamen and Hedicke 1926–1939). Additional details of Rübsaamen's life can be found in Schaffnitt 1928.

Most, if not all, of Rübsaamen's study material is the property of the Zoology Museum, Humboldt University, Berlin. Currently it is located in the Stuttgart Museum of Natural History in the care of E. Möhn. Some specimens are in alcohol, some on slides in glycerin, all arranged by Rübsaamen's collection numbers rather than by taxonomic order. The glycerin on the slides has dried out over the years, so the specimens are in delicate condition and need to be remounted. According to E. Möhn (pers. comm. 1990), the collection requires much attention before it can be made completely available for study.

Joachim da Silva Tavares

J. S. Tavares (1866–1931) was Portuguese and a Jesuit priest. His first permanent post as a priest was a teaching assignment in a secondary school in 1898. Soon after his arrival he began studies on gall makers of Portugal and cofounded the journal *Brotéria*. In 1900 he began publishing on galls and gall makers from the Iberian Peninsula and elsewhere. In 1906 and 1909 Tavares described new Brazilian gall midges collected and sent to him from Brazil by a fellow Jesuit, J. Bruggmann.

In 1910, the new Republican government of Portugal seized the holdings of all religious orders and forced many of the clergy into exile. Tavares had to flee to Spain, leaving behind and losing all his collections, books, and equipment. One month later he was on a ship bound for Argentina, where he stayed for two months, mainly in Córdoba, where he began a gall collection. He then moved to Brazil and stayed for three and a half years. He lived in several of the coastal states before settling in Bahia. His assigned work there included teaching and fruit tree research, but he collected galls extensively and reared the gall makers whenever possible. Tavares collected flowers and fruit of galled plants for later determination because the premier plant taxonomists and collections of Neotropical plants were then in Germany and Austria. He returned to Europe in 1914 shortly before the beginning of World War I and was assigned to a position in Galicia, Spain, near northern Portugal. Tavares would have to wait another five years until the end of the war to have his plants identified, but he immediately began publishing on his South American gall insects, at the same time renewing his studies on Iberian galls. Toward the end of his life, Tavares was finally allowed to return to his native land.

Tavares described a total of 26 genera and 56 species of Argentine and Brazilian cecidomyiids, illustrating his articles with photographs of galls and many anatomical drawings. He published in Portuguese, Spanish, Latin, and French. He was a cultivated and generous man and cooperated effectively with fellow scientists. Of the early cecidomyiid taxonomists, he seems to have had the strongest sense of his readers' needs, providing copious detail and illustrations. Felt received from him many examples of Mediterranean and Brazilian gall midges, which now enrich the Felt Collection. Tavares gave many of his galls to Houard, who used them to good effect (Houard 1933 and elsewhere). These are still well preserved in the Muséum National d'Histoire Naturelle in Paris. Most of the collection that remained with Tavares has not survived. Several years ago, E. Möhn (pers. comm. 1990) discovered in an attic of the Instituto Nun'Alvres, Caldas de Saúde, Portugal, a trunk with the apparent remains of Tavares's collection. These specimens, mainly galls, are presently in the care of E. Möhn in the Stuttgart Museum of Natural History.

Additional details concerning Tavares's life can be found in Tavares's writings, especially Tavares 1915a, and in three obituaries: Leite 1931, Luisier 1932, and Houard 1932.

Ephraim Porter Felt

E. P. Felt (1868–1943) was an American professional entomologist. From 1898 until his retirement in 1928, he was State Entomologist of New York, and from 1928 until his death, Director of Bartlett Tree Laboratories, the research arm of a shade-tree pest control company in Connecticut. Felt's research dealt mainly with North American shade- and forest-tree insects, but he is best known today for his work on galls and gall midges, much of it done as the press of other duties permitted. The crowning achievement of this avocational work was his popular book *Plant Galls and Gall Makers*, which appeared in 1940. In it were keys and illustrations for the known North American galls. Felt was a prodigious worker and active in scientific society activities. He served as editor of the large and influential *Journal of Economic Entomology* from its inception in 1908 until 1936. Additional details of Felt's life and work can be found in Bromley 1944 and Muesebeck and Collins 1944.

Felt named more than 1,150 cecidomyiids, mostly from North America. Of these, 77 valid species and 16 genera were from the Neotropics or have since been found there. His work on gall midge taxonomy was hurried, superficial, and seldom illustrated, and, once done, not seriously reexamined. As a consequence, he described some species several times under different names and even in two or more genera. Felt met Rübsaamen and Kieffer during an official trip to Europe in 1909 (Felt 1911k). He spent a day with Rübsaamen in Remagen, Germany, and visited Kieffer in Bitche, France, for several days. Felt came away from the visit with Kieffer with representatives of about 50 species that are now part of the Felt Collection.

For many years Felt was fortunate to have the services of a very capable laboratory aide, F. T. Hartman. She was responsible for the preparation of most of his types and type series. Her slide mounts, done with Canada balsam, are exceptional and are a model for anyone's efforts. Felt's own slide preparations, made after Hartman retired in 1920, were less well done. Specimens were usually not cleared before placing them in the mounting medium, and critical parts were not often well displayed.

Much of the material Felt described was sent to him by others, and the majority of his Neotropical species were referred to him by the U.S. Department of Agriculture in Washington. Most were reared from known hosts, but others were caught in flight or on spider webs. Fortunately, the unreared species are distinctive. Most of Felt's types are the property of the New York State Museum in Albany. These are presently in the care of the Systematic Entomology Laboratory, USDA, at the U.S. National Museum. Types of some of Felt's Neotropical species are the property of the U.S. National Museum, and one type belongs to Cornell University in Ithaca, New York.

Melville Thurston Cook

An American plant pathologist, M. T. Cook (1869–1952), was fascinated by galls from the point of view of comparative structure. He is of interest here only for having described 12 species of gall midges from Cuba (Cook 1906, 1909). These species were placed in the then catch-all genus *Cecidomyia* and based only on galls that he and a colleague collected in 1904–1906 while Cook was Director of the Agricultural Experiment Station, Santiago de las Vegas, Cuba. Five species can now be placed to genus and seven cannot; of the last seven, three may someday be placed because the plant hosts are known, but the hosts of the remaining four are unknown. I have been unable to locate Cook's gall collection, but perhaps some day similar galls will be found in Cuba and correctly matched with Cook's descriptions and illustrations. It is unfortunate that Cook formally named these species because he never published on them again. More about Cook's life and work can be found in the obituary by LeClerg 1953.

Peter (Pedro) Jörgensen

P. Jörgensen (1870–1937) was a Danish naturalist interested in gall-forming insects. He suffered from tuberculosis, and when he was about 35, he moved to Argentina for his health. There he collected and published on Argentinian insects, particularly Lepidoptera, first as an amateur and later as a professional naturalist. He collected Argentinian insects for many European specialists and sent Kieffer most of the gall insects he collected in and around the province of Mendoza. These were eventually described in Kieffer and Jörgensen 1910. Jörgensen published separately (Jörgensen 1916, 1917) on the galls referred to in the co-authored paper, furnishing the fine illustrations that now make it possible to recognize the galls. Jörgensen continued to suffer from tuberculosis and eventually retired to a friend's farm in Paraguay. There one day he was found murdered from a blow to the head. His killer was never found, but robbery is believed to have been the motive. Additional facts about Jörgensen's life can be found in obituaries by Esben-Petersen 1938 and Henriksen 1936.

Jörgensen's gall collection is apparently lost, although much of the other material he collected during his years in South America was placed in the hands of European specialists during his lifetime. The gall makers he sent to Kieffer for description are also presumed lost.

Jörgensen (1916) wrote that two things were important in order to succeed in rearing gall midges: time and infinite patience. He wrote that one needed at least two years to know the galls of a region. During the first year one could familiarize oneself with the flora and search for galls, trying to rear the gall makers. In the second year one could try again to rear those species that had failed for various reasons to emerge the first year. In my own trips of a few weeks each to Argentina and Brazil, I saw how frustrating it was to be in any one place, often a unique environment, for only one or two days. It was thrilling to find a particular gall I was looking for, but finding only a first-instar larva or that the adults had already flown left me little better off than before.

Juan Brèthes

Born in France, J. Brèthes (1871–1928) moved to Argentina at the age of 19. He completed his university education in Buenos Aires while developing an avid interest in entomology. He became associated with the National Museum in Buenos Aires and eventually earned a position on the staff as the first entomologist to be hired by the museum. Brèthes published 220 papers, mainly taxonomic, and chiefly on Hymenoptera and Diptera. In five papers between 1914 and 1922, he described six genera and nine species of Argentine gall midges. Only four of his species were reared from plants; one other appears to have been reared by a colleague, but Brèthes did not give the name of the host. Most of his cecidomyiid types are in the National Museum in Buenos Aires, mounted uncleared under cover slips on slides, but in a medium that has by now mostly evaporated. These types have allowed me to place some of Brèthes' species. As the only entomologist in his museum, Brèthes was necessarily a generalist. He was also relatively isolated, so had to work without a previously assembled reference collection or full library. In the midst of these disadvantages, it is to Brèthes' credit that some of his species are recognizable today. A detailed account of Brèthes' life can be found in Dallas 1928.

Clodomir Antony Vincent Houard

C. Houard (1873–1943), a Frenchman, was trained in the natural sciences. He developed his life-long interest in galls while earning his university degrees in Paris. After serving a period as an assistant in the Botanical Laboratories in Paris, from 1911 to 1917 he was an inspector for phytopathology and adjunct professor in Caen, and from 1919 to 1934 Director of the Institute of Botany and the Botanical Garden in Strasbourg. He retired officially from the last position in 1934, but left only in 1940 when Germany took Strasbourg at the onset of World War II.

Houard's chief interest was gall structure. He traveled widely in Europe and once to North America to collect galls, both from nature and in herbaria. With these and specimens sent to him by others, he built a large and magnificent collection of galls, still in its original, excellent condition in the Muséum National d'Histoire Naturelle in Paris. Houard's greatest accomplishment was the series of books he produced on galls from various parts of the world (Houard 1908, 1909, 1922, 1923, 1933, and 1940). One projected volume remained uncompleted, that for the galls of North America (exclusive of those on oaks). Most of the notes for the North American manuscript were lost during the occupation of his residence by German soldiers during World War II. Houard died before the war ended.

Houard's 1933 comprehensive work on the galls of the Neotropical Region is still a very useful guide. The book describes, illustrates, and keys galls for which gall makers are still unknown, so should continue to be useful for many years. More details about Houard's life can be found in obituaries by his wife, J. Houard (1948), and by Maresquelle (1944).

Edwin Möhn

E. Möhn (born 1928) is German and began work on Cecidomyiidae with his comprehensive and influential study of the larvae of Palearctic Cecidomyiidae (Möhn 1955a). In 1956, he spent 10 months under the auspices of the Senckenberg Research Institute, Frankfurt-am-Main, as a visiting scientist at the Instituto Tropical de Investigaciones Científicas of the University of El Salvador. While there he collected insects and other organisms, paying particular attention to plant galls from which he reared hundreds of species. Upon his return to Germany in 1957 he accepted a position at the State Museum of Natural History in Stuttgart where he continues his research today. Between 1959 and 1975 in a series of papers on the Asphondyliidi and Lasiopteridi of El Salvador, Möhn described 24 genera and 150 species of Neotropical gall midges, all based on material taken or reared from known hosts. Concurrently he produced a fine, exhaustive work on the classification of the Asphondyliidi, with special reference to the Neotropical fauna (Möhn 1961b). Möhn has since published volume 1 and is working on volume 2 of *System und Phylogenie der Lebewesen*, a projected three-volume classification of all living organisms.

Möhn's cecidomyiid collection is preserved in the SMNS. All specimens are stored in vials filled with alcohol mixed with glycerin and arranged by collection number. The numbers refer to Möhn's notebooks, which contain full collection data and are kept at the museum. To study specimens of a particular species one looks up the name in a book to find the collection numbers under which that species is stored. The numbered vials are located, and the specimens retrieved and temporarily mounted in glycerin under a cover slip on a glass slide. After they have been studied the specimens are replaced in the vials. The collection is well maintained, but using it is cumbersome.

Keith Murray Harris

K. M. Harris (born 1932) is British and began his entomological career as a research entomologist in Nigeria working on insect pests of sorghum and other grains. This early experience with insects of economic importance influenced the direction of his subsequent taxonomic research. Upon his return to the United Kingdom in 1962, he worked on the Barnes Collection of Cecidomyiidae at The Natural History Museum (then the British Museum [Natural History]) in London and subsequently joined the staff of the Royal Horticultural Society at Wisley as entomologist. In 1974, he joined the Commonwealth Institute of Entomology as an insect taxonomist specializing on gall midges and other Diptera of agricultural importance. His comprehensive revision of the coccoid-feeding gall midges of the world (Harris 1968) includes a number of Neotropical species. It was thorough and well illustrated and is still useful for identifying this important group of gall midges. The revision was based on available types and on the considerable amount of material that had accumulated in the British Museum over the course of many years. Harris has since become Director of the International Institute of Entomology and continues to work on gall midges as time permits.

Antonio Wünsch

A. Wünsch was born of German parents in 1949 in Chile. In 1973–1974, while a student of Möhn's in Stuttgart, he spent eight months in the Instituto Colombo-Áleman de Investigationes Científicas, Santa Marta, northern Colombia, collecting and rearing cecidomyiids, particularly in the cactaceous thorn-scrub vegetation in the vicinity of Santa Marta. His thesis (Wünsch 1979) concentrated on the 36 species he found there that belonged to what were then known as the supertribes Lasiopteridi and Asphondyliidi. One new genus and 24 new species were described in his study. His descriptions are detailed and accompanied by excellent illustrations. His collection of Colombian Cecidomyiidae is now in the State Museum of Natural History in Stuttgart and is maintained in the same way as that described for the Möhn Collection. Wünsch now teaches at the Max Born Gymnasium, Backnang, Germany.

Twenty-three other individuals described four or fewer species of Cecidomyiidae originally or subsequently found in the Neotropics. A few of these individuals published considerably more on cecidomyiids of other regions, but most did little more with gall midges. The names of these taxonomists are listed alphabetically, each followed by the number of Neotropical species described and the publication date(s): Ahlberg (1; 1939); Alexander (1; 1936); Barnes (4; 1930, 1932, 1939); Baylac (1; 1987); E. Blanchard (2; 1852); E. E. Blanchard (3; 1938, 1939, 1958); Borgmeier (1; 1931); Byers (1; Gagné and Byers 1985); Cockerell (2; 1892, 1907); Coquillett (2; 1899, 1905); Enderlein (1; 1940); Fyles (1; 1883); Karsch (2; 1877, 1880); Macquart (1; 1826); Mamaev (1; 1967); Molliard (1; 1903); Nijveldt (2; 1967, 1968); Osten Sacken (1; 1878); Philippi (3; 1865, 1873); Pritchard (1; 1951); Rondani (1; 1847); Silvestri (1; 1901); Walker (1; 1856); and Williston (2; 1896). P. Herbst, mentioned in the earlier sketch on Kieffer, was the collector and nominally coauthor with Kieffer of many species described from Chile between 1905 and 1911, but Kieffer was doubtless entirely accountable for the taxonomic expertise of their papers.

2 External Anatomy

This review of the external anatomy of gall midges is more comprehensive than that in Gagné 1989 because the whole family is treated, not only those species associated with plants. The general morphological terminology follows the system used in McAlpine et al. 1981.

Egg

Eggs are smooth and ovoid to elongate-ovoid. In species characterized by long, very attenuate ovipositors, one end of the egg is drawn out to a point (Kieffer 1900, Isidoro and Lucchi 1989). Eggs may be white, yellow, orange, or red. The color may be permanent or may change as the developing larva matures inside (Roskam 1977). Most females bear scores to hundreds of eggs, so each egg is tiny, but females of pedogenetic species bear only two or four larger eggs that completely fill the abdominal cavity.

Larva

Larvae (fig. 1) are legless, cylindrical or depressed-cylindrical, and tapered at both ends. They may be white, yellow, orange, or red depending on the species or age, and the color may vary within a species. Most Cecidomyiidae typically have three instars, but some of the pedogenetic species have one or two (see chapter 3). Instars may be differentiated from one another by head capsule measurements, anatomical differences, and observations over the course of a life cycle (Solinas 1965, Wyatt 1967, Roskam 1977, Isidoro 1987, Gagné and Hatchett 1989, Peña et al.

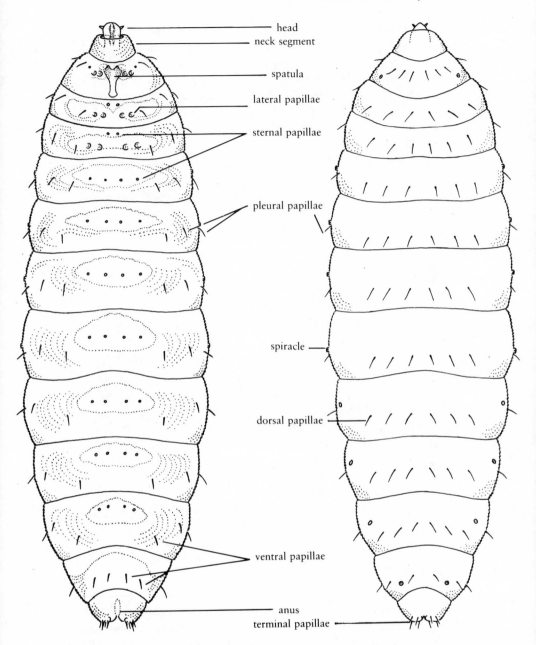

Figure 1. *Dasineura gleditchiae* third instar. Ventral view on left, dorsal on right.

1989). Reports exist of four instars. In some of those cases, the pre-diapausing third instar is counted as a fourth, but it does not molt (e.g., Metcalfe 1933). In other cases, a late second or an early third instar that shows a partially developed spatula is counted as an additional instar (Haridass 1987). One claim for four instars in a Holarctic species (Borkent 1989) is evidently based on samples that contain two separate species.

The last instar of a few phytophagous species develops within the second-instar skin and does not feed. The second-instar skin of these species is called a puparium; it becomes brittle and serves as the cocoon for the developing pupa. This retention of the second-instar skin as a puparium has evolved independently several times in this family (Gagné 1989).

Head

The cone-shaped head capsule is extremely small compared with that of other Mycetophiloidea. It is eyeless but has sensory plates, papillae, and short, cylindrical antennae (Solinas 1968, Solinas et al. 1987). Movable mouthparts are composed of a pair of vertically articulated mandibles and a pair of maxillar plates (Solinas 1971). When the larva feeds, the anterior part of the head capsule is closely appressed to the host tissue, the mandibles pierce, abrade, or grasp the host, salivary secretions are expressed, and strong pharyngeal dilator muscles effect ingestion through the esophagus (Solinas 1968). The head capsule is modified for various modes of life. Predators have long antennae (figs. 179, 180), as do crawling first-instar larvae of many plant-feeding species, but the antennae of most gall midges are less than twice as long as wide. Mandibles are used for scraping, grasping, or piercing (Solinas 1971) and may serve as a conduit for salivary secretions (Hatchett et al. 1990).

Body Segments

Behind the head capsule are a neck segment connecting the head to the thorax, three thoracic segments, and nine abdominal segments including the terminal or caudal segment. First instars have a variable number of spiracles: none; one on the eighth abdominal segment (Wyatt 1967, Peña et al. 1989); two, on the first thoracic and the eighth abdominal segment (Roskam 1977, 1979); or the full complement of nine normally found in the second and third instars, on the first thoracic and first through eighth abdominal segments.

Integument

The integument is smooth, spiny, or rough. Raised, horizontal, rough-ened ridges or welts used in locomotion occur anteroventrally on the abdominal segments, even on many larvae that are otherwise smooth. Some species have ventral pseudopods or dorsal and lateral lobes. The third instar (rarely the second instar also) usually has a spatula, a sclero-tized, deeply pigmented, dermal structure on the venter of the prothorax. It is a structure unique to Cecidomyiidae. The spatula probably origi-nated as an adaptation for digging through soil and was later modified for cutting through hard plant tissue. The spatula has become variously shaped and has been secondarily reduced or lost in some groups. The body has a number of regularly placed sensory papillae with or without setae. Some papillae are lost on sedentary species, particularly those that live in complex galls; such losses are correlated more with convergent larval habits than with phylogenetic affinities (Sylvén 1975). The basic papillar pattern of a third instar is shown in figure 1. Terminal papillae are the most often modified: in particular, many Cecidomyiidi have one or two pairs of short, stubby, slightly recurved setae that are very dif-ferent from the remaining, setiform papillae on that segment (fig. 156). The anus is on the terminal segment. It is caudal in Lestremiinae and pedogenetic Porricondylinae; in remaining cecidomyiids it is usually ven-tral, but it is dorsal on many predators (fig. 178), and dorsal or subdorsal on some gall-forming groups.

Additional references. See Möhn 1955a for an illustrated key to the larvae of Palearctic genera; Mamaev and Krivosheina 1965 for examples of lestremiine and porricondyline larvae; Wyatt 1967 for a review of species with pedogenetic larvae; Mamaev 1968 on larval adaptations to various life strategies; and Gagné 1989 for a key to larvae of Nearctic genera found on plants.

Pupa

The pupal stage offers diagnostic features for taxonomy, including adaptations of the head and abdomen that pupae use in cutting or forcing their way out of plant galls (fig. 108). The face is ventral and may bear various horns or ridges. The antennal bases are the anteriormost parts of a pupa and may be developed into a pair of conical horns. The cephalic sclerite has a pair of elongate setae and the prothorax a pair of spiracles that are usually elongate. Segments 2–7 of most cecidomyiids have a pair of spiracles, but these may be lost on some segments or, in pupae of pedogenetic species, lost completely. Pupae of many groups have spines on the abdominal segments. Most pedogenetic species have a much simplified, larviform pupal stage, called a hemipupa, that gives rise to a generation of larvae (Wyatt 1967). The hemipupa develops within the skin of the preceding larval instar.

Additional references. See Möhn 1961b for a comprehensive study of pupae in the tribe Asphondyliini.

Adult

Representative adults are illustrated in figures 6, 7, 9, and 10 in chapter 5. Despite their simple and fragile appearance, adults have evolved a great number of adaptations to their various ways of life. Most are of uniform light to dark hue, but various groups have developed striking color patterns, either from the base color of the integument or the scale and setal covering. Color patterns are best seen in living or freshly killed specimens. The body of some groups may be partly or entirely covered with very dense scales. The setal vestiture is of practical taxonomic importance.

Head

The head is made up mainly of the two compound eyes that usually join one another at the vertex, or top of the head. Varying degrees of eye and facet size reduction occur. The facets are usually hexagonal when closely appressed, but circular when farther apart. The eyes of cecidomyiids that have a reduced number of eye facets may be separate above. In certain groups the eye facets are placed farther apart laterally than elsewhere. The eyes of some predators are completely divided laterally so that the head appears to have one dorsal and two lateral eyes (fig. 237). Ocelli are present only in the Lestremiinae (figs. 6, 7). Most Cecidomyiidi and some *Ledomyia* species (Lasiopteridi) have an occipital process (fig.

162), which is a narrow, vertical extension of the occiput that bears two strong setae, the "top setae" of Panelius 1965.

Antennae

The antennae primitively have 14 flagellomeres, but some groups have fewer, others more. All subfamilies and supertribes show reductions. Twelve segments seem to be the rule for a large subdivision of Cecidomyiinae. Most species with other than 12 or 14 segments can have a variable number within a species. In the Neotropical Region *Feltomyina polymera* (Alexander) has the most with 63 and *Anarete buscki* (Felt) and the female of *Enallodiplosis discordis* Gagné have the fewest with 7. Male flagellomeres are most generally made up of a large node bearing the setae and various sensoria and, except for the terminal flagellomere, a long apical neck connecting to the next segment (fig. 70). Female flagellomeres are generally cylindrical and typically have much shorter necks (fig. 71) than those of males. In some groups both sexes have reduced necks so that the sexes appear similar (figs. 42, 43). In the supertribe Cecidomyiidi the node of the male flagellomeres is primitively subdivided into two separate spherical or pyriform nodes divided by a smooth internode similar to the distal neck (fig. 141). The male flagellomeres of many subgroups of this supertribe resemble those of females with a simple basal node and neck. This type of gynecoid male antenna is usually stable within species or higher groups where it occurs, but I recently saw a series of *Dicrodiplosis pseudococci* Felt males from Europe, some with normal binodal and some with gynecoid flagellomeres.

Antennal sensoria are primitively setiform, but several other kinds of sensoria have developed, the most common being the transparent sensoria called circumfila that girdle the flagellomeres and that in some Cecidomyiidi form loops longer than the flagellomeres themselves (fig. 203). The circumfila are formed by bifurcate nerve cells whose ends grow together during development (Slifer and Sekhon 1971). Slifer and Sekhon (1971) and Solinas et al. (1987) discuss the function of these and several other kinds of antennal sensoria and illustrate them in fine detail.

Mouthparts

Most cecidomyiids have mouthparts capable of ingesting liquids. They consist of the dorsal labrum, the lateral labella, the ventral hypoproct, and, lateral to the labella, the maxillary palpi (fig. 162). Some species have elongate mouthparts adapted for specialized feeding, and this adaptation is sometimes also associated with a flattened head, long neck, or some other body modifications (Collin 1922, Gagné and Byers 1985, Gagné and Boldt 1989). Other gall midges have almost completely atrophied and probably nonfunctional mouthparts (Jones et al. 1983). Palpi primitively have four segments but may have as few as one segment, and this latter character usually occurs conjointly with general reduction of the remaining mouthparts.

Wing

The most primitive cecidomyiid wings have the complement of veins shown in figure 11. The veins of many groups tend to be concentrated toward the leading edge of the wing, with a corresponding shortening of R_5 (figs. 6, 42). This character is possibly an adaptation for faster or stronger flight, especially in species with short antennae. There are separate tendencies for reduction or loss of the Sc, Rs, M_3, Cu, and Cup. The Lestremiinae are divided into two major groups according to whether they have an M_3 and a simple Cu (fig. 6) or no M_3 and a forked Cu (fig. 7). Elsewhere in the family when Cu is forked, M_3 is evident only as a fold. In some Oligotrophini, the length of R_5 relative to total wing length is positively correlated with body size (Sylvén and Carlbäcker 1981). Some pedogenetic forms have a greatly reduced venation of only two or three long veins. Such wings usually have an elongate fringe of hairs along the wing margin that evidently provide the wing more surface area.

Legs

The legs may be quite long (fig. 10), especially in nonphytophagous groups that roost on spider webs, but they may be secondarily shortened in some groups (fig. 68). Shorter legs may allow adults more rapid flight or the ability to scuttle about on a flat surface such as a leaf. Some sexual dimorphism may occur in leg length; for example, males of some *Rhopalomyia* species have longer legs than females (figs. 70, 71). Tarsi are made up of five tarsomeres except in some pedogenetic forms of Porricondylinae that may have four, three, or two tarsomeres. The most striking change in the leg in this family outside of the Lestremiinae is the great reduction in size of the first tarsomere (fig. 9). The connection between the short first and the much longer second segment is easily broken, allowing long-legged gall midges to sacrifice one or more tarsi caught on a sticky substrate. If gall midges are briskly shaken up in a vial of alcohol, the tarsi will separate at the end of the first tarsomeres. The first tarsomere may have an apicoventral spur (fig. 100). The tarsal claws show a great variety of shapes (figs. 158–161) and may have one or more basal teeth. The empodium, a padlike appendage between the tarsal claws, has a characteristic length relative to the claws, depending on the taxon.

Abdomen

Abdominal terga and sterna are diagnostic in shape and setation. The first through eighth tergites and second through seventh or eighth sternites are usually rectangular, but they may be shaped otherwise or variously subdivided or reduced. The female postabdomen tends to have a conservative form in those groups that simply deposit eggs on various surfaces (fig. 6). Many groups have developed various modifications for laying eggs among bud leaves or into plant tissue (figs. 132, 134). Even a lengthening of the postabdomen causes great changes in the conformation of the individual segments. The progression from a short to a long ovipositor can be seen in several groups, including within genera (Gagné 1985). Cerci of female Lestremiinae and Porricondylinae are two-segmented (with the exception of *Dirhiza*, not yet known from the Neotropics), but are reduced to one segment in all Cecidomyiinae except *Didactylomyia*. The female cerci of most Lasiopteridi are fused into one median lobe (fig. 48), and this modification has occurred separately in various genera of Cecidomyiidi. The male genitalia (figs. 72, 73) are composed of the cerci, the hypoproct, the gonopods, which are divided into the basal gonocoxites and the distal gonostyli, and the aedeagus, which may or may not have associated parameres. The gonopods of Lestremiinae and Porricondylinae are united ventrally (fig. 12) but are separate in the Cecidomyiinae (fig. 72).

Additional references. See McAlpine et al. 1981 on comparative morphology of Diptera in general and Tastás-Duque and Sylvén 1989 on sensory structures of the ovipositor.

3 Biology

Cecidomyiid larvae are responsible for all or most feeding, adults for most dispersal and usually procreation, although reproduction is partly taken over in Heteropezini by larvae and pupae. Larvae and adults have consequently diverged along separate paths. Larvae originally adapted to feed on fungi changed in form and function in various ways to feed on plants and arthropods. From laying eggs on decaying organic matter, females evolved various strategies for finding particular hosts, depositing eggs in or on plant tissue, and adapting to their hosts' biology. The pupa also changed to cope with changing circumstances, such as having to cut its way out of hardened plant tissue. All these modifications through time have resulted in a very large and diverse family.

This chapter reviews the biology of the family, especially as it relates to the Neotropical fauna. It does not repeat all of the review in Gagné 1989, which dealt only with plant-feeding gall midges, but discusses some aspects not treated in that work and updates others.

Larval Feeding Habits

Cecidomyiid larvae can be found in or on fungi, in decaying plant matter, on all parts of living plants, and among colonies of mites and insects. The primitive habit of feeding on fungi in humus and decaying logs is shared by all of the Lestremiinae and Porricondylinae and many Cecidomyiinae. Some Cecidomyiinae (e.g., species of *Mycodiplosis* and *Clinodiplosis*) are found on fungal hyphae or rust spores, and a species of *Ledomyia* forms the only true gall known on fungi (Larew et al. 1987).

Most Cecidomyiinae feed directly on living plants. They may be free-living in buds or flowers or live inside plants and form galls. Plant feeders

22

may be generalists or specialists that are restricted to one or more species of a genus of plants.

Some of the plant-feeding groups have secondarily developed symbiotic associations with fungi. Fungi are found lining the galls made by all *Asphondylia* and other genera of the subtribe Asphondyliini. The association between this subtribe and fungi is obligatory because the fungi, which evidently need the gall to grow in, serve as food for the gall midges (Meyer 1952, 1987). Most Lasiopterini and Alycaulini also have an association with fungi, but in varying degrees. The fungus is a necessary precursor to larval entry into the plant and to subsequent larval feeding of at least some Lasiopterini, even though the larva does not feed exclusively on the fungus (Rohfritsch 1992a,b). In other cases, as in some Alycaulini, the fungus incorporates itself into the gall but is apparently not the larval food (Gagné 1989). Symbiotic fungi may be present or absent in galls made by cecidomyiids of the same genus. While galls of most species of *Asteromyia*, for example, contain a symbiotic fungus, galls of at least *Asteromyia modesta* (Felt) have none.

Many species are obligatory inquilines, or guests, in or on galls made by other cecidomyiids. This habit is widespread among Cecidomyiinae and evidently evolved many times. Some inquilines belong to groups, such as *Trotteria* and Camptoneuromyiini, that are exclusively or almost exclusively inquilinous, but others are in genera, such as *Dasineura* and *Contarinia*, that include mostly primary gall makers. Many inquilines compete directly for food with the gall maker (Parnell 1964, Shorthouse and West 1986); others are generalist fungus feeders that arrive late in the gall cycle, often after the gall maker has departed. Inquilines may increase the time the gall remains on the plant or modify its shape, causing the gall to continue to produce food (Osgood and Gagné 1978, Solinas and Bucci 1982). Many inquilines appear to have a specific target species. Wünsch (1979) regularly found two species, a *Trotteria* and a *Camptoneuromyia*, in galls of a very common *Asphondylia* in northern Colombia but not in galls of other *Asphondylia* species in nearby plants. These same inquilines were associated with the same *Asphondylia* species in El Salvador as well.

A large number of Cecidomyiinae are predators of insects and mites. This is another habit that has arisen many times in cecidomyiids. Predators make up the large tribe Lestodiplosini and the unrelated genera *Aphidoletes*, *Diadiplosis*, *Silvestrina*, and probably *Enallodiplosis*. In addition, some species of the usually fungivorous *Clinodiplosis* and *Hyperdiplosis* are predaceous, perhaps facultatively so. I recently discovered some *Clinodiplosis* larvae with their mouthparts inserted into larvae of a gall-forming *Contarinia* species. Predaceous gall midges, mostly external but some internal (see *Pseudendaphis*), may be specialists on various

Homoptera (aphids, coccoids, psyllids, whiteflies), Hemiptera, Thysanoptera, Acarina (red spider mites, eriophyid mites), or Coleoptera. Some are generalists, among them species of *Lestodiplosis*, that feed on a mixed diet of other cecidomyiids, beetles, lepidopterans, and mites in flower heads (R. Cruttwell, in litt., 1970).

Pedogenesis

Most cecidomyiids begin life as an egg laid by a female adult and progress through three separate larval instars and the pupal and adult stages. The striking exceptions to this general rule are the pedogenetic group, in which the larval or pupal stage assumes the procreative function. This condition evolved independently in Cecidomyiidae at least twice, in one tribe of Lestremiinae and one of Porricondylinae. All pedogenetic species feed on fungi, and several species are pests in commercial mushroom houses. Claims that some gall-forming species of the subfamily Cecidomyiinae are pedogenetic have been shown to be incorrect. Wyatt (1967) summed up previous work on pedogenetic species, reclassified them, and outlined the progressive simplification of both larval and adult stages.

Pedogenetic species have as few as two larval instars. Embryos of one species are already apparent in the larval hemocoel before the first larval molt. Daughter larvae develop within the mother second instar and break out of her cuticle when full-grown. Several to tens of larvae may develop from one larval mother. In other species a greatly simplified, larviform pupa, which may have elaborate spiracles, will develop within a second or third instar and give rise to larvae. A new generation can be produced in 6–14 days. All pedogenetic species also produce normal sexual generations, which in some species appear to be correlated with overcrowding. When the necessary stimulus is present, the last instars, either the second or third, depending on the species, give rise to normal exarate pupae, which then give rise to adults of both sexes. Production of young by the immature stages greatly simplifies and shortens the life cycle.

Initiation of Larval Feeding

Newly hatched larvae of fungus-feeding groups presumably begin feeding immediately if they find themselves on or among their food. Predaceous or plant-feeding larvae that hatch on the surface of a plant must crawl to an appropriate host or feeding site. Upon hatching, larvae of

Diadiplosis coccidivora (Felt) creep under the shell of their mealybug host and begin feeding on the eggs (Parnell 1966). Larvae that hatch from eggs deposited within host tissue may begin feeding directly (Rohfritsch 1990). Once settled on the proper host tissue, larvae begin to feed, unless they enter a period of delayed development (Fernandes et al. 1987).

The first instar of some plant-feeding species, being a crawling form, may have a different appearance from the second, which is usually exclusively a feeding form (Rohfritsch 1980). The third instar may also be different in its shape and in the more rugose and tougher integument, especially if it has to leave the gall and crawl about. During larval development, changes occur in the salivary glands, with different parts active at separate times. An active part of the glands in first-instar larvae produces a factor that causes the galls to grow; a different section is more active in older larvae and produces the salivary secretions used in predigestion and that cause the plant cells to produce food (Stuart and Hatchett 1987). For a comparative study on the structure of the salivary glands and digestive system, see Mamaev 1968.

Stem-infesting species reportedly bore into tissue. Wilson and Heaton (1987) observed larvae of *Neolasioptera brevis* Gagné penetrating the soft tissue of new shoots while gyrating their bodies perpendicular to the stem. Within 8–10 hours, larvae had made their way into the shoot, leaving a small entry channel behind. Pupae later pushed through these channels, allowing the adults to emerge.

Postfeeding Larval Behavior

Upon termination of feeding, most gall-forming Neotropical cecidomyiids remain in the host, but some crawl or force their way out of the gall to drop to the soil. The majority of free-living larvae drop to the soil also, but some (e.g., *Aphidoletes aphidimyza* (Rondani)) may pupate in cocoons on the plant or on the ground. Predators are variable in this regard: larvae of the endoparasitic *Pseudendaphis maculans* Barnes leave the aphid host through the anus and drop to the soil to pupate (Kirkpatrick 1954), but larvae of the coccoid egg-feeding *Diadiplosis coccidivora* remain under the shell of the mother mealybug to pupate (Parnell 1966).

Some full-grown larvae spin white cocoons in which they eventually pupate, but others, notably the Asphondyliini, have completely lost the ability to make cocoons (Möhn 1961b). Certain predators form a cocoon (Harris 1973), while others do not (Kirkpatrick 1954). In some cases only part of a population may make a cocoon (Bishop 1954). Larvae of *Steno-*

diplosis sorghicola form a cocoon only in the diapausing generation (Isidoro 1987).

Pupation in the soil is the primitive habit for gall midges, and larvae of species that do so always have a spatula (fig. 1). Once on the ground, the larvae move to a suitable site by crawling or jumping and dig into the soil with the help of the spatula. A larva jumps by tightly arching its body, the caudal end of the body catching on the anterior end of the spatula, and springing away. A larva may jump more than 10 cm at a time (Kirkpatrick 1954), and the habit is widespread in gall midges (Wyatt 1967, Holz 1970). The probable primitive role of the spatula is for digging through soil (Pitcher 1957). When a larva begins to dig into soil, its head is retracted so that the spatula forms the leading edge of the body. The spatula serves to push grains of soil out of the way to facilitate burrowing (Milne 1961).

Many unrelated groups pupate in their galls. Two main strategies occur. In one, the full-grown larva uses its spatula to cut a channel almost completely through to the exterior. The larva then retreats to its cell. The pupa will later squeeze through the channel and force its way through the remaining plant tissue. In the other method, the pupa cuts its way out to the surface with the help of sclerotized, pointed structures at the anterior end of the body (see fig. 108). Groups using the first strategy always have a well-developed larval spatula (see fig. 49); groups using the second, with the exception of Asphondyliini, usually have lost the spatula or have a weakly developed one.

Full-grown larvae of most Alycaulini and Lasiopterini, which live primarily in woody stem galls, use the spatula to cut a cylindrical passage from their feeding tunnels to the exterior, often leaving only a final thin layer of bark. Larvae of some species dig completely out to the exterior but then retreat into the tunnel after sealing the opening with silk (Wilson and Heaton 1987). Larvae then enter diapause or pupate directly. Eventual passage through the thin remaining bark or silk seal is made by the pupa, which pushes through the remaining weak covering. Many gregarious *Neolasioptera* species in woody stems form a common exit hole.

Larvae of other groups that pupate in their galls (e.g., most *Rhopalomyia*) have secondarily lost the spatula. The task of digging through the sometimes woody tissue is taken over by the pupa, which has well-developed antennal horns and other armature suitable to the purpose. Some pupae of *Asphondylia* species cut a circular channel through their galls by pushing and cutting forward with their antennal horns while rotating their body around a longitudinal axis. Strong abdominal spines help the pupa grip the sides of the channel.

A larva in a complex gall feeds with its head toward the interior of the plant; when full-grown, it reverses direction so that its head is pointed toward the exterior. The pupa that later develops would be unable to perform this function and has only to move anteriorly to exit from the gall. Whether the larva or pupa cuts a channel, the mature pupa exits only about two-thirds of the way to the exterior. The pressure at the end of the tight passage then holds the pupal skin in place while the adult breaks out.

Dormancy

Many species, especially in warm, wet regions of the Neotropics, produce generations continuously (Kirkpatrick 1954, Stiling et al. 1992). Others, particularly in temperate to cold, or dry habitats, have to synchronize adult emergence closely with the seasons or host's cycle so that the proper plant tissue is available for newly hatching larvae. The period of nonactivity is usually spent as a dormant, or disapausing, larva. Most commonly, dormancy occurs in the third-instar larva, either in the ground or in the plant host. Larvae of a *Schizomyia* species in pimento flowers in Jamaica drop to the ground when fully fed, leaving behind the swollen aborted buds, which then rot and fall from the tree (Parnell and Chapman 1966). Because pimento trees in Jamaica flower in March and April, larvae that fall to the ground in October must lie dormant within their cocoons until the following spring, when they will develop into pupae. The diapausing generation of the sorghum midge remains inside the glumes of the host (Isidoro 1987), and third-instar larvae of *Neolasioptera rostrata* Gagné lie dormant in their flower galls on *Baccharis* for many months until the next flowering season. A few cases are known of diapausing first instars in dormant buds. The gall of *Machaeriobia machaerii* Kieffer on *Machaerium aculeatum* in Brazil is initiated in August during the period of vegetative growth, but the first instar remains dormant until November in apparently dry but hypertrophied leaf rachises from which the leaflets have fallen. Further larval and gall development starts only in the following July before the plant's normal vegetative growth cycle begins (Fernandes et al. 1987). Pupal diapause is known from other regions, but it appears to be rare (Gagné 1989); diapause may occur in either a larva or a pupa of one Japanese species (Sunose 1978).

In species with more than one generation per year, dormancy may occur only in the last generation before the plant stops producing new growth (Isidoro 1987), or it may occur in part of each generation (Coutin 1959). The frequency of diapausing larvae increases with successive generations through the season (Readshaw 1966). The first generation of the new plant season will be a composite one, consisting not only of the last generation of the previous year, but also of some individuals from each of the previous year's other generations (Barnes 1956). Extended diapause can benefit a species by ensuring survival of some individuals through lean years or unusual weather (Sunose 1978). Larvae of many plant-feeding Northern Hemisphere cecidomyiids are known to remain in diapause for many years (Gagné 1989).

Emergence of Adults

With the adult fully formed inside, the pupa wriggles to the top of the soil or forces its way through plant tissue. The adult then pushes its way through a split that develops behind the head and along the midline of the pupal skin. The legs, when freed of their pupal sheaths, push against the pupal skin for leverage in pulling the remainder of the body out. Once the adult is completely free, its legs and wings expand rapidly to their full size and harden, after which the adult is able to fly. Individual emergence takes a few minutes to several hours. Dawn and dusk, when conditions are usually more humid than in other parts of the day, are favored times for emergence (Young 1985, Roskam and Nadel 1990). Species with a food source available the year round may emerge at any time of the year (Blahutiak and Alayo 1981, Roskam and Nadel 1990), but plant feeders dependent on a tissue available only during certain times of the year must coordinate their emergence with the biology of their host. In these cases, emergence of a species ranges over a short period, often less than a week, but most of a population may emerge the same day. Male emergence usually peaks slightly before female emergence (Teetes 1985, Yukawa et al. 1976). Earlier emergence allows time for optimal sclerotization of male body parts and proper functioning of the male genitalia before the females emerge.

Flight

Mamaev (1968) outlined four general types of cecidomyiid wing venation that are correlated with the length and function of the legs and with general life habits. The first type is exemplified by the Lestremiidi (fig. 6), which are relatively strong flyers. Their wings are large and long, with the venation more crowded and stronger toward the leading edge. They also have long, cursorial legs, which allow the adults to make rapid progress across an open substrate. In contrast, Micromyiidi (fig. 7) have broader, shorter wings, lending this group to more passive dispersal in flight. These flies have short legs that allow them to scuttle in small cavities amid decaying vegetable matter on the forest floor. A third type of wing, long and with reduced, uncostalized venation, is found in most of the free-living, longer-lived Porricondylinae and Cecidomyiinae with long, thin legs (fig. 10). The fourth type of wing is found in very reduced forms such as the pedogenetic species. These have extremely reduced venation but very long setae along the wing margin that allow for wind dispersal. Adults of pedogenetic gall midges are especially common in amber fossil deposits, possibly because the flies easily drifted into and were trapped by the source resin in much greater numbers than other cecidomyiids.

These four general types of wing venation intergrade considerably throughout the family. Even though gall midges vary in their ability to fly, they are all, relatively speaking, weak fliers. This is especially true for newly emerging females that are so heavy with eggs they cannot take off readily (Freeman and Geoghagen 1987). Passive flight may be a useful strategy for some species in covering distances without great expense of energy. It has been found that once close to their host plants, gall midges actively fly toward their hosts (Sylvén 1979).

Males or both sexes may form swarms (Chiang 1968, Young 1985). Males of some species of *Anarete* are known to swarm above reflective markers in cultivated fields (Chiang and Okubo 1977). Swarms of separate species hover at different heights above the ground. At some point males leave the swarm to mate with passing females, then later rejoin the swarm (Chiang 1968). Adults of many plant-feeding and predaceous gall midges may emerge simultaneously in large numbers and form swarms at the location where males emerge (Yukawa et al. 1976).

Mating

Newly emerged females are extremely attractive to males. Those with protrusible ovipositors exert and wave them, broadcasting a pheromone that may be concentrated at the tip of the ovipositor (McKay and Hatchett 1984, Isidoro 1987). Mating takes place on the ground, on a plant surface, or in the air. In some species copulation is effected without preliminary courtship, with the male simply approaching a female from the rear, attaching, and mating (McKay and Hatchett 1984). In other species courtship may involve wing vibration by males just before copulation (Barnes 1939, Yukawa and Miyamoto 1979) and wing vibration may be accompanied by the male vibrating his antennae and posturing (Baylac 1986). Species that emerge intermittently and in small numbers over the length of a season may require more complex mating behavior than do plant feeders. Once mated, females with protrusible ovipositors retract them and fly away, unattended further by males (McKay and Hatchett 1984, Solinas and Isidoro 1991).

Adult Feeding

Most adults are equipped to feed on liquids. Longer-lived adults probably take water and nectar (Moutia 1958), but most plant-feeding gall midges, especially males that live only long enough to mate, probably do not feed. The labrum, two short lateral labella, and the ventral hypopharynx form a tube for imbibing liquids. In some species these structures are greatly elongated or otherwise modified. Examples of greatly modified mouthparts are those of *Contarinia prolixa* Gagné and Byers, presumably used in taking fluids from floral tubes, and of the African *Farquharsonia rostrata* Collin (Farquharson 1922), used to rob ants of regurgitated food. Species regularly feed at particular sites. Adults, chiefly females, of *Mycodiplosis ligulata* Gagné are consistently found feeding on nectaries in freshly opened cacao flowers (Young 1985). That habit is a regular one of this species because adults have been taken in cacao flowers in both Costa Rica and Bolivia. Adults of species whose larvae are internal parasites of aphids have been observed apparently feeding on the secretions from the aphid cornicles (Kirkpatrick 1954). Some plant-feeding species (e.g., *Rhopalomyia* spp.) have the adult mouthparts so reduced in size as to appear nonfunctional. Their emergence coincides with the availability of the host: adults mate, promptly lay their eggs, and die immediately after. *Didactylomyia longimana* (Felt) feeds on the spiders' prey (Sivinski and Stowe 1981), and females of *Mycodiplosis* species have been seen with swollen abdomens atop large masses of termites taken by spiders (Eberhard 1991).

Roosting

Many Porricondylinae and many nonphytophagous Cecidomyiinae with long, thin legs are commonly seen roosting on spider webs (Sivinski and Stowe 1981, Young 1985). Species that do this are relatively long-lived, probably do not emerge from the pupal stage synchronously, and have many successive generations over a season. I have never found plant-feeding species on spider webs.

Host Seeking

Males of plant-feeding species do not need to fly far from the emergence site; but females, especially if coming out of the soil, may have to travel a considerable distance after mating to find the host. Females are evidently attracted by host chemicals (Sylvén 1979). Females but not males of the Palearctic *Dasineura brassicae* (Winnertz) are attracted to a chemical associated with the host plant (Pettersson 1976). Females of *Aphidoletes aphidimyza*, a predator of aphids, are more attracted to a particular plant species if they were reared from aphids on that species than if they were reared from aphids on a different species (Mansour 1975). Females of *Mayetiola destructor* (Say), a pest of wheat, select particular varieties of the host at least partially on the basis of physical characteristics, such as hairiness (Morrill 1982).

Oviposition

Females must manage to lay their eggs in, on, or near the host. Females of plant-feeding species alight on a plant and find the oviposition site by antennating and palpating the plant surface (Hallberg and Åhman 1987). When a suitable site is found, the female backs into the site by feeling with the tip of her ovipositor (Hallberg and Åhman 1987). Females of most plant-feeding gall midges have long, tapered, telescoped, and pliable ovipositors for inserting eggs into closed flowers and buds. Some species have hard and stiff ovipositors modified for inserting eggs directly into plant tissue, sometimes at particular loci, for example, into stomata (DeClerck and Steeves 1988). Predaceous species lay their eggs singly, close to their prey, with the effect that eggs are scattered around the host colony (Moutia 1958). Species such as *Asphondylia* that pierce plant tissue with their ovipositors apparently deposit only one egg at a time because often only one specimen develops in each flower or bud. But other species that oviposit among flower parts or tight bud leaves may lay several eggs at a time (Solinas and Bucci 1982, Peña et al. 1989). Most cecidomyiid females bear many tiny eggs, but some pedogenetic species bear only two or four relatively large eggs.

Duration of the Life Cycle

The life cycle of gall midges in the warm tropics can be extremely short. As examples, the rust-feeding *Mycodiplosis fungicola* (Felt) in Jamaica took only 10 days (Parnell 1969) from egg to adult, the plant-feeding *Prodiplosis longifila* took about 13 days (Peña et al. 1989), and the predaceous *Diadiplosis coccidivora* in Jamaica took 21 days (Parnell 1966). Where the climate is usually generally favorable or the host constantly produces the proper tissue, generations may be continuous throughout the year (Roskam and Nadel 1990). Freeman and Geoghagan (1989) estimated that *Asphondylia boerhaaviae* Möhn had 17 generations per year in Jamaica. On the other hand, some plant-feeding species may have a limited number of generations per year to coincide with their hosts' cycles. The pimento tree in Jamaica flowers once a year, and the *Schizomyia* species that infests the flowers has but one generation a year, remaining in the soil as a full-grown larva for much of the year between flowering seasons (Parnell and Chapman 1966).

Mortality

Several studies of food webs of Neotropical plant-feeding gall midges have been done. These include Winder 1980, Chen and Appleby 1984a, Cermeño et al. 1985, Fernandes et al. 1987, Freeman and Geoghagen 1989, and Stiling et al. 1992. Larvae within the galls are attacked by parasitic Hymenoptera belonging to several families and also by predaceous Hemiptera, Lepidoptera, and Coleoptera and carnivorous wasps (Vespidae) that feed through the gall tissue. Grazing animals eat whole plants. Free-living gall midge larvae are preyed upon by *Lestodiplosis* species. The coccoid-feeding *Diadiplosis cocci* Felt, is itself severely attacked by mites (Blahutiak and Alayo 1982). Adults emerging from galls are preyed upon by spiders, and Young (1985) lists mantids, spiders, and *Anolis* lizards as predators of cecidomyiid adults feeding in or on flowers. Besides mortality due to parasites and predators, many aspects of accident and weather affect survival: eggs are laid on the wrong host or an unsuitable part of the plant, or larvae die before reaching susceptible tissue. Freeman and Geoghagen (1989) wrote of "endogenous mortality," death from unapparent reasons that accounted for many unsuccessful galls.

4 Collecting, Rearing, and Preparing Specimens for Study

Collecting

Larvae, pupae, and adults are all significant for taxonomy, so it is important to preserve samples of each when collecting and rearing cecidomyiids. For some genera the immature stages have the better diagnostic characters; for others, females or males are more important. Specimens found in a particular niche or host are much more valuable for their associated information than those simply caught in flight. Sexes of flight-caught species often cannot be associated with any confidence. Even when they can be associated by color or other distinguishing marks, the flies lack much meaning outside the context of their hosts, biology, and immature stages. Flight-caught specimens can nevertheless be valuable, particularly for survey work.

Rearing adults requires that larvae be collected when fully grown, after they have finished feeding. Mature larvae can usually be recognized by the presence of the spatula. Larvae that are barely visible or have no spatula are probably in an early instar and will not reach maturity if taken from their food source. Most cecidomyiid galls can be collected in clear plastic bags, but paper bags are an alternative when condensation on the sides of the containers is a problem, especially when the galls cannot be processed until the next day. Clear plastic bags are good rearing containers when the duration of the pupal stage is short. The bags can be opened up occasionally to get rid of condensation that may build up.

Samples of plant damage or galls should be preserved for reference. Hard galls on stems, most bud galls, and the majority of complex galls on leaves need no special treatment and can be kept dry. Any larvae inside the gall tissue will dessicate in place and can be left there for future study.

33

Galls on very old herbarium specimens are still good sources of specimens. Flat galls such as leaf folds or blisters on leaves that might curl or shrivel can be dried between pages of books or in a plant press. The more succulent, large galls are best kept in alcohol to retain their shape. Since plant specimens in alcohol lose their color, it is good to take photographs of the gall or damage.

Rearing

A few fully grown larvae from a collection should be preserved in alcohol or on slide mounts for future reference and identification. The remainder, either free or still within the galls, can be kept for rearing. Many species, particularly if they pupate soon after they are collected, can be reared in the plastic bags in which they were collected or in Petri dishes. Fungal growth can be kept to a minimum by reducing the amount of plant tissue and condensation in the container. Larvae that lie dormant for extended periods need more long-term storage. I have been successful in rearing adults by keeping larvae in a container as shown in figure 2. Specimens are placed in a pot atop a moist mixture of one part clean sand and one part peat moss. The sand keeps the peat moss from binding, and the acidic peat moss impedes fungal growth. The pot should be at least 10 cm in diameter and short enough to stand in a cardboard shoebox with some clearance between the top of the pot and the cover of the box. Plastic pots used by plant growers to raise seedlings are suitable. The soil should be packed more tightly at the very bottom of the pot than at the top to prevent larvae from descending all the way through the soil and out the drainage holes. At the end of the shoebox a hole is made into which a lipped 8- or 10-dram vial will fit snugly. Both the pot and the outer box are labeled as to contents. The box is kept in a sheltered, cool place, and the vial is checked daily for adults. It is occasionally opened to check soil moisture, and water is added as needed. Contamination can be avoided by not reusing boxes and cleaning pots well before reusing them.

When adult cecidomyiids emerge, they will usually fly toward the light into the vial. The vial can quickly be removed and inverted into a killing jar, or the flies can be picked out with an alcohol-moistened artist's brush.

One kind of gall or a flower head will sometimes contain more than one species of cecidomyiid. The number of species can usually be determined by examining the larvae closely, but inquilines sometimes occur in small numbers and can be missed. When larvae leave a host readily, it is wise to retain some of the plant tissue in case another species is present that does not normally leave. For example, *Contarinia* and *Dasineura* in

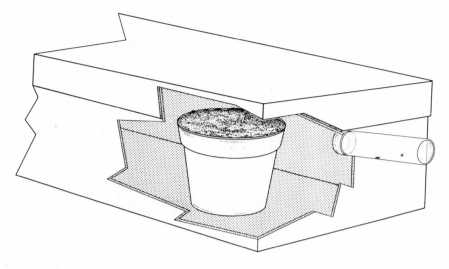

Figure 2. Rearing container.

flower heads of *Baccharis* readily leave the plant when full-grown, but *Neolasioptera* larvae in the same heads remain in the achenes through the winter and could be overlooked.

Preparing Specimens for Study

With experience, one can identify some larvae and adults to generic level with a dissecting microscope. More detailed study, however, will require that specimens be mounted on microscope slides for study under a compound microscope, preferably one with phase contrast. If a slide has to be made, it is best to make a good-quality, permanent one. My technique is outlined below and requires the equipment and materials illustrated in figure 3.

If the specimen is a larva or a pupa, prick it once or twice with a fine insect pin. Place it in a small crucible with a solution of sodium hydroxide made of two pellets of NaOH in about 10 ml of water. The specimen can be heated quickly on a hot plate or left in the solution for a longer period of time at room temperature. If using a hot plate, allow the solution to boil gently for a minute or so. When the specimen is cleared, remove it with an eye dropper and place it in a porcelain spotplate. With a fine forceps transfer the larva or pupa to a drop of acetic acid on the spotplate for a minute to neutralize the sodium hydroxide, and then transfer it successively to 70% and 95% alcohol and clove oil for about five minutes

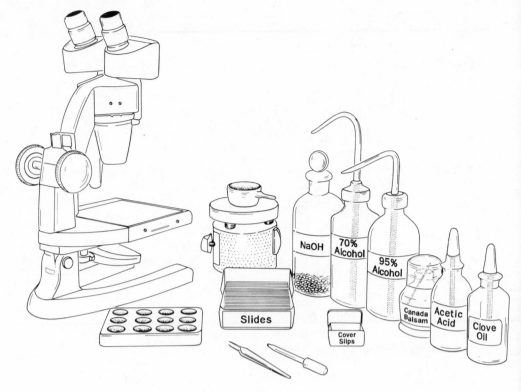

Figure 3. Materials and chemicals for making slide mounts.

in each liquid. Specimens can be kept in the clove oil overnight if necessary. Apply a drop of Canada balsam to the slide by means of a glass rod, transfer the larva or pupa to the drop, and, after arranging the specimen properly with the use of needles, lay a cover slip on it. Pupal skins can be mounted without clearing, but they should be stored in alcohol long enough so that no air remains in the leg or antennal sheaths. A little practice is required to get the Canada balsam just thick enough so that the specimen is not distorted and yet not so thick that one cannot use the high-power microscope objectives. An easier, alternative mounting medium is Euparal, which is miscible with 70% alcohol. With Euparal, two steps can be skipped, the 95% alcohol and the clove oil. The optical properties of Euparal are adequate for most work. Attach a label to each end of the slide: one label to record the host, other collection data, and the mounting medium, the other to record the identification (fig. 4).

Preparation of adults takes a little more care. If a specimen is not already in alcohol, place it in 70% alcohol for five minutes. Remove and

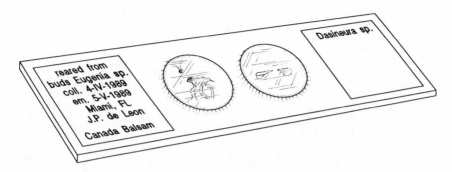

Figure 4. Sample slide mount.

place the wings directly into a drop of clove oil in the spotplate. Puncture the abdomen with a fine insect pin and place the body in the sodium hydroxide solution. When the specimen is cleared, follow the same process as for the larva. The wings can be mounted under one cover slip, the rest of the body under the other (fig. 4). Once the body is on the cover slip in the drop of Canada balsam, remove the head from the body and arrange it for frontal viewing. If the specimen is a male, disconnect the genitalia from the body and mount them dorsoventrally. When several specimens are available, mount some in lateral view with the head and genitalia still attached.

Specimens stored in alcohol deteriorate with time, so mount an adequate sample on slides as soon as possible after collecting or rearing. Because many species have scales that are lost when immersed in alcohol, it is desirable to pin a series on double mounts or on card points. Most adults shrivel slightly, but the scale or color pattern will preserve well. Larvae and pupae can also be kept on card points for long-term storage. These should be kept the same way and with the same care one takes with any pinned specimens.

5 Synopsis of Neotropical Cecidomyiidae

Cecidomyiids are nematocerous Diptera in the infraorder Bibionomorpha and superfamily Sciaroidea (formerly Mycetophiloidea). The superfamily contains the families Cecidomyiidae, Sciaridae, and the former Mycetophilidae, recently divided among six families (Matile 1990). There is some evidence to indicate that the gall midges are most closely related to the Sciaridae (Wood and Borkent 1989). The gall midges are unquestionably monophyletic. Fossils of several tribes are known from early and late Cretaceous amber (Schlee and Dietrich 1970, Gagné 1977a), so the family is evidently very old. Characters unique to Cecidomyiidae within the Bibionomorpha are the lack of tibial spurs (compare figs. 6 and 8); the reduced larval head with modifications, including styliform, elongate mandibles, for a sucking mode of feeding; and the presence in the last instar of the sternal spatula. The family is divided into three subfamilies: the Lestremiinae, the Porricondylinae, and the Cecidomyiinae. These are treated in turn after the key to subfamilies.

Figure 5 shows the relationships among the subfamilies of Cecidomyiidae and the supertribes of Cecidomyiinae. The numbers at the branches refer to the following synapomorphies, which are unique among at least Sciaroidea or are otherwise explained:

1. Tibial spurs lost
2. Larval head reduced in size, specialized in function
3. Sternal spatula present
4. Ocelli lost
5. First tarsomeres greatly shortened, much shorter than second tarsomeres
6. Gonocoxites separate and no longer fused ventrally, and male ninth segment indistinct
7. Flagellomeres reduced in number from 14 to 13 (segment reduction occurs

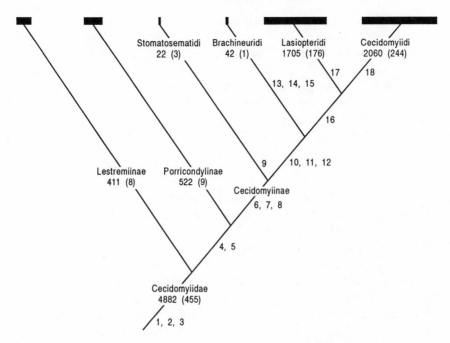

Figure 5. Cladogram showing subfamilies of Cecidomyiidae and supertribes of Cecidomyiinae. The widths of the bars at the top of the cladogram indicate the relative numbers of known species. The numbers immediately below the group names are the total number of described species in the world and, in parentheses, the subset from the Neotropical Region. The numbers under "Cecidomyiidae" are greater than the total for the subgroups, reflecting the fact that some described species are not known well enough to segregate further. Numbers at the branches refer to synapomorphies that are explained in the introduction to chapter 5.

sporadically in all three subfamilies but nowhere else as a regular number in both sexes)

8. Dorsal papillae of the larval eighth abdominal segment reduced from four to two, and ventral papillae of first through seventh abdominal segments reduced from four to two (unique among Cecidomyiidae)

9. Parameres secondarily divided almost to base (convergent with some *Ledomyia* species in Lasiopteridi)

10. Flagellomeres reduced in number from 13 to 12 (segment reduction occurs sporadically in all three subfamilies but nowhere else as a regular number in both sexes)

11. Rs vein weak (occurs in a few Porricondylinae also, but associated there with greatly reduced venation and wing size)

12. Female cerci reduced from two to one segment (unique in Cecidomyiidae except for *Dicerura* in Porricondylinae and *Stomatosema* in Stomatosematidi)

13. Flagellomeres reduced in number to 10 (as a regular number, unique in Cecidomyiidae)
14. Male eighth tergite reduced to a linear, strongly sclerotized, anterior band (unique in Cecidomyiidae)
15. Parameres each reduced to a short, smooth, narrow lobe (occurs also in a tribe of Cecidomyiidi)
16. Occiput with dorsal prominence
17. Parameres clasping aedeagus (unique in Cecidomyiidae)
18. Male flagellomeres binodal with separate circumfila (unique in Cecidomyiidae).

This chapter lists all genera and species known from the Neotropical Region. Information given for each genus includes the original citation and page, the type species, any generic synonyms that have been used for the Neotropical fauna, and a synopsis with such information as the number of species, distribution, diagnostic characters, and any helpful references. Data included for each species are the original citation with page and illustrations (if any), original genus, stages described, kind of type, type locality, deposition, host(s), distribution, and any useful references. Underscored, italicized names are synonyms. Host names followed by an asterisk indicate that the record is from outside the Neotropical Region. Neotropical distribution is by country in alphabetical order, with state, province, or territory given for the larger countries; any extra-territorial distribution follows. A list of abbreviations used appears at the beginning of the book. In the keys figure numbers are in parentheses following the character description.

Key to Subfamilies of Cecidomyiidae

1   Ocelli present; legs with five tarsomeres, the first longer than second (figs. 6, 7) .... Lestremiinae, p. 41
     Ocelli absent; legs with five tarsomeres, the first shorter than second (figs. 9, 10), or with fewer than five tarsomeres 2
2   Rs, if present, as strong as other veins (fig. 11); base of M (M+rm) often sinuous (fig. 11); gonocoxites joined ventrally (fig. 12); female cerci two-segmented (*Didactylomyia* [Cecidomyiinae] has two-segmented cerci, but is unique with its 13 flagellomeres, the last with a terminal nipple) ....
     .... Porricondylinae, p. 45
     Rs absent or weaker than other veins and evanescent at junction with R_1 (fig. 68) (except in Stomatosematidi [fig. 15]); base of M (M+rm) straight or at most weakly sinuous (fig. 85); gonocoxites free ventrally (fig. 72); female cerci one-segmented, often fused mesally (fig. 78) .... Cecidomyiinae, p. 50

Subfamily Lestremiinae

The subfamily Lestremiinae is the oldest of cecidomyiids. Fossils are known from late Cretaceous Canadian amber (Gagné 1977a). Many genera and species are widespread. Some of the species may have spread through commerce. Larvae feed on fungus in decaying organic matter. A few are pests in commercial mushroom houses.

The subfamily is the sister group to the Porricondylinae and Cecidomyiinae (Panelius 1965). It is divided into two distinct supertribes, the Lestremiidi and the Micromyidi, each of which is broken down into several tribes. An example from each supertribe is shown in figures 6 and 7. Wings of Lestremiidi (fig. 6) have a medial fork, a well-developed M_3 vein, and a simple cubitus, while those of Mycromyidi (fig. 7) have a simple medial vein, no M_3 vein, and a forked cubitus. The most relevant studies of the Lestremiinae of the Neotropical Region are Pritchard 1947 and 1951, but see also Kleesattel 1979 and Mamaev 1968 on the fauna of the Palearctic Region. Kleesattel has a full list of available tribal names as used for the Palearctic fauna. Larvae of some European species have been described; the best reviews of larval anatomy, classification, and biology are by Mamaev and Krivosheina (1965) and Mamaev (1968).

This subfamily is especially poorly known in the Neotropics, as indicated in figure 5. So far only 8 described species and 10 genera have been reported. Lestremiinae are nonetheless common and diverse in this region if one may judge from the many undescribed species represented in the U.S. National Museum. All except one of the known Neotropical species belong to cosmopolitan genera. The exception is *Termitomastus*, a genus based on a species associated with termites in Argentina. Were its unique species not brachypterous, it might also be referable to a cosmopolitan genus. Four of the eight species of Lestremiinae are widespread, extending into the Nearctic or Holarctic Regions. I have not presented a key to the few recorded genera, but recommend as a starting point the key to the Nearctic genera in Gagné 1981 and to the Palearctic genera in Kleesattel 1979.

SUPERTRIBE LESTREMIIDI

Tribe Lestremiini

Anarete

Anarete Haliday 1833: 156. Type species, *candidata* Haliday (mon.).
Microcerata Felt 1908b: 309. Type species, *Micromya corni* Felt (orig. des.).

Anarete is a cosmopolitan genus of 24 species. One described species is known from the Neotropics, and I know of several more undescribed species. Refs.: Pritchard 1951, Kim 1967.

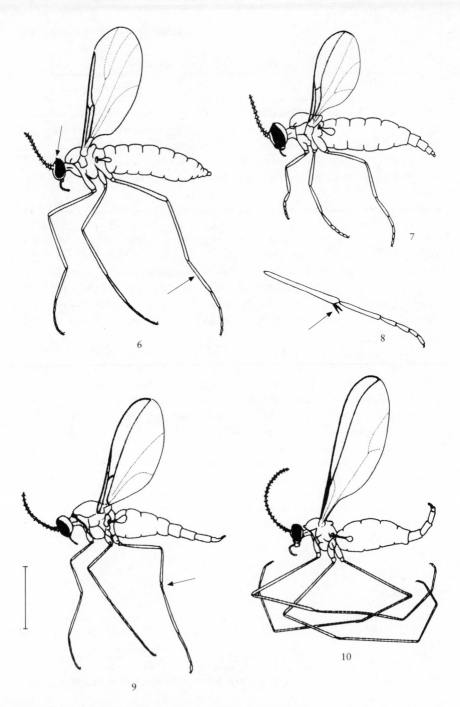

Figures 6, 7, 9, 10. Adult females, Cecidomyiidae. **6,** *Lestremia* sp. (arrow to head: position of ocelli; arrow to leg: elongate first tarsomere). **7,** *Monardia* sp. **9,** *Winnertzia* sp. (arrow: short first tarsomere). **10,** *Camptomyia* sp. **Figure 8.** Leg, *Bradysia* sp. (Sciaridae) (arrow: tibial spurs; scale line = 0.5 mm).

buscki Felt 1915c: 198 (*Microcerata*); ♂. Holotype: Baracoa, Cuba; USNM. Distr.: Cuba, Puerto Rico; USA (widespread). Refs.: Pritchard 1951, Kim 1967.

Lestremia

Lestremia Macquart 1826: 173. Type species, *cinerea* Macquart (mon.).

Lestremia is cosmopolitan and known from 24 species, one of them from Baltic amber. Two species have been reported from the Neotropics. A female is illustrated in figure 6. Ref.: Pritchard 1951.

cinerea Macquart 1826: 173 (*Lestremia*); adult. Syntypes: Europe; depos. unknown. Distr.: Chile; Europe, Hawaii, and North America. Ref.: Pritchard 1951. Note: Edwards (1930) recorded this species from a female collected in Bariloche, Chile, as, "appears indistinguishable from the European *L. fusca*, Mg."
fusca Meigen 1830: 309 (*Lestremia*); ♀. Syntypes: Europe; lost.
nigra Blanchard 1852: 349 (*Lestremia*); adult. Syntypes: Carelmapu and Chiloe, Chile; depos. unknown. Distr.: Chile. Note: Blanchard's description indicates that this species is a lestremiine, but one cannot know what genus it actually belongs to without seeing his specimen.

SUPERTRIBE MICROMYIDI

Tribe Campylomyzini

Corinthomyia

Corinthomyia Felt 1911h: 35. Type species, *Campylomyza hirsuta* Felt (orig. des.) = *Campylomyza brevicornis* Felt.

This genus is known from a single, widespread Holarctic species, which was recorded once from Chile. *Psadaria*, listed separately below, may be a synonym. Ref.: Pritchard 1947.

brevicornis Felt 1907a: 1 (*Campylomyza*); ♀. Holotype: Nassau, New York, USA; Felt Coll. Distr.: Chile; Holarctic.

Psadaria

Psadaria Enderlein 1940: 668. Type species, *pallida* Enderlein (orig. des.).

This genus is known from one Chilean species. Freeman (1953) found the female of the type series to be identical with those of *Corinthomyia brevicornis*.

pallida Enderlein 1940: 669, figs. 36–38 (*Psadaria*); ♂, ♀. Syntypes: Masafuera, Juan Fernandez Is., Chile; ZMHU? Distr.: Chile. Refs.: Freeman 1953, Lagrange 1980 (her fig. 2 of a wing cannot represent that of *Psadaria*).

Tribe Micromyini

Bryomyia

Bryomyia Kieffer 1895d: 78. Type species, *bergrothi* Kieffer (mon.).

Bryomyia is known from 10 Northern Hemisphere species, some of them widespread in the Holarctic Region. The genus has been recorded from Juan Fernandez Is., Chile (Freeman 1953), and I know of other, presumably new species from Chile and Costa Rica. Ref.: Pritchard 1947.

Micromya

Micromya Rondani 1840: 23. Type species, *lucorum* Rondani (mon.).
Micromyia Anonymous 1844: 451, emendation of *Micromya*.
Ceratomyia Felt 1911h: 33. Type species, *johannseni* Felt (orig. des.).

This genus is known from six Old World and two Nearctic species, one of which extends into the Neotropics. Ref.: Pritchard 1947.

johannseni Felt 1911h: 33 (*Ceratomyia*); ♂. Holotype: Ocotlán, Mexico; Felt Coll. Distr.: Costa Rica, Dominica, Mexico; USA (widespread).

Monardia

Monardia Kieffer 1895d: 111. Type species, *stirpium* Kieffer (orig. des.).

This genus is known from 15 species from the Northern Hemisphere, some widespread, and two fossil species, one from Baltic amber and one undescribed species from Mexican amber of Upper Oligocene to Lower Miocene age (Gagné 1973a). A female is illustrated in figure 7. Ref.: Pritchard 1947.

Termitomastus

Termitomastus Silvestri 1901: 1. Type species, *leptoproctus* Silvestri (mon.).

This genus consists of one Neotropical species. It is a commensal in termite galleries and has the wings reduced to stubs and a greatly enlarged abdomen. Only the female is known. Edwards (1929) and Pritchard (1947) agreed that *Termitomastus* is close to *Monardia*. Ref.: Pritchard 1947.

leptoproctus Silvestri 1901: 1 (*Termitomastus*); ♀. Syntypes: S. Ana, Misiones, Argentina; Coxipo, Cuyaba, Brazil; depos. unknown. Habitat: nest of *Anoplotermitis reconditi* (Isoptera). Distr.: Argentina, Brazil. Ref.: Silvestri 1903.

Trichopteromyia

Trichopteromyia Williston 1896: 255. Type species, *modesta* Williston (mon.).

This genus is known from three species, two of them restricted to the Palearctic Region, the third widespread, including the Neotropics. Ref.: Pritchard 1947.

modesta Williston 1896: 255, pl. VIII, figs. 6, 6a, 6b (*Trichopteromyia*); ♀. Syntypes: St. Vincent; BMNH. Distr.: Costa Rica, Dominica, Mexico, St. Vincent; Holarctic.

Tribe Peromyiini

Peromyia

Peromyia Kieffer 1894e: clxxv. Type species, *leveillei* Kieffer (mon.).

Peromyia is a large genus of 54 Holarctic and Oriental species, including one from Baltic amber. The genus has been recorded from Juan Fernandez Is., Chile (Freeman 1953), and an undetermined species, also from Chile, is in the USNM. Ref.: Pritchard 1947.

Subfamily Porricondylinae

Porricondylinae and Cecidomyiinae together are monophyletic (fig. 5) (Panelius 1965). They differ from Lestremiinae in the loss of ocelli and the shortened first tarsomere (figs. 9, 10). This modification may have arisen as an adaptation for roosting on spider webs, where Porricondylinae and Cecidomyiinae are often, but Lestremiinae never, found. Porricondylinae are a very diverse group. Panelius (1965) showed that the subfamily is paraphyletic with relation to the Cecidomyiinae. Porricondylinae have the gonocoxites primitively united ventrally (fig. 12), and most but not all have ringlike antennal circumfila (figs. 13–14) found in all Cecidomyiinae. Wings of Porricondylinae have the Rs as strong as R_3, except in species of Heteropezini and a very few Porricondylini with generally reduced venation. Some Porricondylinae retain part of M_{1+2} (fig. 11).

As in the subfamily Lestremiinae, many genera of this subfamily are widespread. Fossils of one tribe are known from late Cretaceous Canadian amber (Gagné 1977a), and the subfamily is diverse in later amber deposits (Meunier 1904). Most larvae feed on fungus in decaying organic matter. A few occur on plants, particularly under bark, but these are probably feeding on decaying tissue.

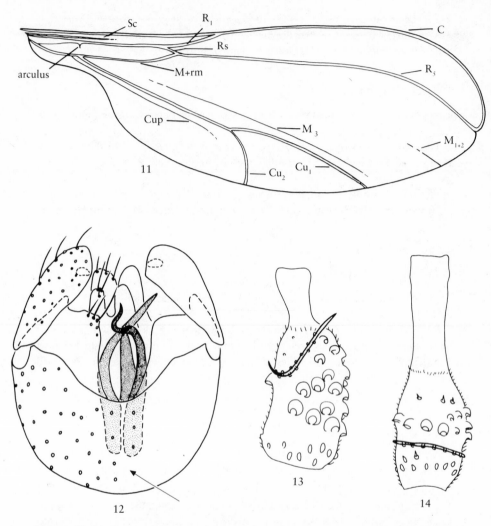

Figures 11–14. Porricondylinae. **11,** Wing, *Porricondyla* sp. **12,** Male genitalia (ventral), *Asynapta mangiferae* (arrow: ventrally fused gonocoxites). **13,** Male third flagellomere, *Winnertzia* sp. **14,** Male third flagellomere, *Camptomyia parrishi*.

Porricondylinae are poorly known in the Neotropics. Only nine species and eight genera have been reported from the region, but sweeps and trap catches contain a great variety of unknown species. I have not presented a key to the few genera that have been recorded, but refer the reader to the key to Nearctic genera in Gagné 1981 and to the revision and key to the Palearctic species in Panelius 1965.

The subfamily is divided into six tribes, of which five are recorded here from the

Neotropics. The most comprehensive study of this subfamily is Panelius 1965. The type of the Nearctic species were reviewed by Parnell (1971) and placed within the classification devised by Panelius (1965). Reviews of larval anatomy and classification of Holarctic genera can be found in Möhn 1955a, Mamaev and Krivosheina 1965, Mamaev 1968, and Panelius 1965.

Tribe Asynaptini

Antennae of the species in the tribe Asynaptini usually have more than 14 flagellomeres, as many as 63 in *Feltomyina*. Adult postabdomens are characteristically recurved dorsoanteriorly (fig. 10). Although few species have so far been described from the Neotropics, this tribe appears to be particularly abundant here.

Asynapta

Asynapta Loew 1850: 21. Type species, *Cecidomyia longicollis* Loew (subseq. des. Karsch 1877: 14).

Asynapta is a cosmopolitan genus of 38 known species. Many species are apparently secondary plant feeders. The three listed below were found under the bark of various plants. Females of a possibly undescribed species repeatedly visit and seem to have a role in pollinating flowers of *Siparuna* spp. (Monimiaceae) (Feil and Renner 1991). The male genitalia of *Asynapta mangiferae* are illustrated in figure 12.

citrinae Felt 1932: 117 (*Asynapta*); ♂, ♀, larva. Syntypes: Isabela, Puerto Rico; Felt Coll. Distr.: Puerto Rico. Host: *Citrus* sp. (grapefruit) (Rutaceae).
gossypii Coquillett 1905: 200 (*Porricondyla*); ♂, ♀. Syntypes: Barbados; USNM. Distr.: Barbados. Host: *Gossypium* sp. (cotton) (Malvaceae).
mangiferae Felt 1909: 299 (*Asynapta*); ♂, ♀. Syntypes: West Indies (specific locality unknown); USNM. Distr.: West Indies. Host: *Mangifera indica* (mango) (Anacardiaceae).

Camptomyia

Camptomyia Kieffer 1894b: 86. Type species, *erythromma* Kieffer (orig. des.).

This large cosmopolitan genus is known from 62 species, including one from Baltic amber. Besides the single species listed below for this region, I know of several other, undescribed species, including some from Dominican amber. A female is illustrated in figure 10 and a male flagellomere in figure 14.

parrishi Felt 1915a: 152 (*Porricondyla*); ♂. Holotype: Bartica, Guyana; Felt Coll. Distr.: Guyana.

Feltomyina

Feltomyia Alexander 1936: 12 (preocc. Kieffer 1913a). Type species, *polymera* Alexander (orig. des.).

Feltomyina Alexander 1937: 60; new name for *Feltomyia* Alexander. Type species, *Feltomyia polymera* Alexander (aut.)

This genus is known from one described Neotropical species and at least one undescribed species from North America (Gagné 1981).

polymera Alexander 1936: 14 (*Feltomyia*); ♀. Holotype: Potrerillos, Chiriquí, Panama; USNM. Distr.: Panama.

Tribe Diallactini

The wing of this tribe is distinctive, with its distally situated Rs and diagonal R_5. The latter is fairly straight but meets C well posteriad of the wing apex. Unlike the antennae of most other Porricondylinae, diallactine antennae bear no circumfila or other extrasetal sensoria.

Haplusia

Haplusia Karsch 1877: 15. Type species, *plumipes* Karsch (orig. des.).
Johnsonomyia Felt 1908b: 417. Type species, *rubra* Felt (orig. des.).

Haplusia is a cosmopolitan genus of 18 described species, including one from Baltic amber. These species are among the most elegantly marked cecidomyiids. In addition to the three Neotropical species listed below, Williston (1896) identified the genus from St. Vincent, and several undescribed species are represented in the USNM. Ref.: Gagné 1978b.

braziliensis Felt 1915a: 153 (*Johnsonomyia*); ♀. Holotype: Igarapé Açu, Pará, Brazil; CUI. Distr.: Brazil.
cincta Felt 1912c: 103 (*Johnsonomyia*); ♀. Holotype: Polochic River, Guatemala; USNM. Distr.: Guatemala.
plumipes Karsch 1877: 16 (*Haplusia*); ♀. Holotype: Bahia, Brazil; ZMHU. Distr.: Brazil.

Tribe Heteropezini

The tribe Heteropezini contains most of the pedogenetic cecidomyiids. Adults are characterized by the loss of some tarsomeres, reduced antennae with 10 or fewer flagellomeres, and wing venation that may be reduced to two or three veins. The larval stage of some species has a reduced number of instars. The pupal stage of a pedogenetic generation of some species develops a much simplified "hemipupa" that may or may not be enclosed within the ultimate larval integument. Wyatt (1967) offers a comprehensive review of this tribe.

Henria

Henria Wyatt 1959: 175. Type species, *psalliotae* Wyatt (orig. des.).

This genus is known from two Nearctic, one Palearctic, and some undescribed species from Mexican (Gagné 1973a) and Dominican amber.

Miastor

Miastor Meinert 1864: 42. Type species, *metroloas* Meinert (mon.).

Four species of *Miastor* are known from the Holarctic and Australian Regions, and Williston (1896) recorded the genus for an undetermined species from St. Vincent.

Tribe Porricondylini

Species in the tribe Porricondylini have 14 or fewer antennal flagellomeres, but always a constant number within a species or sex. It is clear from the variety of undescribed Neotropical species represented in the USNM that this tribe is rich in species. The wing of a *Porricondyla* sp. is illustrated in figure 11.

Holoneurus

Holoneura Kieffer 1894b: 84 (preocc. Tetens 1891). Type species, *cincta* Kieffer (subseq. des. Kieffer 1894c: 316).
Holoneurus Kieffer 1895b: 115; new name for *Holoneura* Kieffer. Type species, *Holoneura cincta* Kieffer (aut.).

This genus is a mainly Holarctic, probably catchall genus of 18 described species. The one Neotropical species was reared from a rotting fig branch. Ref.: Parnell 1971.

occidentalis Felt 1911i: 190 (*Holoneurus*); ♂, ♀, larva. Syntypes: Paraíso, Panama; USNM. Distr.: Panama.

Tribe Winnertziini

Members of the tribe Winnertziini have a characteristic wing with a straight M base and a simple cubitus (fig. 9). Antennae have U-shaped (fig. 13) or bladelike sensoria instead of the ringlike circumfila seen in most porricondylines.

Winnertzia

Winnertzia Rondani 1860: 290. Type species, *Cecidomyia lugubris* Winnertz (orig. des.).

This large cosmopolitan genus of 60 described species was recorded from St. Vincent by Williston (1896) for an undetermined species. Several undescribed species from the Caribbean and South America are represented in the USNM. A female is illustrated in figure 9 and a male flagellomere in figure 13.

Subfamily Cecidomyiinae

The Cecidomyiinae are the youngest and much the largest of the three subfamilies of Cecidomyiidae (fig. 5). This subfamily is not known from Cretaceous fossils, although it is well represented in fossil amber from the Eocene and later. The group is much more diverse in its habits than the other two subfamilies and is of greater interest to humans because of its association with plants. Cecidomyiinae also contain primitive, generalist fungus feeders, but various lines have evolved into specific fungus feeders, plant feeders, or predators of insects and mites. It is apparent that the Neotropical Cecidomyiinae are collectively quite different from those in the rest of the world. Of the 137 genera represented here, 96 are endemic and many of them have little apparent relationship to those elsewhere. Three, perhaps four, genera are recent immigrants, and the remaining 37 are widespread or cosmopolitan.

The actual number of species of Cecidomyiinae that must occur in the Neotropics is at present incalculable. I know of undescribed species in almost every genus and of many for which new genera must be erected. In the discussion that follows, numbers of species given for various tribes, subtribes, and genera are of described species only.

Unique, shared characters of the subfamily are the ventrally separated gonocoxites (fig. 72), the reduced number of flagellomeres from the primitive 14 typical of the Porricondylinae to 13, 12, or fewer, or the augmented number of more than 12 that is always irregular within a species, and the presence of only two dorsal papillae on the eighth larval segment (fig. 1).

The subfamily is divided here into four supertribes (fig. 5) treated here in the following order: the relatively species-poor Stomatosematidi; the slightly larger Brachineuridi, reelevated here to a rank equal to the other supertribes; and the very large supertribes Lasiopteridi and Cecidomyiidi, between which all the remaining species of the subfamily are almost equally divided. The parameres, the mesobasal lobes between the base of the gonocoxites and the base of the aedeagus (fig. 72), are helpful in working out this classification. The parameres, as defined here, should be equivalent to the structure that Panelius (1965) termed the tegmen in the Porricondylinae and Mamaev (1968) the paramere in Lestremiinae.

The Stomatosematidi share two derived characters unique to the family, the division of the parameres into two approximately equally large parts (fig. 20) and the membranous female eighth tergite (fig. 24). The remaining three supertribes of Cecidomyiinae differ from the Stomatosematidi in the weak Rs and the reduction of the regular number of flagellomeres to 12. The Brachineuridi have the antenna reduced to a regular 10 flagellomeres, completely lack the Rs, have the male eighth abdominal tergite modified into a very short, strongly sclerotized band, and have the parameres reduced to a mostly smooth stub with two or three apical setae (fig. 29). The Lasiopteridi and Cecidomyiidi together share the cephalic occipital process (fig. 162) on which are situated a pair of strong setae ("top setae" of Panelius 1965). This cephalic process is widespread among Cecidomyiidi but rare in Lasiopteridi, where I have found it only in some species of the tribe Ledomyiini. It is the only synapomorphy I know of for both Lasiopteridi and Cecidomyiidi. The parameres of the Lasiopteridi usually clasp the aedeagus, at least partially (fig. 72). This arrangement apparently lends rigidity to the aedeagus, which appears to have a weaker wall in Lasiopteridi than in Cecidomyiidi. In the Cecidomyiidi the parameres are usually lacking (fig. 143), but when present do not sheath the aedeagus (fig. 123). Additional characters of Cecidomyiidi not known elsewhere in the Cecidomyiidae are the binodal male flagellomeres with separate circumfila (fig. 141) (unless secondarily gynecoid), and the more cylindrical than bulbous female flagellomeres that lack the strongly recurved setae of Lasiopteridi (fig. 172).

Key to Supertribes of Cecidomyiinae

1   Antenna with 12 flagellomeres, usually binodal in males (fig. 141), cylindrical in females and some males (figs. 94, 95), the setae not strongly recurved near base; parameres lacking (fig. 143) or, if present, not clasping aedeagus (fig. 123) . . . . Cecidomyiidi, p. 99
     Antenna with 7–42 flagellomeres; if with 12, flagellomeres with some setae strongly recurved at their bases (fig. 41) and parameres sheathing aedeagus (fig. 72) 2

2   Antenna with 7 flagellomeres; eyes divided at vertex and medially (fig. 237) . . . . ♀ *Enallodiplosis*; see under Cecidomyiidi, p. 99
     Antenna with 7 (rarely) to 42 flagellomeres; eyes not divided at vertex and medially 3

3   Antenna with 13 flagellomeres, the last with elongate apical process (fig. 18); Rs as strong as R_5 (fig. 15); gonocoxites splayed (figs. 19, 21), parameres free, not clasping aedeagus . . . . Stomatosematidi, p. 52
     Antenna with 7–42 flagellomeres; if with 13, the last without apical process; Rs lacking or weaker than R_5 (fig. 68); gonocoxites usually parallel, parameres either rudimentary (fig. 29) or at least partially clasping aedeagus (fig. 72) 4

4   Antenna with 10 flagellomeres; male eighth and usually seventh tergites reduced to linear, strongly sclerotized horizontal bands (fig. 39); parameres

rudimentary, glabrous, each with two to three apical setae (fig. 29); female postabdomen not protrusible (figs. 30, 31) . . . . Brachineuridi, p. 54
Antenna with 7–42 flagellomeres; male tergites, if short, not so strongly linear (fig. 53) (except in *Meunieriella*); parameres well developed, mostly setulose, and at least partially sheathing aedeagus (figs. 55, 72) (except in *Calmonia*); female postabdomen at least slightly protrusible (figs. 43, 48) . .
. . . . Lasiopteridi, p. 58

SUPERTRIBE STOMATOSEMATIDI

The supertribe Stomatosematidi contains two cosmopolitan genera, both present in the Neotropics. Some of the species are widespread. Characters that distinguish this group from other Cecidomyiinae are the presence of 13 flagellomeres, the last with an elongate apical process (fig. 18), and the strong Rs wing vein (fig. 15). *Didactylomyia* has a very short aedeagus, hypoproct, and cerci in relation to the elongated gonopods (figs. 19, 20) and two-segmented female cerci (fig. 25), while *Stomatosema* has normally proportioned genitalia (fig. 21) and one-segmented female cerci (fig. 24).

Didactylomyia

Didactylomyia Felt 1911h: 39. Type species, *Colpodia longimana* Felt (orig. des.).

This genus contains one cosmopolitan species that is a kleptoparasite in spider webs (Sivinski and Stowe 1981). Distinguishing characters are shown in figures 15, 16, 19, 20, 25. Ref.: Gagné 1975a.

longimana Felt 1908b: 416 (*Colpodia*); ♂. Holotype: Auburndale, Massachusetts, USA; Felt Coll. Distr.: Colombia, Dominica, Mexico, USA (Florida); cosmopolitan.

Stomatosema

Stomatosema Kieffer 1904: 380. Type species, *nemorum* Kieffer (mon.).

Figures 15–25. Stomatosematidi. **15, 16,** *Didactylomyia longimana*: **15,** wing; **16,** tarsal claw and empodium. **17, 18,** *Stomatosema dominicensis*: **17,** male third flagellomere; **18,** male 13th flagellomere. **19, 20,** *Didactylomyia longimana*: **19,** male genitalia (dorsal); **20,** aedeagus and parameres (enlarged). **21–23,** *Stomatosema dominicensis*: **21,** male genitalia (dorsal); **22,** aedeagus and parameres (ventral; arrow: two-lobed paramere flanking aedeagus); **23,** male hypoproct (ventral). **24,** *Stomatosema* sp., female postabdomen (dorsolateral). **25,** *Didactylomyia longimana*, female postabdomen (dorsolateral).

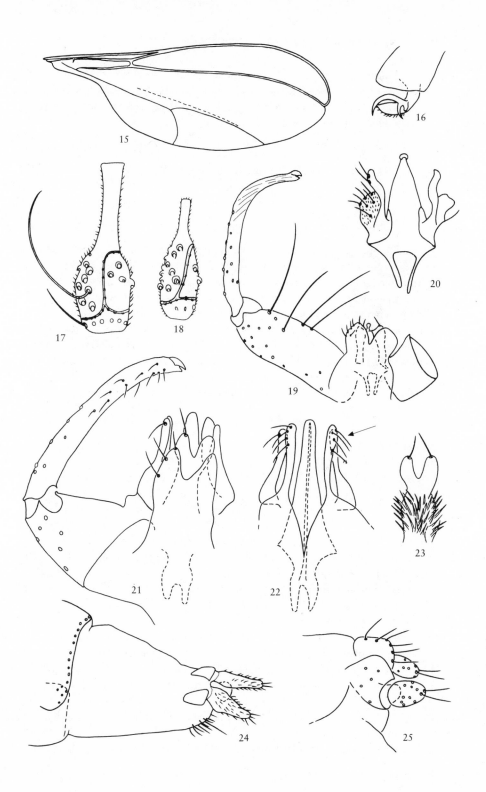

15

16

17

18

19

20

21

22

23

24

25

One described species of this cosmopolitan genus is listed here that extends into Mexico from the Holarctic Region. I know of several other species from the Neotropics, one of which is described below. A representative female postabdomen is shown in figure 24. Ref.: Gagné 1975a.

obscura Mamaev 1967: 877 (*Vanchidiplosis*); ♂. Holotype: Volkovo, Yarslavl Province, Russia; ASM. Distr.: Mexico; USA, Russia. Refs.: Mamaev 1968, Gagné 1975a.

Stomatosema dominicensis Gagné, new species

Description: Wing 1.5 mm long. Third flagellomere (fig. 17) with completely setulose neck; 13th flagellomere as in figure 18. Male genitalia (figs. 21–23) distinct from other known species by the parameres, each divided from its base into two subequal, elongate, glabrous lobes, one with one seta, the other with several. Hypoproct covered ventrobasally with spiniform setulae.

Holotype: ♂; 1.7 mi E Pont Cassé, Dominica, 12-III-1965, W. W. Wirth; USNM.

Etymology: The name *dominicensis* is a Latin adjective meaning from Dominica.

Remarks: This species differs from *S. obscura* in the shape of the parameres. In *S. obscura* the two lobes are dissimilar, one linear and glabrous, but the other triangular and almost completely covered with recurved setulae (Gagné 1975a).

SUPERTRIBE BRACHINEURIDI

The cosmopolitan supertribe Brachineuridi is represented in the Neotropics by *Chrybaneura* and four other genera newly recorded from this region, *Brachineura*, *Coccidomyia*, *Epimyia*, and *Rhizomyia*. These five genera are associated in the same taxon for the first time. The close relationship between *Epimyia* and *Rhizomyia* is demonstrated by the short, strongly sclerotized male seventh and eighth abdominal tergites, the peculiar parameres, and the presence of a regular number of 10 flagellomeres. Their affinity was previously obscured by the cylindrical flagellomeres, short R_5 vein, the bizarre genitalia, and thick scale covering of *Epimyia*. An undescribed Neotropical species of Brachineuridi in the USNM that is not covered with scales and has bulbous, necked flagellomeres similar to *Rhizomyia* has the bizarre male genitalia of *Epimyia* (figs. 36–38).

Brachineuridi are distinguished as follows: antennae with 10 sexually dimorphic flagellomeres, the nodes and necks longer in males, the nodes cylindrical and bulbous; R_5 two-thirds as long to as long as the wing (figs. 26, 32); Cu forked or simple, if forked, the tines reaching wing edge, if simple, the vein evanescent apically (fig. 26); tarsal claws narrow, curved near midlength (fig. 35), toothed, at least on the foreclaws; male seventh (except in some *Brachineura*) and eighth abdominal tergites reduced to a strongly sclerotized, linear band (fig. 39), and the corresponding sternites

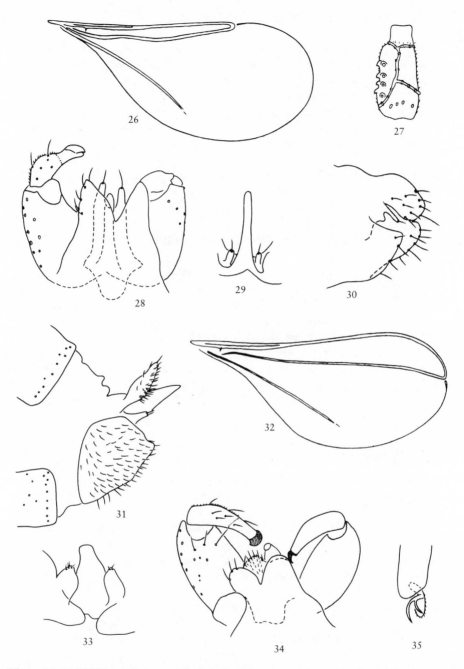

Figures 26–35. Brachineuridi. **26–30,** *Chrybaneura harrisoni*: **26,** wing; **27,** male third flagellomere; **28,** male genitalia (dorsal); **29,** aedeagus and parameres (ventral); **30,** female postabdomen (lateral). **31,** *Coccidomyia pennsylvanica*, female postabdomen (lateral). **32–34,** *Coccidomyia* sp., from epiphyte in Brazil: **32,** wing; **33,** aedeagus and flanking parameres (dorsal); **34,** male genitalia (dorsal); **35,** *Rhizomyia* sp., tarsal apex.

shorter and more weakly developed than preceding sternites; male genitalia are variable but always with free, short to long, cylindrical or stout (in *Coccidomyia* only), glabrous parameres with one to three terminal setae (figs. 29, 33); the female eighth tergite and sternite are weakly or not sclerotized, have no associated setae, and the postabdomen is not protrusible (figs. 30, 31); the female cerci are separate and the hypoproct wide and usually bilobed, and the ninth sternum is divided longitudinally, except in *Coccidomyia* in which it is entire and enlarged. *Brachineura* and *Epimyia* are unique in several ways: their flagellomeres are usually cylindrical and closely covered with longitudinal rows of scales (fig. 40), and the scutum is completely covered with scales between the regular setal rows.

Key to Neotropical Genera of Brachineuridi

1   R_5 much shorter than wing (fig. 26)................................. 2
     R_5 as long as wing (fig. 32) 4
2   Scutum with extensive areas devoid of scales.............. *Chrybaneura*
     Scutum completely covered with scales between setal rows 3
3   Aedeagus greatly modified, bulbous at base (figs. 37–38)........ *Epimyia*
     Aedeagus evenly cylindrical (fig. 29) .... *Brachineura*
4   Cu simple, evanescent before wing edge (fig. 26); tarsal claws toothed only on foreleg; palpus two segmented .... *Coccidomyia*
     Cu forked, attaining wing edge; all tarsal claws toothed; palpus four segmented .... *Rhizomyia*

Brachineura

Brachineura Rondani 1840: 16. Type species, *fuscogrisea* Rondani (mon.).
Brachyneura Kieffer 1894d: 201. Emendation.

Brachineura is known from 17 species outside of the Neotropics. The USNM has several undetermined specimens from this region. In this genus R_5 joins C at two-thirds to three-quarters wing length (as in fig. 26), Cu is simple and evanescent before the wing margin, and the body is extensively covered with scales.

Chrybaneura

Chrybaneura Gagné 1968b: 33. Type species, *harrisoni* Gagné (orig. des.).

I know *Chrybaneura* from several Neotropical species, only one of them described, and one undescribed Nearctic species. The species are associated with arthropod remains. Distinguishing characters are as follows: R_5 joins C at two-thirds to three-quarters wing length, and Cu is simple and evanescent before the wing margin (fig. 26); the antenna is sexually dimorphic, the nodes and necks longer in males (fig. 27) than in females; tarsal claws are all toothed; and the parameres are short or long and cylindrical (figs. 28, 29). A female postabdomen is illustrated in figure 30. *Chrybaneura* is distinguished from *Rhizomyia* by the shorter R_5 and Cu veins.

harrisoni Gagné 1968b: 34, figs. 1–6 (*Chrybaneura*); ♂, ♀. Holotype: ♂; Changuinola, Panama; USNM. Host: associated with spider egg cases and lepidopteran pupal exuviae. Distr.: Costa Rica, Ecuador, Panama.

Coccidomyia

Coccidomyia Felt 1911h: 45. Type species, *pennsylvanica* Felt (orig. des.).

This genus is known from one species, the Nearctic *Coccidomyia pennsylvanica*, associated with scale insects, but possibly feeding only on their remains. In addition, I know of an undescribed species associated with an epiphytic bromeliad in Bahia, Brazil. The wing of this genus (fig. 32) has the long R_5 of *Rhizomyia* but the simple and apically evanescent Cu of *Chrybaneura*. The tarsal claws and antennae of *Coccidomyia* are similar to both of those general, as is the postabdomen of both sexes, except for the enlarged female ninth sternum and triangular cerci (fig. 31) and the larger parameres (figs. 33, 34). Ref.: Harris 1968.

Epimyia

Epimyia Felt 1911h: 38. Type species, *carolina* Felt (orig. des.).

Epimyia is known from one described North American species. I know also of several undescribed Nearctic and Neotropical species represented in the USNM. These species have the same thick covering of scales and reduced wing venation of *Brachineura*, but the male genitalia are highly modified (figs. 36–38), especially the aedeagus, which is much enlarged and recurved basally.

Rhizomyia

Rhizomyia Kieffer 1898: 56. Type species, *perplexa* Kieffer (mon.).

This genus is known from 17 species in the Holarctic and Oriental Regions. The USNM collection has a specimen of an undetermined species from Mexico. *Rhizomyia* and *Chrybaneura* appear similar except in their very different wing venation, *Rhizomyia* with R_5 as long as the wing and Cu forked, *Chrybaneura* with its short R_5 and simple Cu.

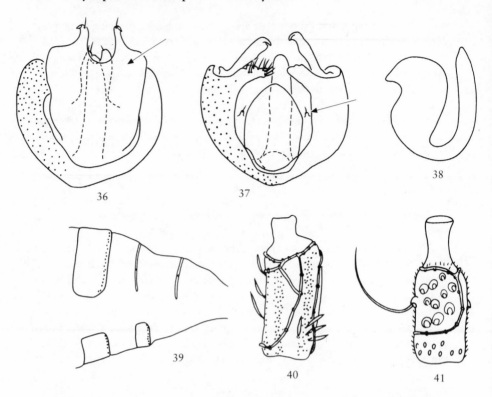

Figures 36–41. Brachineuridi. 36–40, *Epimyia* sp.: 36, male genitalia (dorsal; arrow: cerci); 37, male genitalia (ventral; arrow: a paramere); 38, aedeagus (lateral); 39, male sixth to eighth abdominal segments; 40, male third flagellomere. 41, *Coccidomyia pennsylvanica*, male third flagellomere.

SUPERTRIBE LASIOPTERIDI

Lasiopteridi are characterized by the presence in the males of parameres that clasp the aedeagus (fig. 72). The exception to this is *Calmonia* with its parameres splayed away from the aedeagus (fig. 77), but I consider this a secondary modification. The clasping parameres may lend rigidity to the aedeagus, which appears thin-walled in Lasiopteridi. The number of antennal flagellomeres ranges between 7 and 43 and is variable within a species and between sexes, except in some Ledomyiini, many of which have a regular number of 12 flagellomeres.

Lasiopteridi are divided here into six tribes, the Alycaulini, Camptoneuromyiini, Lasiopterini, Ledomyiini, Oligotrophini, and Trotteriini. They are treated below in alphabetical order. The Ledomyiini are possibly the sister group of the remainder of the supertribe. All the other tribes are monophyletic except the Oligotrophini, which are a large, paraphyletic or polyphyletic category that may be ancestral to each of the remaining four tribes.

The short R_5 (figs. 56, 61) often seen in this supertribe appears to have evolved separately many times. It occurs in each of the six tribes, primitively in four, and occurs also in the supertribes Brachineuridi and Cecidomyiidi, as well as in both supertribes of the Lestremiinae. A short R_5 appears to be correlated with short legs and antennae. The combined modifications may allow for stronger, more directed flight or for better maneuverability when walking.

Key to Tribes and Genera of Neotropical Lasiopteridi

Many of the genera keyed here are known from only one sex, others from sketchy original descriptions. Host information and larval characters, used in this key when helpful, are sometimes the only information available to separate genera. Too little is known about larvae generally to offer a key to larvae, but the generic synopses give available information and some illustrations of larvae and pupae also.

Angeiomyia is not included in the key. This genus was based only on a larva from leaf galls of *Hydrangea* (Saxifragaceae); although traditionally placed with Lasiopteridi and listed under that tribe here, *Angeiomyia* could belong instead to the Cecidomyiidi.

1   R_5 much shorter than wing, nearly parallel with R_1, and separated from the adjacent R_1 or C by less to little more than vein diameter (fig. 42) 2
        R_5 shorter to longer than wing (figs. 56, 70); if shorter, then bowed away from R_1, separated from the adjacent R_1 or C by appreciably more than vein diameter (fig. 56). 18

2(1)   Gonopods extremely elongate and narrow (fig. 60); distal half of ovipositor with dorsal, wedge-shaped, sclerotized processes dorsally, the lateral setulae usually enlarged and grasping (fig. 59); mostly inquilines . . . .
       . . . . Camptoneuromyiini, in part, *Meunieriella*
       Gonopods not extremely elongate and narrow (fig. 54); distal half of ovipositor clear, without dorsal wedge-shaped structrues, and lateral setulae not modified 3

3(2)   Male seventh tergite, occasionally eighth, setose along caudal margin and sclerotized at least posteriorly (figs. 52, 53); basal half of ovipositor with some dorsal striae interrupted and enlarged to form series of sclerotized, pigmented bumps (fig. 48) . . . . Alycaulini, 4
       Male seventh and eighth tergites unsclerotized posteriorly, setae and scales lacking or present only laterally on seventh tergite (fig. 65); ovipositor on basal half with uninterrupted posterodorsal striae or without striae. 16

4(3)   Female cerci separate or only partly fused (fig. 47) 5
       Female cerci fused, occasionally slightly notched apically (fig. 48) 7

5(4)   Larva with eight terminal papillae; in leaf vein swellings of *Mikania* (Asteraceae). . . . . *Alycaulus* in part
       Larva with two or four terminal papillae. 6

6(5)   Female cerci with short, thick sensory setae; larva with median tooth of spatula serrate and two terminal papillae; in leaf spot on *Smilax* (Smilacaceae). . . . . *Smilasioptera*
       Female cerci without such setae; larva with median tooth of spatula simple and four terminal papillae; in swollen veins of *Pseudocalymma* (Bignoniaceae). . . . . *Lobolasioptera*

7(4)   Pupal integument brown, brittle; in leaf blisters of various plants, including *Baccharis* and *Eupatorium* (Asteraceae) . . . . *Geraldesia*
       Pupal integument whitish, pliable. 8

8(7)   In leaf blister galls of Astereae (Asteraceae); adult palpi reduced to one to three segments . . . . *Asteromyia*
       If in leaf blister galls, not on Asteraceae; palpi various 9

9(8)   In stem galls of *Baccharis* (Asteraceae); pupa with long antennal horns; larval spatula reduced, with one lateral papilla per side. . . . . *Baccharomyia*
       If on *Baccharis*, not on stems; pupa usually with short antennal horns; larval spatula, if reduced, with two or more lateral papillae 10

10(9) In swollen leaf veins of *Mikania* (Asteraceae); spatula with three teeth, the middle longest . . . . *Alycaulus* in part
       If in swollen leaf veins, on other hosts; spatula various. 11

11(10) In leaf spot on *Achatocarpus* (Phytolaccaceae); larva with short, wide spatula and four terminal papillae; female fused cerci wide, bulbous. . . . .
       . . . . *Atolasioptera*
       On other hosts; larva and female various 12

12(11) Leaf gall on *Coccoloba*, inquiline of an undescribed gall maker; larval spatula three-toothed, the median tooth wide, slightly notched . . . . . *Marilasioptera*
       On other hosts; larval spatula various. 13

13(12) In stem swellings of *Iresine* (Amaranthaceae); larva with short, wide spatula, two lateral papillae per side, and four terminal papillae. . . . . . . *Epilasioptera*
       Not on *Iresine*; larva various 14

14(13) In stem, petiole and leaf midrib swellings of *Gilibertia* (Araliaceae) and *Pseudocalymma* (Bignoniaceae); female fused cerci broad, slightly convex distally . . . . *Brachylasioptera*
       In other hosts; female various. 15

15(14) From grass stems; terminal segment of larva with caudal pair of lobes . . . . .
       . . . . *Chilophaga*
       Not from grass stems; terminal segment of larva without caudal lobes . . . . .
       . . . . *Neolasioptera*

16(3) Basal half of ovipositor with lateral group of numerous, strong, modified setae, the fused cerci with hooked setae (fig. 64); larvae with three or four pairs of terminal papillae, the setae equally short; gall makers . . . .
       . . . . Lasiopterini, 17
       Basal half of ovipositor without such setae; larvae with four pairs of terminal papillae, one pair noticeably shorter than the other three; inquilines 18

17(16) R_5 more than three-quarters wing length (fig. 63); palpus one-segmented
.... *Isolasioptera*

R_5 less than two-thirds wing length; palpus four-segmented .... *Lasioptera*

18(1, R_5 shorter than wing, joining C noticeably in front of wing apex (fig. 68)
16) 19

R_5 as long as wing, joining C at or behind wing apex (fig. 70) 28

19(18) Mid- and hindfemora successively larger than forefemur (fig. 79)..........
.... Trotteriini, *Trotteria*

All femora of approximately equal size (fig. 68)...................... 20

20(19) Scutum completely covered with scales between setal rows 21

Scutum without scales outside vicinity of setal rows 26

21(20) First two flagellomeres not connate; ovipositor bilaterally flattened, cerci
fused; in flower buds of *Kielmeyera* (Clusiaceae).......................
.... Oligotrophini in part, *Arcivena*

First two flagellomeres connate (fig. 57); ovipositor cylindrical, cerci fused or
separate; inquilines .... Camptoneuromyiini, 22

22(21) R_5 usually conspicuously bowed (fig. 56); gonocoxites tapering from wide
base to narrower apex; larva with six lateral papillae on each side of spatula
(fig. 2)... 23

R_5 usually nearly straight, parallel with R_1; gonocoxites cylindrical; larva
with three or four lateral papillae on each side of spatula.............. 24

23(22) Some setae of male flagellomeres at least twice length of flagellomere ......
.... *Camptoneuromyia*

Setae of male flagellomeres not appreciably longer than length of flagellomere
.... *Salvalasioptera*

24(22) Female fused cerci with sparse and unevenly distributed setae; male unknown; larval spatula with one tooth and thoracic pleural papillae with setae
.... *Incolasioptera*

Female fused cerci with many and evenly distributed setae; larval spatula with
two teeth and thoracic pleural papilla without setae 25

25(24) Larva with three lateral papillae on each side of spatula... *Domolasioptera*

Larva with four lateral papillae on each side of spatula... *Copinolasioptera*

26(20) Claws thin, curved at midlength (fig. 66); paramere partially divided, one part
smooth, the other covered with setulae (fig. 67); female cerci separate .....
.... Ledomyiini, *Ledomyia* in part

Claws robust, curved beyond midlength; paramere simple, completely covered with setulae (fig. 72); female cerci fused (fig. 78) or separate (fig. 75) ..
.... Oligotrophini in part, 27

27(26) Parameres not clasping aedeagus (fig. 77); female cerci separate but tiny (fig.
75); palpus one-segmented; from leaf galls on *Ficus* (Moraceae) ....
.... *Calmonia*

Parameres clasping aedeagus (fig. 72); female cerci fused (fig. 69); palpus
usually four-segmented; on other plants ....

. . . . *Dasineura* (see also *Austrolauthia*)

28(18) Antenna with 8–10 or 12 flagellomeres, usually regular in number; parameres subdivided into two lobes, one glabrous the other setulose; female cerci
separate. . . . . Ledomyiini, *Ledomyia* in part
Antenna with 7–42 flagellomeres, usually more than 10 and irregular in
number within genus and species; if with 7–13, parameres almost completely
setulose and undivided; female cerci separate or fused . . . .
. . . . Oligotrophini remainder, 29

29(28) Tarsal claws simple. 30
Tarsal claws toothed. 34

30(29) Female cerci separate; male unknown; pupa with elongate antennal horns;
from spherical leaf galls on *Psychotria* (Rubiaceae) . . . . *Apodiplosis*
Female cerci fused (fig. 78); pupa various; from other hosts 31

31(30) Female flagellomeres each girdled with six circumfila; male reportedly without parameres; from spherical stem galls on *Lycium* (Solanaceae) . . . .
. . . . *Lyciomyia*
Female flagellomeres each girdled with no more than two circumfila; male
with parameres; on other hosts. 32

32(31) Ovipositor short, barely protrusible; from bud galls on *Clusia* (Clusiaceae)
. . . . *Uleia*
Ovipositor elongate-protrusible; from other plants 33

33(32) From elongate, complex leaf and stem galls on *Lippia* (Verbenaceae) . . . . . .
. . . . *Pseudomikiola*
From other plants . . . . *Rhopalomyia* in part

34(29) Palpus one- to three-segmented; if three, third segment much shorter than
second. 35
Palpus three- to four-segmented, last segment about as long as the preceding
. . . . 39

35(34) Female cerci separate; male unknown; from cylindrical leaf gall on *Serjanea*
(Sapindaceae). . . . . *Haplopalpus*
Female cerci fused; from other hosts 40

36(35) Antenna with more than 30 flagellomeres; Cu forked near basal third; first
tarsomeres with spur; from fruit of *Ficus* (Moraceae). . . . . *Ficiomyia*
Antenna with fewer than 20 flagellomeres; Cu forked beyond half its length;
first tarsomeres without spur; from other hosts 37

37(36) Ovipositor short, barely protrusible; from *Guarea* (Meliaceae). . . . .
. . . . *Guarephila*
Ovipositor elongate; from other hosts. 38

38(37) Palpus one-segmented; from *Baccharis* (Asteraceae) . . . . *Scheueria*
Palpus two-segmented; from *Chrysanthmum* (Asteraceae). . . . .
. . . . *Rhopalomyia* in part

39(34) Female with separate cerci; from *Colliguaya* (Euphorbiaceae) . . *Promikiola*
Female with fused cerci; from various hosts 40

40(39) From galls on *Pernettya* (Ericaceae) 41

Tribe Alycaulini

The large tribe Alycaulini is restricted to the Americas. Of 19 genera, nine are endemic to the Neotropics, seven to the Nearctic, and three occur in both regions. The Neotropical Region has a total of 86 species known for this tribe, while the Nearctic has 97. A few species span both regions, extending from eastern North America to Central America, and one even from southern North America to Argentina. Most are primary plant feeders causing simple swellings on stems, tendrils, or petioles, or live in flowers or fruit of Asteraceae. Many evidently feed on symbiotic fungi. A few are inquilines in galls of other cecidomyiids. Except for *Neolasioptera*, most of the genera have only one to several species and are restricted to a particular host genus.

Most of the 86 species of Neotropical Alycaulini are in *Neolasioptera*, a catchall genus. *Baccharomyia* has four species, *Geraldesia* three, *Alycaulus* and *Brachylasioptera* two each, and the remaining seven genera include one species each. In addition to the described species, most of the 39 undetermined species from simple stem and petiole galls referred to in the host list probably belong to the Alycaulini.

Unique, derived characters of Alycaulini involve modifications of the ovipositor and male posterior abdominal tergites not found elsewhere in Cecidomyiinae. The dorsal, longitudinal ridges of the basal half of the ovipositor are interrupted and form rows of sclerotized processes (fig. 48). These may be used to abrade plant tissue when laying eggs. The male seventh tergite and occasionally the eighth have the posterior margin sclerotized and setose (figs. 52, 53). Reduction in these two tergites begins at the anterior border, whereas in the Lasiopterini, Camptoneuromyiini, and Oligotrophini, any reduction proceeds from the posterior border of the tergites (fig. 62). A further possibly apomorphic character in this tribe is the presence of a tiny tooth below the large basal tooth of the tarsal claws. The additional tooth is not present in all species, possibly due to secondary loss. The female cerci of at least one *Alycaulus* species are separate, the apparently primitive state for the tribe. Characters that are shared with other tribes of Lasiopteridi are the short antennae of both sexes (figs. 42, 43), the short R_5 that is also straight and almost parallel with C (fig. 42), and the robust, toothed claws. Most larvae show adaptations for cutting through wood: these are the cylindrical body shape and strongly sclerotized spatula (fig. 49). The larval lateral papillae are reduced to four or fewer per side (fig. 50), and the terminal papillae may be reduced from eight (fig. 51) to six, four, or two. The robust spatula and reduced lateral papillae occur also in other species that live in woody stem galls,

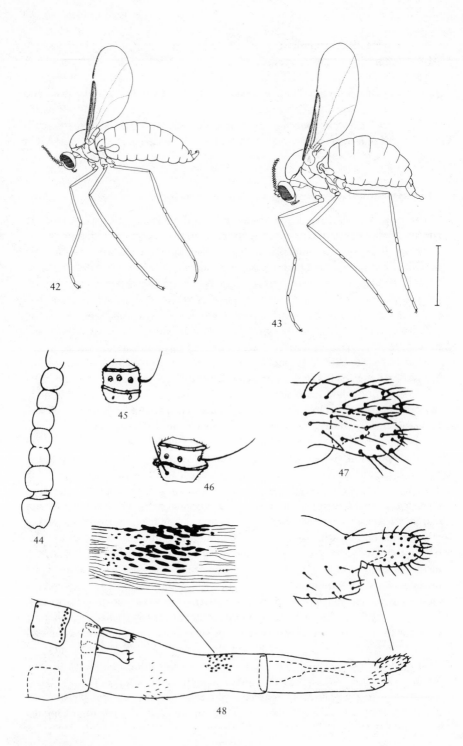

42

43

44

45

46

47

48

including the Lasiopterini (Möhn 1955a), the Oligotrophini (Sylvén 1975), and the genus *Bruggmanniella* of the supertribe Cecidomyiidi (q. v.)

Alycaulus

Alycaulus Rübsaamen 1916a: 476. Type species, *mikaniae* Rübsaamen (mon.).

The two described species of this Neotropical genus cause leaf vein swellings on *Mikania* (Asteraceae). They have similar larvae with a robust, wide, three- to several-pointed spatula and eight short terminal papillae. The pupa of *A. trilobatus* has strong antennal horns, but that of *A. mikaniae* is unknown. The female cerci of the two species are very different: they are separate in *A. mikaniae*, but fused in *A. trilobatus*. Either that difference is unimportant or the two species should not belong in the same genus. The separate cerci are certainly the primitive character state for the tribe. I have seen cerci similar to those of *A. mikaniae* on an apparently new species of *Alycaulus* that I reared from leaf galls on *Andira parviflora* in Brazil (fig. 47). The male of the type species differs from other alycaulines in its long and narrow gonostyli, but the males of *A. trilobatus* and the species from *Andira* are still unknown.

mikaniae Rübsaamen 1916a: 476, figs. 54–59 (*Alycaulus*); ♂, ♀, pupa, larva. Syntypes: Auristella on Rio Acre, Amazonas, Brazil; ZMHU. Host: *Mikania* sp. (Asteraceae). Distr.: Brazil.
trilobatus Möhn 1964b: 583, figs. 145, 146 (*Alycaulus*); larva. Holotype: NE Cojutepeque, Dept. Cuscatlán, El Salvador; SMNS. Hosts: *Mikania cordifolia*, *M. micrantha*, and *Mikania* sp. (Asteraceae). Distr.: Colombia, Costa Rica, El Salvador. Ref.: Wünsch 1979.

Asteromyia

Asteromyia Felt 1910c: 348. Type species, *Cecidomyia carbonifera* Felt (orig. des.) = *Asteromyia euthamiae* Gagné.

This genus is known from eight Nearctic species, one of which also occurs in South America on a widespread weed, *Conyza canadensis* (Asteraceae). All form blister galls on leaves and green stems of Astereae (Asteraceae). Adults have reduced palpi of

---

Figures 42–48. Alycaulini. **42, 43,** *Neolasioptera lathami* (scale line = 0.5 mm): **42,** male; **43,** female. **44, 45,** *Neolasioptera portulacae*: **44,** male scape, pedicel, and base of flagellum; **45,** male third flagellomere. **46,** *Baccharomyia ornaticornis*, male third flagellomere. **47,** *Alycaulus* sp. from *Andira*, Brazil, female cerci and hypoproct (dorsolateral). **48,** *Neolasioptera cruttwellae*, female postabdomen with details of dorsal ridges of eighth segment and of fused cerci (dorsolateral).

one to three segments and a short ovipositor with short, broad, fused cerci. The spatula is variably shaped in the genus, but is short and wide in *A. modesta*. Larvae have only three lateral papillae on each side of the spatula and four terminal papillae. Refs.: Gagné 1968c, 1969b.

modesta Felt 1907b: 163 (*Choristoneura*); ♂, ♀. Syntypes: Albany, New York, USA; Felt Coll. Host: *Conyza canadensis* (Asteraceae). Distr.: Argentina, Brazil; USA (widespread). Ref.: Gagné 1968c.

Atolasioptera

Atolasioptera Möhn 1975: 8. Type species, *calida* Möhn (orig. des.).

The one known species lives in circular blister galls of *Achatocarpus* (Phytolaccaceae). Adults have a three-segmented palpus, and the female a short ovipositor with short, broad, fused cerci. The larva has a short, broad, four-toothed spatula, four lateral papillae per side, and four terminal papillae. Differences between this genus and *Asteromyia* are minor: known *Asteromyia* species have only three instead of four lateral papillae per side and occur on Asteraceae.

calida Möhn 1975: 8, pl. 2, figs. 10–19 (*Atolasioptera*); ♀, pupa, larva. Holotype: ♀; San Diego, E La Libertad, La Libertad, El Salvador; SMNS. Host: *Achatocarpus nigricans* (Phytolaccaceae). Distr.: El Salvador.

Baccharomyia

Baccharomyia Tavares 1917b: 129. Type species, *ramosina* Tavares (orig. des.).

This genus is known from four described species from Argentina and southern Brazil, three of them newly removed from *Neolasioptera*. All live in stem swellings of *Baccharis* (Asteraceae). Adults have one- or two-segmented palpi, enlarged antennal pedicels, and flagellomeres that are characteristically widened below the circumfila (fig. 46). The female has a longitudinally divided eighth tergite, a long ovipositor, and sensory setae on the bulbous, fused cerci. Pupae have very long antennal horns. Larvae have a reduced spatula or none, one or two lateral setae per side, and four terminal papillae. Differences between two recently collected series I have determined as *B. ornaticornis* and *B. ramosina* indicate that this genus is very diverse and the species are closely host-specific.

cordobensis Kieffer and Jörgensen 1910: 363 (*Lasioptera*); larva. New combination. Syntypes: Arias and La Carlota, Córdoba, Argentina; presumed lost. Host: *Baccharis coridifolia* (Asteraceae). Distr.: Argentina.
interrupta Kieffer and Jörgensen 1910: 375 (*Lasioptera*); ♂, ♀, pupa, larva. New

combination. Syntypes: Chacras de Coria, Mendoza, Argentina; presumed lost. Host: *Baccharis subulata* (Asteraceae). Distr.: Argentina.

ornaticornis Kieffer and Jörgensen 1910: 368 (*Lasioptera*); ♂, ♀, pupa, egg. New combination. Syntypes: Chacras de Coria, Mendoza, Argentina; presumed lost. Host: *Baccharis salicifolia* (Asteraceae). Distr.: Argentina.

ramosina Tavares 1917b: 130, fig. B (1–6) (*Baccharomyia*) ♀, pupa, larva. Syntypes: Novo Friburgo, Rio de Janeiro, Brazil; presumed lost. Host: *Baccharis trimera*. Distr.: Argentina, Brazil.

Brachylasioptera

Brachylasioptera Möhn 1964b: 576. Type species, *rotunda* Möhn (orig. des.).

This genus is known from two Neotropical species that form galls on plants of two different host families. *Brachylasioptera gilibertiae*, based on the larva only, was only provisionally placed here because of the similarity of the spatula to that of the type species. Adult palpi are two- or three-segmented, the ovipositor is short and its fused cerci broad, slightly concave caudally. The pupal antennal horns are long and sclerotized. The larva has a robust spatula, three lateral papillae per side, and six terminal papillae. It is noteworthy that three other species of Alycaulini besides *B. rotunda* occur in various simple swellings of the same host.

gilibertiae Möhn 1964b: 579, fig. 129 (*Brachylasioptera*); larva. Holotype: vic. Los Chorros, La Libertad, El Salvador; SMNS. Host: *Gilibertia arborea* (Araliaceae). Distr.: El Salvador.

rotunda Möhn 1964b: 576, figs. 113–128 (*Brachylasioptera*); ♂, ♀, pupa, larva. Holotype: ♂; San Diego, E La Libertad, La Libertad, El Salvador; SMNS. Host: *Pseudocalymma macrocarpa* (Bignoniaceae). Distr.: El Salvador.

Chilophaga

Chilophaga Gagné 1969b: 1,358. Type species, *Lasioptera colorati* (orig. des.).

This genus is known from four North American species, one of them from southern Florida. All feed in grass culms. Adult legs are conspicuously banded with alternating light and dark bands. The palpi are four-segmented, the ovipositor is long and its fused cerci elongate and narrow. The larva has a pair of conical processes on the terminal segment.

gyrantis Gagné *in* Gagné and Stegmaier 1971: 335, figs. 1–8 (*Chilophaga*); ♂, ♀, pupa, larva. Holotype: ♂; Miami, Florida, USA; USNM. Host: *Aristida gyrans* (Poaceae). Distr.: USA (Florida).

Epilasioptera

Epilasioptera Möhn 1964b: 573. Type species, *iresinis* Möhn (orig. des.).

The single known species lives in stem galls of *Iresine* (Amaranthaceae). Adults have two-segmented palpi and enlarged antennal scapes. The larva has a broad, short spatula with two widely separated teeth, two lateral papillae per side, and four terminal papillae.

iresinis Möhn 1964b: 573, figs. 98–112 (*Epilasioptera*); ♂, ♀, pupa, larva. Holotype: ♂; San Salvador, El Salvador; SMNS. Host: *Iresine calea* (Amaranthaceae). Distr.: El Salvador.

Geraldesia

Geraldesia Tavares 1917b: 134. Type species, *eupatorii* Tavares (orig. des.).

This Neotropical genus is known from three described species from blister leaf galls, one each on *Baccharis* and *Eupatorium* in Asteraceae and one on Polygonaceae. I know of at least one other undescribed species from *Baccharis* leaves in Argentina. Palpi are three-segmented, and the antennal pedicel is enlarged. The ovipositor is short, and its fused cerci are short and broad. Pupae are unique among Alycaulini because they are brown and brittle. The pupal antennal horns are elongate and pointed, and the last abdominal segment is two-lobed. Larvae may or may not have a spatula and have only two or three lateral papillae per side and four terminal papillae. Ref.: Möhn 1975.

cumbrensis Möhn 1975: 5, pl. 1, figs. 19–26, pl. 2, figs. 1–9 (*Geraldesia*): ♂, ♀, pupa, larva. Holotype: ♂; Hacienda San José, vic. San José, Santa Ana, El Salvador; SMNS. Host: *Baccharis trinervis* (Asteraceae). Distr.: El Salvador.
eupatorii Tavares 1917b: 134, fig. C (1–5) (*Geraldesia*); ♀, pupa. Syntypes: Rio de Janeiro, Brazil; presumed lost. Host: *Eupatorium* sp. (Asteraceae). Distr.: Brazil, El Salvador. Ref.: Möhn 1975.
polygonarum Wünsch 1979: 111, pl. 36 (*Geraldesia*); ♂, ♀, pupa, larva. Holotype: ♂; Cerro San Fernando, Colombia; SMNS. Host: *Triplaris/Symmeria/Ruprechtia* sp. (Polygonaceae). Distr.: Colombia.

Lobolasioptera

Lobolasioptera Möhn 1964b: 580. Type species, *media* Möhn (orig. des.).

The single known species causes swellings in leaf veins of *Pseudocalymma* (Bignoniaceae), as do three other species of Alycaulini. Adults have four-segmented palpi, and the female has a short ovipositor with a deep division between the fused cerci. The larva has a robust spatula with three lateral papillae and four terminal papillae.

media Möhn 1964b: 581, figs. 130–144 (*Lobolasioptera*); ♂, ♀, pupa, larva. Holotype: ♂; vic. San Diego, E La Libertad, La Libertad, El Salvador; SMNS. Host: *Pseudocalymma macrocarpa* (Bignoniaceae). Distr.: El Salvador.

Marilasioptera

Marilasioptera Möhn 1975: 62. Type species, *tripartita* Möhn (orig. des.).

This genus contains one species, an inquiline in leaf galls of another cecidomyiid on *Coccoloba* (Polygonaceae). *Marilasioptera* is keyed and placed here with the Alycaulini, although Möhn (1975) compared it with *Meunieriella* (Camptoneuromyiini) and relatives because the species is an inquiline. Adults have four-segmented palpi, and the female has elongate fused cerci. The larva does not resemble other Camptoneuromyiini because of its more robust, three-toothed spatula and the presence of only six terminal papillae. All other genera of that tribe have a clove-shaped spatula (except for *Incolasioptera* with its single-toothed spatula) and eight terminal papillae.

tripartita Möhn 1975: 63, pl. 9, figs. 18–25 (*Marilasioptera*); ♀, larva. Holotype: ♀; vic. La Libertad, La Libertad, El Salvador; SMNS. Host: ex leaf galls of an undetermined cecidomyiid on *Coccoloba* sp. (Polygonaceae). Distr.: El Salvador.

Neolasioptera

Neolasioptera Felt 1908b: 330. Type species, *Lasioptera vitinea* Felt (subseq. des. Coquillett 1910: 575).
Physalidicola Brèthes 1917a: 240. Type species, *argentata* Brèthes (orig. des.). New synonymy.
Luisieria Tavares 1922: 43. Type species, *fariae* Tavares (orig. des.).
Neurolasioptera Brèthes 1922: 138. Type species, *baezi* Brèthes (orig. des.). New synonymy.
Dilasioptera Möhn 1964b: 570, as subgenus of *Neolasioptera*. Type species, *Neolasioptera serrata* Möhn (orig. des.).

This genus is known from 68 Neotropical and 64 Nearctic species. Four of the species span both regions. *Neolasioptera* is a diverse, catchall group of species that lacks the modifications distinguishing the other, more narrowly defined genera of Alycaulini. Figures 42–45, 48–55 show adults, a larva, and various anatomical parts of *Neolasioptera* species.

Physalidicola Brèthes and *Neurolasioptera* Brèthes show nothing to distinguish them generically from many species of *Neolasioptera*. I have seen Brèthes' specimens

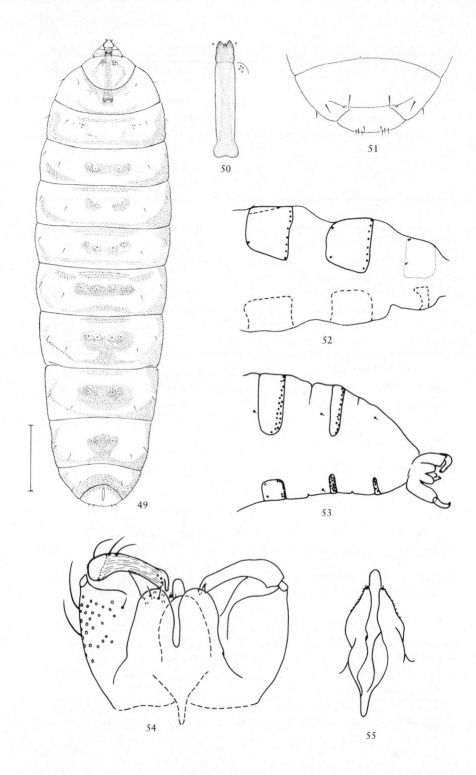

49

50

51

52

53

54

55

of both genera. *Neurolasioptera* would be distinguishable from *Neolasioptera* if it had simple tarsal claws as originally described, but no legs now remain on any of the three syntypes of *N. baezi.* If the tarsal claws were untoothed, they would be the only untoothed claws in Alycaulini. Brèthes described the empodia of *Physalidicola* as almost absent, but his syntypes have empodia that are as long as the claws.

aeschynomenis Brèthes 1918: 312 (*Lasioptera*); ♀. Syntypes: Buenos Aires, Argentina; INTAC. Host: *Aeschynomene montevidensis* (Fabaceae). Distr.: Argentina.

amaranthi Möhn 1964a: 47, figs. 1–14 (*Neolasioptera*); ♂, ♀, pupa, larva. Holotype: ♂; San Salvador, El Salvador; SMNS. Host: *Amaranthus spinosus* (Amaranthaceae). Distr.: El Salvador.

aphelandrae Möhn 1964a: 90, figs. 220–221 (*Neolasioptera*); larva. Holotype: larva; N El Cimarrón, La Libertad, El Salvador; SMNS. Host: *Aphelandra deppeana* (Acanthaceae). Distr.: El Salvador.

argentata Brèthes 1917a: 240 (*Physalidicola*); ♂, ♀. New combination. Syntypes: Buenos Aires, Argentina; Instituto Nacional de Tecnologia Agropecuaria, Castellar, Buenos Aires. Host: *Physalis viscosa* and *Physalis* sp. (Solanaceae). Distr.: Argentina.

baezi Brèthes 1922: 138 (*Neurolasioptera*); ♂, ♀. New combination. Syntypes: Paraná, Entre Ríos, Argentina; INTAC. Host: *Teucrium inflatum* (Lamiaceae). Distr.: Argentina.

borreriae Möhn 1964a: 93, figs. 230–238 (*Neolasioptera*); ♀, pupa, larva. Holotype: ♀; E San Marcos, San Salvador, El Salvador; SMNS. Host: *Borreria verticillata* (Rubiaceae). Distr.: El Salvador.

brickelliae Möhn 1964b: 560, figs. 42–52 (*Neolasioptera*); ♂, ♀, pupa, larva. Holotype: ♂; road to La Palma, ca. 15 km before La Palma, Chalatenango, El Salvador; SMNS. Host: *Brickellia pacayensis* (Asteraceae). Distr.: El Salvador.

caleae Möhn 1964b: 556, figs. 15–29 (*Neolasioptera*): ♂, ♀, pupa, larva. Holotype: ♂; near Los Chorros, La Libertad, El Salvador; SMNS. Host: *Calea integrifolia* (Asteraceae). Distr.: El Salvador.

camarae Möhn 1964b: 565, figs. 66–73 (*Neolasioptera*); ♂, larva. Holotype: ♂; San Diego, E La Libertad, La Libertad, El Salvador; SMNS. Host: *Lantana camara* (Verbenaceae). Distr.: Colombia, El Salvador. Ref.: Wünsch 1979.

capsici Möhn 1964a: 78, figs. 161–170 (*Neolasioptera*); ♀, pupa, larva. Holotype: ♀; near San Diego, E La Libertad, La Libertad, El Salvador; SMNS. Host: *Capsicum baccatum* (Solanaceae). Distr.: El Salvador.

Figures 49–55. Alycaulini. **49–51,** *Neolasioptera fontagrensis*: **49,** larva (ventral; scale line = 0.5 mm); **50,** spatula and associated papillae; **51,** terminal segments (dorsal). **52,** *Neolasioptera martelli*, male sixth through eighth abdominal segments (dorsolateral). **53–55,** *Neolasioptera vitinea*: **53,** male postabdomen from sixth abdominal segment (dorsolateral except genitalia dorsal); **54,** male genitalia (dorsal); **55,** aedeagus and flanking parameres (ventral).

celtis Möhn 1964b: 553, figs. 1–14 (*Neolasioptera*); ♂, ♀, pupa, larva. Holotype: ♂; near San Diego, E La Libertad, La Libertad, El Salvador; SMNS. Host: *Celtis iguanea* (Ulmaceae). Distr.: El Salvador.

cerei Rübsaamen 1905: 81 (*Lasioptera*); ♀. Syntypes: Cabo Frio, Rio de Janeiro, Brazil; ZMHU. Host: *Cereus setaceus* (Cactaceae). Distr.: Brazil.

cestri Möhn 1964a: 80, figs. 171–184 (*Neolasioptera*); ♂, ♀, pupa, larva. Holotype: ♂; Hacienda San José, vic. San José, Santa Ana, El Salvador; SMNS. Hosts: *Cestrum auriantiacum* and *C. lanatum* (Solanaceae). Distr.: El Salvador.

cimmaronensis Möhn 1975: 62, pl. 11, figs. 3–12 (*Neolasioptera*); ♀, pupa, larva. Holotype: ♂; N El Cimmarón, La Libertad, El Salvador; SMNS. Host: a *Vernonia*-like Asteraceae. Distr.: El Salvador.

cissampeli Möhn 1964a: 55, figs. 44–54 (*Neolasioptera*); ♀, pupa, larva. Holotype: ♀; San Salvador, El Salvador; SMNS. Host: *Cissampelos pareira* (Menispermaceae). Distr.: El Salvador.

clematicola Möhn 1964a: 52, figs. 29–43 (*Neolasioptera*); ♂, ♀, pupa, larva. Holotype: ♂; NW Ciudad Arce, Santa Ana, El Salvador; SMNS. Host: *Clematis dioica* (Ranunculaceae). Distr.: El Salvador.

combreti Möhn 1964a: 68, figs. 112–124 (*Neolasioptera*); ♂, ♀, pupa, larva. Holotype: ♂; N El Cimarrón, La Libertad, El Salvador; SMNS. Host: *Combretum farinosum* (Combretaceae). Distr.: El Salvador.

compositarum Möhn 1964b: 563, figs. 53–65 (*Neolasioptera*); ♂, ♀, pupa, larva. Holotype: ♂; San Salvador, El Salvador; SMNS. Hosts: *Hymenostephium cordatum* and *Tridax procumbens* (Asteraceae). Distr.: El Salvador, Guatemala.

cordiae Möhn 1964a: 71, figs. 127–134 (*Neolasioptera*); ♀, pupa, larva. Holotype: ♀; SE Ciudad Arce, La Libertad, El Salvador; SMNS. Host: *Cordia cana* (Boraginaceae). Distr.: El Salvador.

cruttwellae Gagné 1977b: 115, figs. 9–14 (*Neolasioptera*); ♂, ♀, pupa, larva. Holotype: ♀; Simla, Trinidad; USNM. Hosts: *Chromolaena odorata* and *C. ivaefolia* (Asteraceae). Distr.: Colombia, Trinidad.

odorati Wünsch 1979; 100, pl. 34 (*Neolasioptera*); ♂, pupa, larva. New synonym. Holotype: ♂; Colombia; SMNS. Host: *Chromolaena odorata*.

cusani Wünsch 1979: 104, pl. 34 (*Neolasioptera*); larva. Holotype: larva; Donama, Colombia; SMNS. Host: an undetermined Asteraceae. Distr.: Colombia.

dentata Möhn 1964b: 567, figs. 74–75 (*Neolasioptera*); larva. Holotype: larva; vic. San José, Santa Ana, El Salvador; SMNS. Host: *Eupatorium* sp. (Asteraceae). Distr.: El Salvador.

diclipterae Wünsch 1979: 78, pls. 27, 28 (*Neolasioptera*); ♂, pupa, larva. Holotype: ♂; Rio Manzanares, Colombia; SMNS. Host: *Dicliptera assurgens* (Acanthaceae). Distr.: Colombia.

donamae Wünsch 1979: 97, pl. 33 (*Neolasioptera*); larva. Holotype: larva; Donama, Colombia; SMNS. Host: *Iresine angustifolia* (Amarantaceae). Distr.: Colombia.

erigerontis Felt 1907a: 163 (*Choristoneura*); ♂, ♀. Syntypes: Albany, New York, USA; Felt Coll. Host: *Conyza canadensis* (Asteraceae). Distr.: El Salvador; USA (widespread). Ref.: Möhn 1964b.

erythroxyli Möhn 1964a: 62, figs. 86–94 (*Neolasioptera*); ♀, pupa, larva. Holotype: ♀; N Los Cóbanos, Sonsonate, El Salvador; SMNS. Host: *Erythroxylum mexicanum* (Erythroxylaceae). Distr.: El Salvador.

exeupatorii Gagné. New name for *eupatorii* Cook. Host: *Eupatorium villosum* (Asteraceae). Distr.: Cuba.

eupatorii Cook 1909: 144 (*Cecidomyia*); gall. New combination. Preoccupied by *Cecidomyia eupatorii* Felt 1907. Syntypes: gall; Cuba; depos. unknown.

exigua Möhn 1964a: 82, figs. 185–193 (*Neolasioptera*); ♀, pupa, larva. Holotype: ♀; N El Cimarrón, La Libertad, El Salvador; SMNS. Host: *Solanum umbellatum* (Solanaceae). Distr.: El Salvador.

fariae Tavares 1922: 44, figs. 11–16 (*Luisieria*); ♂, ♀, pupa, larva. Syntypes: Retiro, near Bahia, Brazil; presumed lost. Host: double-convex leaf gall of unknown plant. Distr.: Brazil. Ref.: Houard 1933.

frugivora Gagné 1977b: 116, figs. 15–19 (*Neolasioptera*); ♂, ♀, pupa, larva. Holotype: ♀; Simla, Trinidad; USNM. Hosts: *Chromolaena odorata* and *Fleischmannia microstemon* (Asteraceae). Distr.: Trinidad.

grandis Möhn 1964a: 87, figs. 209, 210 (*Neolasioptera*); larva. Holotype: larva; near Rio San Felipe, San Vicente, El Salvador; SMNS. Host: *Pseudocalymma macrocarpa* (Bignoniaceae). Distr.: El Salvador.

heliocarpi Möhn 1964a: 66, pl. 9, fig. 109 (*Neolasioptera*); larva. Holotype: San Juan Higenio, Cumbre, La Libertad, El Salvador; SMNS. Host: *Heliocarpus donnellsmithii* (Tiliaceae). Distr.: El Salvador.

heterothalami Kieffer and Jörgensen 1910: 399 (*Lasioptera*); ♂, ♀, pupa, larva, egg. Syntypes: LaPaz, Mendoza, and Cancete, San Juan, Argentina; presumed lost. Host: *Baccharis artemisioides* and *B. spartioides* (Asteraceae). Distr.: Argentina.

hibisci Felt 1907b: 155 (*Choristoneura*); ♂, ♀. Syntypes: Staten I., New York, USA; Felt Coll. Hosts: *Hibiscus moscheutos** (Malvaceae) and *Cucumis melo* (Cucurbitaceae). Distr.: Honduras; USA (southeastern).

hyptis Möhn 1964a: 74, figs. 137–151 (*Neolasioptera*); ♂, ♀, pupa, larva. Holotype: ♂; NE Sitio del Niño, La Libertad, El Salvador; SMNS. Host: *Hyptis suaveolens* (Lamiaceae). Distr.: El Salvador. Ref.: Möhn 1975.

indigoferae Möhn 1964a: 55, figs. 70–85 (*Neolasioptera*); ♂, ♀, pupa, larva. Holotype: ♂; San Salvador, El Salvador; SMNS. Host: *Indigofera suffruticosa* (Fabaceae). Distr.: El Salvador.

ingae Möhn 1964a: 57, figs. 55–69 (*Neolasioptera*); ♂, ♀, pupa, larva. Holotype: ♀; Hacienda San Julián, SE San Julián, Cumbre, La Libertad, El Salvador; SMNS. Host: *Inga leptoloba* and *I. spuria* (Fabaceae). Distr.: El Salvador.

iresinis Möhn 1964a: 50, figs. 15–28 (*Neolasioptera*); ♂, ♀, pupa, larva. Holotype: ♂; San Salvador, El Salvador; SMNS. Host: *Iresine celosia* (Amaranthaceae). Distr.: El Salvador.

lantanae Tavares 1922: 48 (*Luisieria*); larva. Syntypes: Roça da Madre de Deus, near Bahia, Brazil; presumed lost. Host: *Lantana* sp. (Verbenaceae). Distr.: Brazil. Refs.: Tavares 1917b, pl. X, fig. 8 (given as 9 in error); Tavares 1918a: 29 (his item #12).

lapalmae Möhn 1964a: 94, figs. 239–252 (*Neolasioptera*); ♂, ♀, pupa, larva.

Holotype: ♂, SE La Palma, Chalatenango, El Salvador; SMNS. Host: *Calea zaca-techichi* (Asteraceae). Distr.: El Salvador.

lathami Gagné 1971: 154, figs. 1–6 (*Neolasioptera*); ♂, ♀, pupa, larva. Holotype: ♂; Yemassee, South Carolina, USA; USNM. Hosts: *Baccharis halimifolia* and *Baccharis* spp. Distr.: USA (Florida; SE USA). Ref.: Diatloff and Palmer 1987.

malvavisci Möhn 1975: 72, pl. 10, fig. 14 (*Neolasioptera*); larva. Holotype: larva; N El Cimarrón, La Libertad, El Salvador; SMNS. Host: *Malvaviscus arboreus* (Malvaceae). Distr.: El Salvador.

martelli Nijveldt 1967: 125, figs. 2–10 (*Neolasioptera*); ♂, ♀, pupa, larva. Holotype: ♂; Chapingo, Distrito Federal, Mexico; Nijveldt Coll. Host: *Agave* sp. (Amaryllidaceae). Distr.: Mexico.

melantherae Möhn 1964a: 97, figs. 253, 254 (*Neolasioptera*); larva. Holotype: NW Salcoatitlán, Sonsonate, El Salvador; SMNS. *Melanthera nivea* (Asteraceae). Distr.: El Salvador.

merremiae Möhn 1964a: 70, figs. 125, 126 (*Neolasioptera*); larva. Holotype: near San Diego, E La Libertad, La Libertad, El Salvador; SMNS. Host: *Merremia umbellata* (Convolvulaceae). Distr.: El Salvador.

mincae Wünsch 1979: 95, pl. 33 (*Neolasioptera*); pupa, larva. Holotype: larva; Minca, Colombia; SMNS. Host: *Melanthera aspera* (Asteraceae). Distr.: Colombia.

monticola Kieffer and Herbst 1909: 124 (*Lasioptera*); ♂, ♀, pupa, larva. Syntypes: Pennalolen, Santiago, Chile; presumed lost. Host: *Gymnophyton polycephalum* (Apiaceae). Distr.: Chile.

odontonemae Möhn 1964a: 91, figs. 222–229 (*Neolasioptera*); ♂, pupa, larva. Holotype: ♂; km 48 on road to Sonsonate, Sonsonate, El Salvador; SMNS. Host: *Odontonema stricta* (Acanthaceae). Distr.: El Salvador.

olivae Wünsch 1979: 92, pls. 32–33 (*Neolasioptera*); ♀, pupa, larva. Holotype: ♀; Donama, Colombia; SMNS. Host: *Calea caracasana* (Asteraceae). Distr.: El Salvador.

parvula Möhn 1964b: 558, figs. 30–41 (*Neolasioptera*); ♂, ♀, larva. Holotype: ♂; Vulcan Boquerón, 1650 m elev., La Libertad, El Salvador; SMNS. Host: *Calea integrifolia* (Asteraceae). Distr.: El Salvador.

phaseoli Möhn 1975: 71, pl. 10, figs. 6–13 (*Neolasioptera*); ♀, pupa. Holotype: ♀; Hacienda San José, vic. San José, La Libertad, El Salvador; SMNS. Host: *Phaseolus* sp. (Fabaceae). Distr.: El Salvador.

piriqueta Felt 1917a: 193 (*Lasioptera*); ♂, ♀, pupa. Syntypes: Mayaguez, Puerto Rico; Felt Coll. Host: *Piriqueta ovata* (Turneraceae). Distr.: Puerto Rico.

portulacae Cook 1906: 251 (*Cecidomyia*); gall. Syntypes: Cuba; depos. unknown. Host: *Portulaca oleracea* (Portulaccaceae). Distr.: Colombia, Cuba, El Salvador, Jamaica, Mexico, USA (Florida), West Indies (Montserrat, Nevis, St. Kitts, St. Vincent). Refs.: Möhn 1975: 72; Wünsch 1979: 74.

<u>*portulacae*</u> Cook 1909: 145 (*Cecidomyia*); gall. Syntypes: Cuba; ? New synonym. Note: Cook 1909 described *portulacae* as a new species separate from the one of 1906. He explained in the introduction to his 1909 paper that "The numbering [his

species were given consecutive numbers] is continuous with that in the first [1906] paper." Host: *Portulaca* sp.

portulacae Felt 1911j: 84 (*Lasioptera*); ♂, ♀, pupa, larva. Syntypes: St. Vincent, Felt Coll. Host: *Portulaca oleracea.*

pseudocalymmae Möhn 1964a: 84, figs. 194–208 (*Neolasioptera*); ♂, ♀, pupa, larva. Holotype: ♂; near San Diego, E La Libertad, La Libertad, El Salvador; SMNS. Host: *Pseudocalymma macrocarpa* (Bignoniaceae). Distr.: El Salvador.

rostrata Gagné *in* Gagné and Boldt 1989: 170, figs. 1–10 (*Neolasioptera*); ♂, ♀, larva. Holotype: ♂; Dodge I., Miami, Florida, USA: USNM. Hosts: *Baccharis halimifolia**, *B. myrsinites*, *B. pingraea*, and *B. spartioides*. Distr.: Argentina, Dominican Republic; USA (southern).

salvadorensis Möhn 1964a: 88, figs. 211–219 (*Neolasioptera*); ♂, pupa, larva. Holotype: ♂; S El Delirio, San Miguel, El Salvador; SMNS. Hosts: *Pithecoctenium echinatum* (Bignoniaceae). Distr.: El Salvador. Ref.: Möhn 1975.

salviae Möhn 1964a: 76, figs. 152–160 (*Neolasioptera*); ♀, pupa, larva. Holotype: ♀; San Salvador, El Salvador; SMNS. Host: *Salvia occidentalis* (Lamiaceae). Distr.: El Salvador.

samariae Wünsch 1979: 89, pl. 32 (*Neolasioptera*); ♂, pupa, larva. Holotype: ♂; Tigreria or Donama, Colombia; SMNS. Host: *Adenocalymma dugandii* (Bignoniaceae). Distr.: Colombia.

senecionis Möhn 1964a: 98, figs. 255–269 (*Neolasioptera*); ♂, ♀, pupa, larva. Holotype: ♂; N El Cimarrón, La Libertad, El Salvador; SMNS. Host: *Senecio kermesinus* (Asteraceae). Distr.: El Salvador.

serjaneae Möhn 1964a: 64, figs. 95–108 (*Neolasioptera*); ♂, ♀, pupa, larva. Holotype: ♂; NW Quezaltepeque, forest of La Toma, La Libertad, El Salvador; SMNS. Host: *Serjania racemosa* (Sapindaceae). Distr.: El Salvador.

serrata Möhn 1964b: 571, figs. 89–97 (*Neolasioptera*); ♀, pupa, larva. Holotype: ♀; SE La Palma, Chalatenango, El Salvador; SMNS. Host: *Calea zacatechichi* (Asteraceae). Distr.: El Salvador.

sidae Möhn 1964a: 67, figs. 110, 111 (*Neolasioptera*); larva. Holotype: San Salvador, El Salvador; SMNS. Host: *Sida rhombifolia* (Malvaceae). Distr.: El Salvador.

tournefortiae Möhn 1964a: 73, figs. 135, 136 (*Neolasioptera*); larva. Holotype: NE Sitio del Niño, La Libertad, El Salvador; SMNS. Host: *Tournefortia volubilis* (Boraginaceae). Distr.: El Salvador.

tribuli Wünsch 1979: 81, pls. 29, 30 (*Neolasioptera*); ♂, ♀, pupa, larva. Holotype: ♂, Santa Marta or Mamatoco, Colombia; SMNS. Host: *Tribulus cistoides* (Zygophyllaceae). Distr.: Colombia.

tridentifera Kieffer and Jörgensen 1910: 398 (*Lasioptera*); ♀, pupa, larva. Syntypes: La Paz, Mendoza, and Cancete, San Juan, Argentina; presumed lost. Host: *Heliotropium curassavicum* (Boraginaceae). Distr.: Argentina.

urvilleae Tavares 1909: 25 (*Lasioptera*); ♂, ♀. Syntypes: Rio Grande do Sul, Brazil; presumed lost. Host: *Urvillea uniloba* (Sapindaceae). Distr.: Brazil.

verbesinae Möhn 1964a: 100, figs. 270–283 (*Neolasioptera*); ♂, ♀, pupa, larva.

Holotype: ♂; road to La Palma, ca. 15 km before La Palma, Chalatenango, El Salvador; SMNS. Hosts: *Verbesina alternifolia**, *V. fraseri*, *V. sublobata*, and *V. turbacensis* (Asteraceae). Distr.: El Salvador; USA (Maryland).

vernoniensis Möhn 1964a: 103, figs. 284, 285 (*Neolasioptera*); larva. Holotype: SE Ciudad Arce, La Libertad, El Salvador; SMNS. Host: *Vernonia* sp. (Asteraceae). Distr.: Brazil, El Salvador.

Smilasioptera

Smilasioptera Möhn 1975: 10. Type species, candelariae Möhn (orig. des.).

This genus is known from one species that forms leaf galls on *Smilax* (Smilacaceae). Adults have a three-segmented palpus and an enlarged antennal pedicel. The female has a short ovipositor with the fused cerci secondarily divided into two lobes bearing sensory setae. Larvae have a robust, three-toothed spatula, four lateral papillae per side, and only two terminal papillae.

candelariae Möhn 1975: 11, pl. 2, figs. 20–28, pl. 3, figs. 1–9 (*Smilasioptera*); ♂, ♀, pupa, larva. Holotype: ♂; SE Candelaria, Santa Ana, El Salvador; SMNS. Host: *Smilax mexicana* (Smilacaceae). Distr.: El Salvador.

Tribe Camptoneuromyiini

Camptoneuromyiini are restricted to the Americas. Thirty-eight species are known from the Neotropics, seven from the Nearctic. Five of the seven known genera are endemic to Central and South America, and the remaining two are shared between the Neotropical and Nearctic Regions. All Neotropical species are inquilines in galls formed by other cecidomyiids, although the two reared Nearctic species of *Meunieriella* are known gall makers (Gagné and Valley 1984 and Gagné unpub.). Many of the species in the larger genera, *Camptoneuromyia*, *Domolasioptera*, and *Meunieriella*, are differentiated on the basis of slight differences, so identification will depend on host association.

Camptoneuromyia petioli Mamaeva (1964), a Palearctic species, does not belong to this tribe. The original figures of the male genitalia and wing of that species indicate that the species belongs to *Ledomyia* in the strict sense (Gagné 1985), where it will be a new combination.

The only character unique to this tribe is the unequal larval terminal papillae, with one pair shorter than the remaining three. Other distinguishing characters of the tribe are the short R_5 (fig. 56), which is only one-third to one-half the length of the wing and usually bowed but sometimes nearly straight and the short antennae of both sexes (fig. 57), which show only slight sexual dimorphism. The ovipositor is elongate but otherwise unmodified, except in *Meunieriella* with its modified setulae. Larvae have a clove-shaped spatula (except reduced to one tooth in *Incolasioptera*) and primitively six but sometimes four lateral papillae on each side of the spatula. Refs.: Gagné 1969b, Möhn 1975.

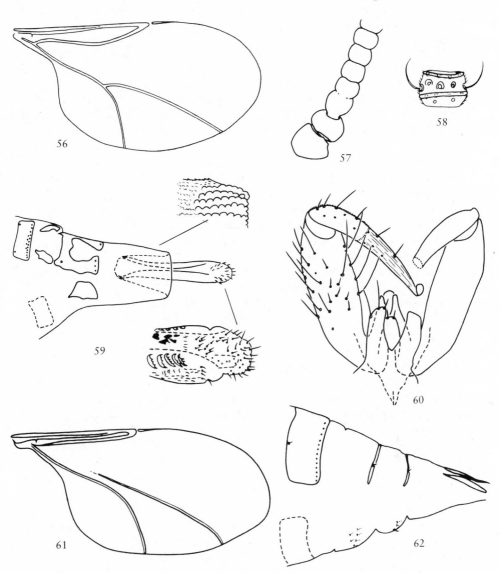

Figures 56–62. Camptoneuromyiini. **56,** *Camptoneuromyia* sp., Trinidad, wing. **57–62,** *Meunieriella graciliforceps*: **57,** male scape, pedicel, and base of flagellum; **58,** male third flagellomere; **59,** female postabdomen with details of striae of eighth segment and of falciform setulae and cerci (dorsolateral); **60,** male genitalia (dorsal); **61,** wing; **62,** male sixth through eighth abdominal segments, showing pseudotergites beyond eighth tergite (dorsolateral).

Camptoneuromyia

Camptoneuromyia Felt 1908b: 334. Type species, *Dasineura virginica* Felt (subseq. des. Coquillett 1910: 518).

This genus is known from nine Neotropical and six Nearctic species. The R_5 wing vein is curved well away from R_1 (fig. 56), the male gonocoxites gradually narrow from the base, and the larva has six lateral papillae on each side of the spatula. Refs.: Möhn 1975, Wünsch 1979.

boerhaaviae Möhn 1975: 22, pl. 4, figs. 25–32 (*Camptoneuromyia*); ♂, larva. Holotype: ♂; forest km 81, vic. Rio San Felipe, San Vicente, El Salvador; SMNS. Hosts: ex galls of *Asphondylia boerhaaviae* on *Boerhavia erecta* and *B. caribaea* (Nyctaginaceae). Distr.: Colombia, El Salvador. Ref.: Wünsch 1979.

burserae Möhn 1975: 16, pl. 3, figs. 18–29, pl. 4, figs. 1–5 (*Camptoneuromyia*); ♂, ♀, pupa, larva. Holotype: ♂; W lava fields, NE Sitio del Niño, La Libertad, El Salvador; SMNS. Host: ex galls of *Burseramyia burserae* on *Bursera simiruba* (Burseraceae). Distr.: El Salvador.

calliandrae Möhn 1975: 26, pl. 5, figs. 8–15 (*Camptoneuromyia*); ♀, larva. Holotype: ♀; SE La Palma, Chalatenango, El Salvador; SMNS. Host: ex flower galls of an unknown Cecidomyiidi on *Calliandra houstoniana* (Fabaceae). Distr.: El Salvador.

crotalariae Möhn 1975: 24, pl. 5, figs. 1–7 (*Camptoneuromyia*); ♀, larva. Holotype: ♂; ravine near ITIC, San Salvador, San Salvador, El Salvador; SMNS. Host: ex flower galls of an unknown Cecidomyiidi on *Crotalaria longirostrata* (Fabaceae) and of *Asphondylia indigiferae* on *Indigofera suffruticosa* (Fabaceae). Distr.: El Salvador.

exigua Möhn 1975: 23, pl. 4, figs. 33–38 (*Camptoneuromyia*); ♀. Holotype: forest km 81, vic. Rio San Felipe, San Vicente, El Salvador; SMNS. Host: with *Camptoneuromyia boerhaaviae* ex galls of *Asphondylia boerhaaviae* on *Boerhaavia erecta* (Nyctaginaceae). Distr.: El Salvador.

meridionalis Felt 1910b: 269 (*Camptoneuromyia*); ♀. Syntypes: St. Vincent; Felt Coll. Host: ex galls of *Schizomyia ipomoeae* on *Ipomoea* sp. (Convolvulaceae). Distr.: St. Vincent, Trinidad. Ref.: Gagné 1977b.

siliqua Möhn 1975: 20, pl. 4, figs. 14–24 (*Camptoneuromyia*); ♀, pupa, larva. Holotype: ♀; km 5, road to Los Cóbanos, Sonsonate, El Salvador; SMNS. Host: ex galls of *Asphondylia vincenti* on *Jussiaea suffruticosa* (Onagraceae). Distr.: El Salvador.

waltheriae Möhn 1975: 14, pl. 3, figs. 10–17 (*Camptoneuromyia*); ♀, larva. Holotype: ♀; E Tablón del Coco, Santa Ana, El Salvador; SMNS. Host: ex galls of *Asphondylia waltheriae* on *Waltheria americana* (Sterculiaceae). Distr.: El Salvador.

xylosmatis Möhn 1975: 18, pl. 4, figs. 6–13 (*Camptoneuromyia*); ♂, larva. Holotype: ♂; SE Candelaria, Hacienda Cortez, Santa Ana, El Salvador; SMNS. Host: ex galls of *Asphondylia xylosmatis* on *Xylosma flexuosa* (Flacourtiaceae). Distr.: El Salvador.

Copinolasioptera

Copinolasioptera Möhn 1975: 39. Type species, *salvadorensis* Möhn (orig. des.).

This Neotropical genus contains one species. It differs from *Camptoneuromyia* in the following ways: R_5 lies close and is almost parallel to R_1, the gonocoxites are cylindrical and narrow throughout, and the larva has only four lateral papillae per side.

salvadorensis Möhn 1975: 40, pl. 6, figs. 14–26 (*Copinolasioptera*); ♂, ♀, larva. Holotype: ♂; SE Ciudad Arce, La Libertad, El Salvador; SMNS. Host: ex leaf galls of a Cecidomyiidi on *Hymenaea courbaril* (Fabaceae). Distr.: El Salvador.

Dialeria

Dialeria Tavares 1918b: 78. Type species, *styracis* Tavares (orig. des.).

This genus is based on a single female reared from galls caused by another cecidomyiid. The original description is sketchy but does include a drawing of a wing that shows an R_5 about two-thirds the length of the wing and bowed away from C, reminiscent of a *Camptoneuromyia* wing (fig. 56). I place this genus with the Camptoneuromyiini because of that resemblance and because the species is apparently an inquiline. The species should be found again in association with the gall.

styracis Tavares 1918b: 78, pl. III, fig. 13 (*Dialeria*); ♀. Syntypes: Caeteté, Bahia, Brazil; presumed lost. Host: ex galls of *Styraxdiplosis caetitensis* on *Styrax* sp. (Styracaceae). Distr.: Brazil.

Domolasioptera

Domolasioptera Möhn 1975: 31. Type species, *adversaria* Möhn (orig. des.).

This Neotropical genus contains seven described species. R_5 is straight in most species but bowed in *curatellae*. Otherwise it differs from *Camptoneuromyia* by the narrow, cylindrical gonocoxites and from all other Camptoneuromyiini in the reduction of lateral larval papillae to three on each side of the spatula and of ventral papillae of the eighth abdominal segment to one pair.

acuario Wünsch 1979: 115, pl. 37 (*Domolasioptera*); ♂, ♀, pupa. Holotype: ♂; Rio Buritaca, Colombia; SMNS. Host: ex galls of *Alycaulus trilobatus* on *Mikania cordifolia* (Asteraceae). Distr.: Colombia. Note: This species could be a *Neolasioptera*; larvae are needed for verification.

adversaria Möhn 1975: 32, pl. 5, figs. 32–39 (*Domolasioptera*); ♂, larva. Holotype: ♂; Hacienda Cortez, SE Candelaria, Santa Ana, El Salvador; SMNS. Host: ex empty galls of *Asphondylia xylosmatis* or *Xylosma flexuosa* (Flacourtiaceae). Distr.: El Salvador.

ayeniae Möhn 1975: 35, pl. 6, figs. 3, 4 (*Domolasioptera*); larva. Holotype: road to Metapán, NE Santa Ana, Santa Ana, El Salvador; SMNS. Host: ex fruit galls of *Asphondylia ayeniae* on *Ayenia pusilla* (Sterculiaceae). Distr.: El Salvador.

baca Möhn 1975: 38, pl. 6, figs. 12, 13 (*Domolasioptera*); larva. Holotype: ravine near ITIC, San Salvador, San Salvador, El Salvador; SMNS. Host: ex galls of *Asphondylia indigoferae* and dried flowers on *Indigofera suffruticosa* (Fabaceae). Distr.: El Salvador.

curatellae Möhn 1975: 37, pl. 6, figs. 5–11 (*Domolasioptera*); ♀, larva. Holotype: ♀; SE La Palma, Chalatenango, El Salvador; SMNS. Host: ex closed flowers infested with a Cecidomyiidi on *Curatella americana* (Dilleniaceae). Distr.: El Salvador.

securidacae Möhn 1975: 33, pl. 5, figs. 40–45 (*Domolasioptera*); ♂. Holotype: ravine near ITIC, San Salvador, San Salvador, El Salvador; SMNS. Host: ex closed flowers infested with a Cecidomyiidi on *Securidaca sylvestris* (Polygalaceae). Distr.: El Salvador.

thevetiae Möhn 1975: 34 (*Domolasioptera*); larva. Holotype: SE Santo Domingo, km 54, San Vicente, El Salvador; SMNS. Host: ex flower galls of *Asphondylia thevetiae* on *Thevetia plumeriaefolia* (Apocynaceae). Distr.: El Salvador.

Incolasioptera

Incolasioptera Möhn 1975: 64. Type-species, *siccida* Möhn (orig. des.).

This Neotropical genus contains one species. The male is unknown. The female differs from the remainder of the tribe in its sparsely setose, elongate, fused cerci, and the larva by its single toothed spatula and the presence of setae on the thoracic ventral papillae. The spatula has only four lateral papillae on each side.

siccida Möhn 1975: 65, pl. 9, figs. 26–30 (*Incolasioptera*); ♀, larva. Holotype: larva; NE San Diego, Santa Ana, El Salvador; SMNS. Host: ex leaf galls of an undetermined Cecidomyiidi on *Celtis iguanea* (Ulmaceae). Distr.: El Salvador.

Meunieriella

Meunieria Rübsaamen 1905: 137. Type-species, *dalechampiae* Rübsaamen (orig. des., as n. g., n. sp). Preocc. Kieffer 1904.

Meunieriella Kieffer 1909: 35, new name for *Meunieria* Rübs. Type species, *Meunieria dalechampiae* Rübsaamen (aut.).

Dolicholabis Tavares 1918b: 72. Type species, *lantanae* Tavares (orig. des.).

This genus is known from 18 Neotropical and one Nearctic species. While all the Neotropical species are inquilines, the Nearctic species is responsible for blister leaf galls on *Gleditsia* (Fabaceae) (Gagné and Valley 1984). An undescribed Nearctic species forms a blister gall on leaves of *Smilax* (Smilacaceae). The male and female postabdomens (figs. 59, 62) are remarkably distinct from all other genera in the tribe and supertribe. The gonocoxites and gonostyli are extremely long (fig. 60), and the postabdomen usually shows two pseudotergites beyond the eighth tergite. The distal half of the ovipositor has falcate setulae (fig. 59). The genus otherwise differs from *Camptoneuromyia* in having a nearly straight R_5 and only four lateral papillae on each side of the spatula. Ref.: Möhn 1975.

acaciae Möhn 1975: 52, pl. 7, figs. 32–33 (*Meunieriella*); larva. Holotype: SW Turin, road to Ahuachapán, Ahuachapán, El Salvador; SMNS. Host: ex stem swelling of an undetermined Cecidomyiidi on *Acacia riparioides* (Fabaceae). Distr.: El Salvador.

acalyphae Möhn 1975: 45, pl. 7, figs. 3–13 (*Meunieriella*); ♂, pupa, larva. Holotype: ♂; ravine near ITIC, San Salvador, San Salvador, El Salvador; SMNS. Host: ex leaf gall of an undetermined Cecidomyiidi on *Acalypha unibracteata* (Euphorbiaceae). Distr.: El Salvador.

armeniae Möhn 1975: 50, pl. 7, figs. 24–29 (*Meunieriella*); ♂. Holotype: W Ateos, La Libertad, El Salvador; SMNS. Host: with *Meunieriella lonchocarpi* ex leaf vein gall of an undetermined Cecidomyiidi on *Lonchocarpus* sp. (Fabaceae). Distr.: El Salvador.

cordiae Möhn 1975: 43, pl. 6, figs. 27–42, pl. 7, figs. 1, 2 (Meunieriella); ♂, ♀, pupa, larva. Holotype: ♂; SE Candelaria, Santa Ana, El Salvador; SMNS. Host: ex leaf gall of an undetermined Cecidomyiidi on *Cordia alliodora* (Boraginaceae). Distr.: El Salvador.

dalechampiae Rübsaamen 1905: 137 (*Meunieria*); ♂, ♀, pupa. Syntypes: Palmeira, Rio de Janeiro, Brazil; ZMHU. Host: *Dalechampia filicifolia* (Euphorbiaceae). Distr.: Brazil.

datxeli Wünsch 1979: 123, pl. 39 (*Meunieriella*): ♂, ♀, pupa, larva. Holotype: ♂; Bahia Nenguange, Colombia; SMNS. Host: ex galls of *Geraldesia polygonarum* on *Triplaris/Symmeria/Ruprechtia* sp. (Polygonaceae). Distr.: Colombia.

eugeniae Möhn 1975: 51, pl. 7, figs. 30, 31 (*Meunieriella*); larva. Holotype: NE El Cimmarón, La Libertad, El Salvador; SMNS. Host: ex leaf blister gall of an undetermined Cecidomyiidi on *Eugenia* sp. (Myrtaceae). Distr.: El Salvador.

gairae Wünsch 1979: 127, pls. 40, 41 (*Meunieriella*); ♂, ♀, pupa, larva. Holotype: ♂; Bella Vista, Colombia; SMNS. Host: ex gall of an undetermined Cecidomyiidi on *Cordia subtruncata* (Boraginaceae). Distr.: Colombia.

graciliforceps Kieffer and Jörgensen 1910: 429 (*Lasioptera*); ♂, ♀, pupa. New combination. Syntypes: Prov. Mendoza, Argentina; presumed lost. Host: *Prosopis strombulifera* (Fabaceae). Distr.: Argentina.

ingae Möhn 1975: 59, pl. 8, figs. 29–34 (*Meunieriella*); ♂. Holotype: NE El Cimmarón, La Libertad, El Salvador; SMNS. Host: ex stem gall of undetermined cecidomyiid on *Inga leptoloba* (Fabaceae). Distr.: El Salvador.

insignis Tavares 1922: 13, figs. 2–6 (*Dolicholabis*); ♂, ♀. Syntypes: vic. Bahia, Brazil; part presumed lost, 3 ♂, 3 ♀, possibly syntypes, sent to Felt by Tavares and deposited in Felt Coll. Host: ex gall of *Dasineura braziliensis* on *Protium heptaphyllum* (Burseraceae). Distr.: Brazil.

lantanae Tavares 1918a: 73, pl. III, figs. 1–6, pl. IV, fig. 1 (*Dolicholabis*); ♂. Syntypes: Nova Friburgo, Rio de Janeiro, Brazil; presumed lost. Host: ex gall of *Schismatodiplosis lantanae* on *Lantana* sp. (Verbenaceae). Distr.: Brazil. Ref.: Nijveldt 1968, a tentative identification of this species from leaf galls on *Manihot utilissima* (Euphorbiaceae) in Surinam.

lonchocarpi Möhn 1975: 49, pl. 7, figs. 16–23 (*Meunieriella*); ♀, larva. Holotype: ♀; Los Chorros, La Libertad, El Salvador; SMNS. Host: with *Meunieriella armeniae* ex various leaf rolls and swellings on *Lonchocarpus* (Fabaceae), and ex leaf blister gall of an unknown Cecidomyiidi on *Mimosa* sp. Distr.: El Salvador.

lucida Möhn 1975: 54, pl. 7, figs. 34–37, pl. 8, figs. 1–13 (*Meunieriella*); ♂, ♀, pupa, larva. Holotype: ♂; NE Sitio del Niño, W of lava fields, La Libertad, El Salvador; SMNS. Host: ex blister leaf gall of an undetermined Cecidomyiidi on *Ficus ovalis* (Moraceae). Distr.: El Salvador.

machaerii Möhn 1975: 58, pl. 8, figs. 21–28 (*Meunieriella*); ♀, larva. Holotype: ♀; ravine near ITIC, San Salvador, San Salvador, El Salvador; SMNS. Host: ex cecidomyiid gall on *Machaerium biovulatum* (Fabaceae). Distr.: El Salvador.

magdalenae Wünsch 1979: 119, pl. 38 (*Meunieriella*); ♂, ♀, pupa, larva. Holotype: ♂; on road to Bahia Nenguange, Colombia; SMNS. Host: ex galls of an undetermined cecidomyiid on *Coccoloba candolleana* (Polygonaceae). Distr.: Colombia.

meridiana Möhn 1975: 60, pl. 9, figs. 1–17 (*Meunieriella*); ♀, larva. Holotype: ♀; NE El Cimmarón, La Libertad, El Salvador; SMNS. Host: ex gall of *Neolasioptera cimarronensis* on a *Vernonia*-like Asteraceae. Distr.: El Salvador.

pisoniae Möhn 1975: 47, pl. 7, figs. 14, 15 (*Meunieriella*); larva. Holotype: larva; NW Las Chinamas, Ahuachapán, El Salvador; SMNS. Host: ex gall of *Bruggmannia pustulans* on *Pisonia micranthocarpa* (Nyctaginaceae). Distr.: El Salvador.

randiae Möhn 1975: 56, pl. 8, figs. 14–20 (*Meunieriella*); ♀, pupa, larva. Holotype: larva; Los Chorros, La Libertad, El Salvador; SMNS. Host: ex gall of *Bruggmannia randiae* on *Randia spinosa* (Rubiaceae). Distr.: El Salvador.

Salvalasioptera

Salvalasioptera Möhn 1975: 28. Type species, *merremiae* Möhn (orig. des.).

This genus contains a single species reared from larvae that dropped to the ground from flowers. It differs from *Camptoneuromyia* in having more setae on the antennal flagellomeres, but none much longer than the length of a flagellomere.

merremiae Möhn 1975: 29, pl. 5, figs. 16–31 (*Salvalasioptera*); ♂, ♀, pupa, larva. Holotype: ♂; W Armenia, Sonsonate, El Salvador; SMNS. Host: *Merremia umbellata* (Convolvulaceae). Distr.: El Salvador.

Tribe Lasiopterini

Lasiopterini are a large, chiefly Old World tribe. About 230 species have been described from the Old World, but only 37 from the Nearctic Region and three from the Neotropics. One of the three Neotropical species ranges from the southern United States to Nicaragua, the other two are in an endemic genus known only from Central America. Most Lasiopterini, including the three Neotropical species, form simple swellings on stems, petioles, and leaf ribs of various plants, but some live in galls or galleries formed by other insects. Most are associated with a symbiotic fungus.

The synapomorphic characters of the Lasiopterini are in the ovipositor (fig. 64): a lateral group of enlarged and flattened setae are present just distal of the eighth tergite and hooked setae and a glabrous area are present on the fused cerci. The tribe superficially resembles Alycaulini in the short antennae that are similar in both sexes, the short R_5, the thick covering of scales, including the white spot at the juncture of R_5 and the costa, and generally similar larvae, particularly in the presence of four lateral setae on each side of the spatula. Males of Lasiopterini differ from those of Alycaulini in the manner of reduction in length of the male seventh tergite, which loses sclerotization along its posterior margin (fig. 65), unlike in Alycaulini, where reduction proceeds from the anterior margin (fig. 53).

Isolasioptera

Isolasioptera Möhn 1964b: 584. Type species, *eupatoriensis* Möhn (orig. des.).

Isolasioptera, known from two Neotropical species, is unique in this tribe for its one-segmented palpus and its long R_5 vein (fig. 63), about three-fourths rather than two-thirds the length of the wing.

eupatoriensis Möhn 1964b: 585, pl. 10, figs. 147–153, pl. 11, figs. 154–162 (*Isolasioptera*); ♂, ♀, pupa, larva. Holotype: ♂; km 48 on road to Sonsonate, Sonsonate, El Salvador; SMNS. Hosts: *Eupatorium morifolium* and *Eupatorium* sp. (Asteraceae). Distr.: El Salvador, Mexico.

palmae Möhn 1975: 78, pl. 11, figs. 13–22 (*Isolasioptera*); ♂, pupa, larva. Holotype: ♂; SE La Palma, Chalatenango, El Salvador; SMNS. Host: *Eupatorium collinum* (Asteraceae). Distr.: El Salvador.

Lasioptera

Lasioptera Meigen 1818: 88. Type species, *picta* Meigen (des. Karsch 1877: 14 (validated by International Commission on Zoological Nomenclature 1970: 95) = *Lasioptera rubi* (Schrank).

This is a catchall genus with about 125 species from all over the world. It is relatively poor in number of species in the Nearctic Region, and dwindles to only one species in Central America. Figures 64 and 65 show female and male postabdomens.

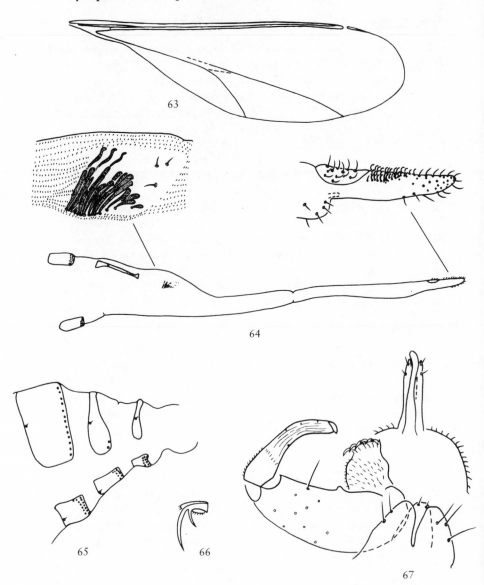

Figures 63–65. Lasiopterini. **63,** *Isolasioptera* sp. ex *Eupatorium adenophorum*, Mexico, wing. **64, 65,** *Lasioptera kallstroemia*: **64,** female postabdomen with details of lateral setae on eighth segment and of modifications on the fused cerci (lateral); **65,** male sixth through eighth abdominal segments (lateral). **Figures 66, 67.** Ledomyiini, *Ledomyia manihot*. **66,** Foretarsal claws. **67,** Male genitalia (dorsal).

kallstroemia Felt 1935: 6 (*Lasioptera*); ♀. Syntypes: Austin, Texas, USA; Felt Coll.
  Hosts: *Kallstroemia intermedia**, *Kallstroemia* sp. (Zygophyllaceae), and *Solanum*
  spp. (Solanaceae). Distr.: Mexico, Guatemala, Nicaragua; USA (Texas).

Tribe Ledomyiini

This cosmopolitan tribe is extremely diverse. While several problematic genera are
used for the Palearctic fauna, all of the New World fauna is included in the catchall
genus *Ledomyia* (Gagné 1985). The genus is also common in Mexican (Gagné 1973a)
and Dominican amber. Three Neotropical species have been described, and I know of
at least 15 undescribed species from this region. Data accompanying Neotropical
specimens in the USNM indicate that some species are associated with fungal fruiting
bodies or rotting plant tissue, some with insects or insect remains. Several Nearctic
species have been reared from xylem vessels in logs (Rock and Jackson 1985). This
tribe has one synapomorphy, the divided parameres (fig. 67). One part is setulose, the
other, usually longer, is glabrous except for apical setae. The antenna has 8–10 or a
regular number of 12 sexually dimorphic flagellomeres. The head of a few species has
a dorsal occipital prominence, unlike the remainder of the Lasiopteridi. The tarsal
claws (fig. 66) are thin, toothed on the forelegs, toothed or simple on the mid- and
hindlegs. The wing is quite variable, with R_5 much shorter to longer than the wing.
The ovipositor is short to elongate and the cerci are separate. Refs.: Gagné 1976,
1985.

Ledomyia

Lepidomyia Kieffer 1894d: 201. Type species, *lugens* Kieffer (mon.). Preocc. Loew
  1864.
Ledomyia Kieffer 1895e: cccxx, new name for *Lepidomyia* Kieffer. Type species,
  Lepidomyia lugens Kieffer (aut.).
Lasiopteryx of Felt 1911h: 44.
Phaenolauthia Kieffer 1912a: 2. Type species, *Ledomyia obscuripennis* Kieffer (orig.
  des.).

collarata Gagné *in* Larew et al. 1987: 502, figs. 1–5 (*Ledomyia*); larva. Holotype:
  Cerro de la Neblina, Amazonian Federal Territory, Venezuela; IZAM. Host: *Xylaria enterogena* (Ascomycetes). Distr.: Venezuela.
manihot Felt 1912e: 144 (*Lasiopteryx*); ♂, ♀. Syntypes: St. Vincent; Felt Coll. Host:
  Manihot utilissima (Euphorbiaceae). Distr.: St. Vincent.
schwarzi Felt 1911i: 191 (*Lasiopteryx*); ♂, ♀. Syntypes: Paraíso, Panama; USNM.
  Host: rotting fig branch. Distr.: Panama.

Tribe Oligotrophini

This tribe is a large, heterogeneous assemblage of primary plant feeders. It is the
largest tribe of Lasiopteridi in the Holarctic Region, but it is too soon to gauge how

Table 1. Character matrix for the tribe Oligotrophini. See text for explanation of characters. For character 1 the matrix lists the lowest number of antennal flagellomeres for the genus in the Neotropics. Numbers followed by an apostrophe indicate a separate morphocline from the same numbers without an apostrophe. A dash indicates that the state could not be determined from either specimens or descriptions.

Character	Angeiomyia	Apodiplosis	Arcivena	Austrolauthia	Calmonia	Calopedilla	Dasineura	Ficiomyia	Guarephila	Haplopalpus	Liciomyia	Pernettyella	Promikiola	Pseudomikiola	Rhopalomyia	Riveraella	Scheueria	Uleia	Xyloperrisia
1	–	17	17	–	10	–	15	29	15	24	15	20	17	17	15	15	18	20	19
2	–	1	0	2	3	1	0	1	3	3	1	0	0	1	2	0	3	2	0
3	–	1	0	0	0	0	0	0	0	0	1	0	0	1	0,1	0	0	1	0
4	–	1	2	–	2	1	2	0	1	0	0	1	1	1	0,1	1	1	1	1
5	–	–	1	–	0	–	1	0	–	–	–	–	–	–	0	–	–	–	–
6	–	–	1	–	–	–	1	0	0	–	–	–	–	0	0	–	–	–	–
7	–	1	1	–	1	1	1	1	0	0	1	1	0	1	1	1	1	0	–
8	–	0	1	–	0	1	1	1	1	0	1	1	0	1	1	1	1	1	–
9	1	–	0	–	0	2′	0	1′	–	–	–	2′	–	–	1′,2′	1	–	1	1
10	–	–	0	–	–	–	0	2	–	–	–	–	–	–	0	–	–	–	–
11	–	–	1	–	–	–	0	1	–	–	–	–	–	–	1	–	–	–	–

numerous it is in the Neotropics. Nineteen genera are known from this region, and, except for the cosmopolitan, catchall genera *Dasineura* and *Rhopalomyia*, are strictly Neotropical. All 42 species recorded here are endemic. While most are responsible for plant galls, larvae of some live freely in flowers. Some species are known only from sketchy original descriptions and the types of others are lost, but most species should be found again through host associations.

The Oligotrophini share no known apomorphic character and contain all of the genera remaining after the Alycaulini, Camptoneuromyiini, Lasiopterini, Ledomyiini, and Trotteriini are removed. Table 1 shows several character states of each genus to preclude repeating them in the separate generic diagnoses. The character states in the grid are presented as morphoclines from 0 to 1, 2, or 3. The morphoclines do not necessarily imply polarity. The numbers to the left of the matrix correspond to the following characters:

1. Number of flagellomeres
2. Palpus with four (0), three (1), two (2), or one (3) segments
3. Tarsal claws toothed (0) or simple (1)
4. R_5 longer (0), as long as (1), or shorter (2) than wing

5. Gonocoxites completely setulose (0) or partially bare (1)
6. Female eighth tergite entire (0) or longitudinally divided (1)
7. Ovipositor not or barely protrusible (0) or elongate-protrusible (1)
8. Female cerci separate (0) or fused (1)
9. Spatula clove-shaped (0), enlarged (1), reduced (1'), or absent (2')
10. Larva with six (0), four (1), or two lateral papillae on each side of the spatula
11. Larva with eight (0) or six (1) terminal papillae

Angeiomyia

Angeiomyia Kieffer and Herbst 1909: 124. Type species, *spinulosa* Kieffer and Herbst (orig. des. as n. g., n. sp.).

This genus is known only from the larva of a species found in spherical leaf galls on *Hydrangea*. It is not necessarily an oligotrophine but has historically been placed in this tribe. The spatula is robust with four teeth of approximately equal length and reminiscent of some in *Neolasioptera* and *Asphondylia*, which belong to other tribes. The larva is reportedly covered with strong yellow spicules. Ref.: Kieffer 1913a.

spinulosa Kieffer and Herbst 1909: 124, fig. 5 (*Angeiomyia*); larva. Syntypes: Chile; presumed lost. Host: *Hydrangea scandens* (Saxifragaceae). Distr.: Chile.

Apodiplosis

Apodiplosis Tavares 1922: 37. Type species, *praecox* Tavares (orig. des.).

This genus is known only from the female and pupa of a species taken from spherical, thick-walled leaf galls on *Psychotria*. The female is noteworthy for the separate cerci at the end of an elongate, protrusible ovipositor. I once placed the genus with the Cecidomyiidi (Gagné 1968a), but it belongs with the Lasiopteridi because of its 17 or 18 flagellomeres.

praecox Tavares 1922: 35, figs. 7, 8 (*Apodiplosis*); ♀, pupa. Syntypes: Nova Friburgo, Rio de Janeiro, Brazil; presumed lost. Host: *Psychotria* sp. (Rubiaceae). Distr.: Brazil.

Arcivena

Arcivena Gagné 1984: 129 (*Arcivena*). Type species, *kielmeyerae* Gagné (orig. des.).

This genus is known from one species that lives in buds of *Kielmeyera*. The very short R_5 (fig. 74) and the lack of necks on the male flagellomeres separate this genus from all other Neotropical Oligotrophini. The ovipositor is strongly sclerotized and bilaterally flattened.

kielmeyerae Gagné 1984: 130, figs. 31–42 (*Arcivena*); ♂, ♀, pupa, larva. Holotype: ♂; Mogi-Guaçu, São Paulo, Brazil; MZSP. Host: *Kielmeyera* spp. (Clusiaceae). Distr.: Brazil.

Austrolauthia

Austrolauthia Brèthes 1914: 153. Type species, *spegazzini* Brèthes (orig. des.).

This genus is based on a species whose host is unknown. There is nothing in the original description to distinguish *Austrolauthia* from *Dasineura* in the broad sense, and it can probably not be identified unless the types are located.

spegazzini Brèthes 1914: 153, fig. 3 (*Austrolauthia*); ♂, ♀. Syntypes: La Plata, Argentina; possibly lost, not found in MACN. Host: unknown, but evidently reared because mention is made of parasites. Distr.: Argentina.

Calmonia

Calmonia Tavares 1917b: 173. Type species, *urostigmatis* Tavares (orig. des.).

This Neotropical genus contains two species, both from complex leaf galls on *Ficus* (Moraceae) in Brazil. The male genitalia are remarkable among Oligotrophini because the parameres are held well separate of the aedeagus (fig. 77). Although Cu was described as evanescent and shown as missing in an illustration accompanying the original description, it is present on Tavares's specimens of *C. urostigmatis* in the Felt Collection (fig. 76). The ovipositor is protrusible and has tiny, separate cerci (fig. 75). The pupa has elongated antennal horns.

fici Gagné. New name for *Asteromyia urostigmatis* Tavares, secondary homonym. Host: *Ficus* sp. (Moraceae). Distr.: Brazil.
  urostigmatis Tavares 1917b: 169, pl. XI, figs. 7, 8 (*Asteromyia*); ♀, pupa. Syntypes: Itaparica (Bahia) and S. Antonio de Barra (Bahia), Brazil; presumed lost. Note: This species seems best placed in *Calmonia* because it forms a gall similar to and on the same host genus as the type species.
urostigmatis Tavares 1917b: 173, pl. X, figs. 10–12, pl. XI, fig. 9 (*Calmonia*); ♂, ♀, pupa, larva. Syntypes: Nova Friburgo, Rio de Janeiro, Brazil; part presumed lost, but ♂♂ and ♀♀, possible syntypes sent to Felt by Tavares, are in the Felt Collection. Host: *Ficus* sp. (Moraceae). Distr.: Brazil.

Calopedilla

Calopedilla Kieffer 1913b: 49. Type species, *Rhopalomyia herbsti* Kieffer (orig. des.).

The one known species of *Calopedilla* causes complex, ovoid bud galls on *Baccharis*. It might belong to subgenus *Diarthronomyia* of *Rhopalomyia* for its toothed claws, but, unlike *Diarthronomyia*, the palpus has three instead of two segments. Ref.: Kieffer 1913a.

herbsti Kieffer 1903: 226 (*Rhopalomyia*); ♂, ♀, pupa, larva. Syntypes: Chile (vic. Concepción?); presumed lost. Host: *Baccharis rosmarinifolia* (Asteraceae). Distr.: Chile.

Dasineura

Dasineura Rondani 1840: 17. Type species, *obscura* Rondani (subseq. des. Rondani 1856: 200).
Dasyneura Agassiz and Loew 1846: 11, emendation (preocc. Saunders 1842).
Perrisia Rondani 1846: 371. Type species, *Cecidomyia urticae* Perris (orig. des.).

Dasineura is a large, catchall genus with hundreds of species described in the world. Most of the seven Neotropical species listed probably do not belong in the genus. The traditional definition of the genus fits the characters in Table 1. Illustrated here are representative adults (figs. 68, 69), larva (fig. 1), and male genitalia (figs. 72, 73).

braziliensis Tavares 1922: 12 (*Perrisia*); ♀, larva. Syntypes: vic. Bahia, Brazil; presumably lost. Host: *Protium heptaphyllum* (Burseraceae). Distr.: Brazil.
chilensis Kieffer and Herbst 1909: 121, fig. 3 (*Perrisia*); ♂, ♀, pupa. Syntypes: Concepción, Chile; presumed lost. Host: *Baccharis rosmarinifolia* (Asteraceae). Distr.: Chile.
corollae Gagné 1977b: 118, figs. 23–29 (*Dasineura*); ♂, ♀, pupa, larva. Holotype: ♂; Curepe, Trinidad; USNM. Host: *Chromolaena odorata* (Asteraceae). Distr.: Trinidad.
eugeniae Felt 1912c: 106 (*Dasyneura*); ♂, ♀. Syntypes: Key West, Florida, USA; USNM. Host: *Eugenia buxifolia* (Myrtaceae). Distr.: Puerto Rico, USA (Florida).
gardoquiae Kieffer and Herbst 1905: 65 (*Dasyneura*); larva. Syntypes: Concepción, Chile; presumed lost. Host: *Gardoquia gilliesii* (Lamiaceae). Distr.: Chile. Ref.: Kieffer and Herbst 1906.
gracilicornis Kieffer and Herbst 1905: 64 (*Perrisia*); gall. Syntypes: vic. Concepción, Chile; presumed lost. Host: a shrub. Distr.: Chile. Ref.: Kieffer and Herbst 1906.
subinermis Kieffer and Herbst 1911: 696 (*Perrisia*); adult, pupa. Syntypes: Valparaíso, Chile; presumed lost. Host: *Baccharis rosmarinifolia* (Asteraceae). Distr.: Chile.

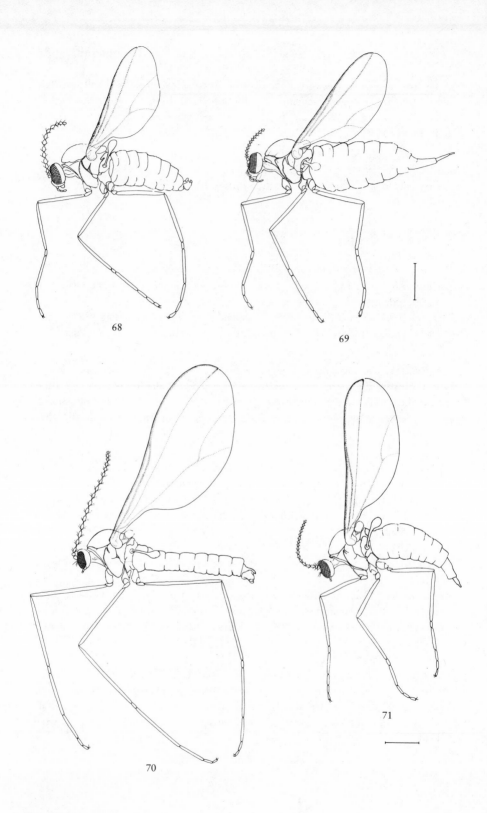

68

69

70

71

Ficiomyia

Ficiomyia Felt 1922b: 5. Type species, *perarticulata* Felt (orig. des.).

This genus is known from one species from southern Florida and a possibly undescribed species from Dominica. The Florida species lives in syconia of *Ficus*. This genus is truly remarkable and stands alone among the Oligotrophini for its large eyes, strongly marked wings with Cu forked near the basal third of the wing, the presence of spurs on the first tarsomeres, and the elongate ventroapical lobe of the gonocoxites. The larva has a short, wide, two-toothed spatula. Ref.: Roskam and Nadel 1990.

perarticulata Felt 1922b: 5 (*Ficiomyia*); ♂, ♀. Lectotype (desig. Roskam *in* Roskam and Nadel 1990): ♂; Miami, Florida, USA; Felt Coll. Host: *Ficus citrifolia* (Moraceae). Distr.: USA (Florida).
birdi Felt 1934: 132 (*Ficiomyia*); ♂, ♀. Lectotype (desig. Roskam *in* Roskam and Nadel 1990): ♂; Royal Palm State Park, Dade Co., Florida, USA; USNM. Host: *Ficus citrifolia* (Moraceae).

Guarephila

Guarephila Tavares 1909: 18. Type species, *albida* Tavares (mon.).

The single known species of *Guarephila* forms spherical leaf galls on *Guarea*.

albida Tavares 1909: 18, pl. II, figs. 7–10 (*Guarephila*); ♂, ♀. Syntypes: Rio Grande do Sul, Brazil; presumed lost. Host: *Guarea* sp., possibly *trichilioides* (Meliaceae). Distr.: Brazil.

Haplopalpus

Haplopalpus Rübsaamen 1916a: 471. Type species, *serjaneae* Rübsaamen (mon.).
Hablopalpus, error, Rübsaamen 1916a: 474.

The single species known of this genus forms complex leaf galls on *Serjania*. The genus is striking for its bowed, two-toothed tarsal claws and its nonprotrusible ovipositor with separate cerci. The pupa has elaborate antennal horns. Except for the antennae with its 24 flagellomeres, this species would pass for a Cecidomyiidi.

serjaneae Rübsaamen 1916a: 473, figs. 49–52 (*Haplopalpus*); ♀, pupa. Syntypes:

Figures 68–71. Oligotrophini (scale line = 0.5 mm). **68, 69,** *Dasineura gleditchiae*: **68,** male; **69,** female. **70, 71,** *Rhopalomyia pomum* (partially redrawn from Jones et al. 1983): **70,** male; **71,** female.

Auristella am Rio Acre, Amazonas, Brazil; ZMHU. Host: *Serjania* sp. (Sapindaceae). Distr.: Brazil.

Lyciomyia

Lyciomyia Kieffer and Jörgensen 1910: 412. Type species, *gracilis* Kieffer and Jörgensen (mon.).

The single species in this genus may be an inquiline. It was reared from stem swellings apparently made by *Jorgensenia falcigera*. Male antennae were not described, so were probably missing originally, but the female flagellomeres are most unusual, each ringed with six separate circumfila. The male reportedly has no parameres, but they may have been overlooked. Ref.: Kieffer 1913a.

gracilis Kieffer and Jörgensen 1910: 412, fig. 40 (*Lyciomyia*); ♂, ♀, egg. Syntypes: Mendoza, Argentina; presumed lost. Host: *Lycium gracile* (Solanaceae). Distr.: Argentina.

Pernettyella

Pernettyella Kieffer and Herbst 1909: 126. Type species, *longicornis* Kieffer and Herbst (orig. des. as n. g., n. sp.).

The one species known in this genus causes an enlargement of the buds of its host. The lack of a tooth on the male gonostyli is unique among the Oligotrophini. The tarsal claws are a little shorter than the empodia. The pupa has no antennal or frontal horns and the larva no spatula. Ref.: Kieffer 1913a.

longicornis Kieffer and Herbst 1909: 126, fig. 7 (*Pernettyella*); ♂, ♀, pupa, larva. Syntypes: Concepción, Chile; presumed lost. Host: *Pernettya furens* (Ericaceae). Distr.: Chile.

Promikiola

Promikiola Kieffer and Herbst 1911: 700. Type species: *rubra* Kieffer and Herbst (orig. des. as n. g., n. sp.).

The one known species was reared from the same type of stem swelling on *Colliguaja* made by *Riveraella colliguayae*. Ref.: Kieffer 1913a.

rubra Kieffer and Herbst 1911: 700, fig. 5 (*Promikiola*); ♀, pupa (a ♂ pupa, presumably, from reference in Kieffer 1913a). Syntypes: Valparaíso, Chile; presumed lost. Host: *Colliguaja odorifera* (Euphorbiaceae). Distr.: Chile.

Pseudomikiola

Pseudomikiola Brèthes 1917b: 411. Type species, *lippiae* Brèthes (orig. des.).

This genus, a probable synonym of *Rhopalomyia*, is known from one species reared from an unspecified gall on *Lippia*. The types are not in good condition, but it should be possible to find this species again. A female postabdomen is shown in figure 78.

lippiae Brèthes 1917b: 153 (*Pseudomikiola*); ♂, ♀. Syntypes: Concepción de l'Uruguay, Entre Ríos, Argentina; MACN. Hosts: *Lippia turbinata* and *Lippia* sp. (Verbenaceae).

Rhopalomyia

Rhopalomyia Rübsaamen 1892: 370. Type species, *Oligotrophus tanaceticola* Karsch (subseq. des. Kieffer 1896: 89).
Diarthronomyia Felt 1908b: 339. Type species, *artemisiae* Felt (orig. des.), = *Rhopalomyia pomum* Gagné.
Misospatha Kieffer 1913b: 48. Type species, *Rhopalomyia globifex* Kieffer and Jörgensen (orig. des.).
Panteliola Kieffer 1913b: 49. Type species, *Rhopalomyia haasi* Kieffer (orig. des.).

This is a catchall genus with 135 species known from the Holarctic Region, where almost all form galls on Asteraceae. Nine Neotropical species are currently placed here, seven of them described by Kieffer and Jörgensen (1910). These were reared from several families of plants, only two from Asteraceae. *Diarthronomyia* can be treated as a subgenus of *Rhopalomyia*, comprising those species with toothed claws that occur on Anthemideae (Asteraceae).

Rhopalomyia species have a protrusible ovipositor, but the eighth tergite, instead of being longitudinally divided, as, for example, in *Dasineura*, is merely elongate. Larvae may or may not have a spatula; when they do not, the pupae usually have well-developed antennal horns. Galls may be simple to complex, but pupation always occurs in galls. A representative male and female are shown in figures 70 and 71. Ref.: Gagné 1975b.

ambrosiae Gagné 1975c: 54 (*Rhopalomyia*); ♂, ♀, pupa. Holotype: ♂; Hialeah, Florida, USA; USNM. Hosts: *Ambrosia artemisiifolia** and *Ambrosia* spp. (Asteraceae). Distr.: USA (Florida; Texas).
bedeguaris Kieffer and Jörgensen 1910: 403, fig. 32 (*Rhopalomyia*); ♂, ♀, pupa. Syntypes: Cordillera de Mendoza, Mendoza, Argentina; presumed lost. Host: *Lycium chilense* (Solanaceae). Distr.: Argentina.
chrysanthemi Ahlberg 1939: 60 (*Diarthronomyia*); ♂, ♀. Syntypes: Sweden; depos.

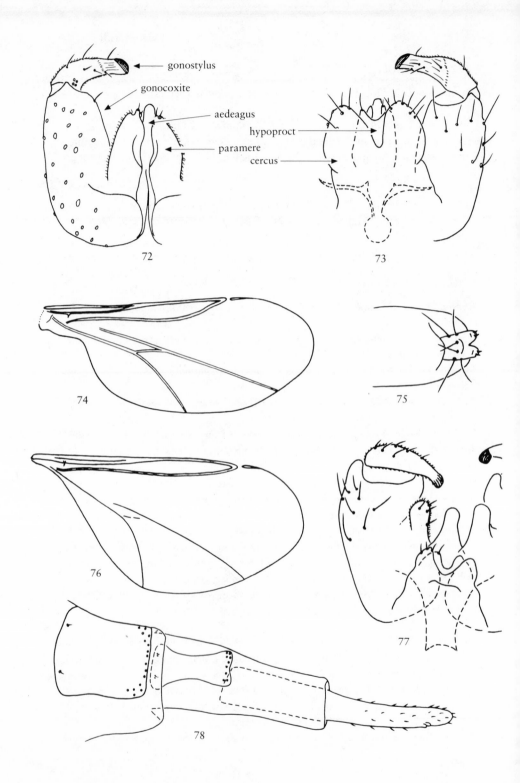

gonostylus

gonocoxite

aedeagus

hypoproct

paramere

cercus

72

73

74

75

76

77

78

unknown. Host: *Chrysanthemum* spp. (Asteraceae). Distr.: Argentina, Colombia; North America, Europe, Japan. Refs.: Barnes 1948, Mallea et al. 1983.

globifex Kieffer and Jörgensen 1910: 364, figs. 2, 2a (*Rhopalomyia*); ♂, ♀, pupa, larva. Syntypes: Chacras de Coria, Pedregal, San Ignacio, and Potrevillos, all in Mendoza Prov., and Cancete, San Juan, Argentina; presumed lost. Host: *Baccharis salicifolia* (Asteraceae). Distr.: Argentina.

lippiae Kieffer and Jörgensen 1910: 401, fig. 31 (*Rhopalomyia*); ♂, ♀, pupa, egg. Syntypes: Cordillera de Mendoza, Mendoza, Argentina; presumed lost. Host: *Lippia foliolosa* (Verbenaceae). Distr.: Argentina.

nothofagi Gagné 1974: 447, figs. 1–7 (*Rhopalomyia*); ♂, ♀, pupa. Holotype: ♂; near Valdivia, Chile; USNM. Host: *Nothofagus obliqua* (Fagaceae). Distr.: Chile.

oreiplana Kieffer and Jörgensen 1910: 441 (*Rhopalomyia*); ♀, pupa. Syntypes: Blanco Encallada, San Juan, Argentina; presumed lost. Host: *Verbena seriphioides* (Verbenaceae). Distr.: Argentina.

prosopidis Kieffer and Jörgensen 1910: 427, fig. 51 (*Rhopalomyia*); pupa, larva. Syntypes: Chacras de Coria and La Paz, Mendoza, and Alto Pencoso, San Luis, Argentina; presumed lost. Hosts: *Prosopis alpataco*, *P. campestris*, and *P. flexuosa* (Fabaceae). Distr.: Argentina.

tricyclae Kieffer and Jörgensen 1910: 440, fig. 61 (*Rhopalomyia*); pupa. Syntypes: Blanco Encalado and San Ignacio, San Juan, Argentina; presumed lost. Host: *Tricycla spinosa* (Nyctaginaceae). Distr.: Argentina.

verbenae Kieffer and Jörgensen 1910: 441 (*Rhopalomyia*); pupa. Syntypes: San Juan Prov., Argentina; presumed lost. Host: *Verbena aspera* (Verbenaceae). Distr.: Argentina.

Riveraella

Riveraella Kieffer and Herbst 1911: 699. Type species, *colliguayae* Kieffer and Herbst (orig. des. as n. g., n. sp.).

This genus consists of one described species and one other listed only as *Riveraella* sp. from a different gall on the same host (Kieffer and Herbst 1911). *Promikiola rubra* and *R. colliguayae* were reared from similar galls on *Colliguaja*. Little is known of this genus to distinguish it from others with four palpal segments and toothed claws. Ref.: Kieffer 1913a.

colliguayae Kieffer and Herbst 1911: 698, figs. 3, 4 (*Riveraella*); ♀, pupa, larva.

Figures 72–78. Oligotrophini. 72, 73, *Dasineura corollae*: 72, male genitalia (ventral); 73, male genitalia (dorsal). 74, *Arcivena kielmeyerae*, wing. 75–77, *Calmonia urostigmatis*: 75, ovipositor apex (dorsal); 76, wing; 77, male genitalia (dorsal). 78, *Pseudomikiola lippiae*, female postabdomen (dorsolateral).

Syntypes: Valparaiso, Chile; presumed lost. Host: *Colliguaja odorifera* (Euphorbiaceae). Distr.: Chile.

Scheueria

Scheueria Kieffer and Herbst 1909: 120. Type species, *longicornis* Kieffer and Herbst (orig. des., as n. g., n. sp.).

This genus is known from two species reared from bud galls on *Baccharis* (Asteraceae). In addition to the characters listed in Table 1, the female flagellomeres are surrounded with numerous, sinuous circumfila, the tarsal claws are thick, and the empodia are longer than the tarsal claws. Ref.: Kieffer 1913a.

agglomerata Kieffer 1913a: 39 (cites description of gall in Kieffer and Herbst 1909: 120, fig. 2) (*Scheueria*); gall. Syntypes: Concepción, Chile; presumed lost. Host: *Baccharis eupatorioides* (Asteraceae). Distr.: Chile.
longicornis Kieffer and Herbst 1909: 120, fig. 1 (*Scheueria*); ♂, ♀, pupa. Syntypes: Concepción, Chile; presumed lost. Host: *Baccharis eupatorioides* (Asteraceae). Distr.: Chile.

Uleia

Uleia Rübsaamen 1905: 85. Type species, *clusiae* Rübsaamen (mon.).
Ulea Houard 1933: 249, error or emendation.

This genus is known from one species reared from an aggregate bud gall on *Clusia*. This genus reportedly has a large scape and pedicel that are much wider than the flagellum. The larva has a two-toothed spatula that is greatly enlarged anteriorly.

clusiae Rübsaamen 1905: 85 (*Uleia*); ♂, ♀, larva. Syntypes: Santa Clara and Bom Fim on Juruá River, Amazonas, Brazil; ZMHU. Host: *Clusia* sp. (Clusiaceae). Distr.: Brazil.

Xyloperrisia

Xyloperrisia Kieffer 1913a: 68. Type species, *Perissia azarae* Kieffer (mon.).

This genus is known from one species reared from stem swellings of *Pernettya*. The tarsal claws are a little shorter than the empodia, and the male gonostyli are enlarged at midlength. The pupa has strongly developed antennal horns and the larva a strongly developed spatula. Ref.: Kieffer 1913a.

azarae Kieffer 1903: 227 (*Perrisia*); ♂, pupa, larva. Syntypes: Chile (vic. Concepción?); presumed lost. Host: *Pernettya furens* (Ericaceae). Distr.: Chile. Ref.: Kieffer and Herbst 1906. Note: The name *azarae* derives from an initial misidentification of the host as *Azara integrifolia* (Kieffer and Herbst 1905).

azarai, emendation for *azarae*, Kieffer, p. 64, *in* Kieffer and Herbst 1905 (a patronymic to honor Azara [Kieffer and Herbst 1906]).

Unplaced Species of Oligotrophini

The following species were originally placed in *Cecidomyia, Janetiella,* or *Oligotrophus,* all of which at one time were used as catchall genera. *Janetiella* still is an omnibus genus for Holarctic groups of species (Gagné 1989), but *Oligotrophus* is now restricted to Holarctic species that live in Cupressaceae and *Cecidomyia* to species that live in conifer resin. All except one of the species listed below should be found again because their hosts were named and galls described.

acuticauda Kieffer and Herbst 1906: 227 (*Janetiella*); ♀. Syntypes: vic. Concepción, Chile; presumed lost. Host: undetermined shrub. Distr.: Chile. Note: Nomen nudum in Kieffer and Herbst 1905: 64.

avicenniae Cook 1909: 145 (*Cecidomyia*); gall. Syntypes: Cuba; depos. unknown. Hosts: *Avicennia germinans, A. nitida,* and *A. tomentosa* (Verbenaceae). Distr.: Belize, Brazil, Cuba, USA (Florida).

eugeniae Kieffer and Herbst 1909: 125, fig. 6 (*Oligotrophus*); larva. Syntypes: Concepción, Chile; presumed lost. Host: *Myrceugenia stenophylla* (Myrtaceae). Distr.: Chile.

lyciicola Kieffer and Jörgensen 1910: 409, 414 (*Oligotrophus*); gall. Syntypes: Mendoza Prov., Argentina; presumed lost. Host: *Lycium chilense* and *Lycium gracile* (Solanaceae). Distr.: Argentina.

montivaga Kieffer and Jörgensen 1910: 432 (*Janetiella*); ♂, ♀, pupa. Syntypes: Mendoza Prov., Argentina; presumed lost. Host: *Senecio mendocinus* (Solanaceae). Distr.: Argentina.

nectandrae Kieffer 1910b: 442 (*Oligotrophus*); adult (sex unspecified), pupa. Syntypes: Paraguay; presumed lost. Host: *Nectandra megapotamica* (Lauraceae). Distr.: Paraguay.

Tribe Trotteriini

The tribe Trotteriini consists of one cosmopolitan genus of 21 species. All are inquilines in galls of other cecidomyiids, chiefly Asphondyliini. Four species have been described from the Neotropics, and I know of two additional, undescribed species. Adults (fig. 79) are so modified that the tribe's closest relative among Lasiopteridi is not apparent. The postocciput is long and covers the dorsal part of the head, so that the eyes and antennae are moved ventrad. The antennal scape is elongate, as is the pronotum, and the hind femora are greatly enlarged. Möhn (1963b) noted that the hind legs are carried outstretched and do not touch the ground when the fly is at rest. The wing (fig. 80) is narrowed, especially behind Cu, and Cu is simple; R_5 is

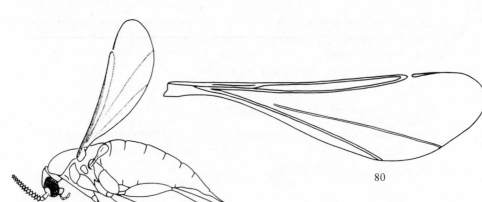

Figures 79, 80. *Trotteria* sp., Dominica: 79, female (scale line = 0.5 mm); 80, wing.

short and bowed. The ovipositor is shaped generally as in *Dasineura* but is much more attenuate; the cerci are fused and, in most species, needlelike. Tarsal claws are toothed. Larvae have a clove-shaped spatula with a full complement of six lateral papillae per side and eight terminal papillae of equal length. Ref.: Möhn 1975.

Trotteria

Choristoneura Rübsaamen 1892: 342. Type species, *Cecidomyia obtusa* Loew (mon.). Preocc. Lederer 1859.

Trotteria Kieffer 1901: 561, new name for *Choristoneura* Rübsaamen. Type species, *Cecidomyia obtusa* Loew (aut.).

Four species are known from this region. In addition, there is a record of *Trotteria caryae* Felt from Dominica (Gagné 1969b), but it may be a misidentification (see comments in Möhn 1975).

lapalmae Möhn 1975: 68, pl. 9, figs. 31–37, pl. 10, figs. 1–5 (*Trotteria*); ♂, pupa, larva. Holotype: ♂; SE La Palma, Chalatenango, El Salvador; SMNS. Host: ex galls of *Perasphondylia reticulata* on *Eupatorium* sp. (Asteraceae). Distr.: El Salvador.

lindneri Möhn 1963b: 1, figs. 1–8 (*Trotteria*); ♂, pupa, larva. Holotype: ♂; San Salvador, El Salvador; SMNS. Host: ex galls of *Asphondylia indigoferae* on *Indigofera suffruticosa* (Fabaceae). Distr.: El Salvador.

rivinae Wünsch 1979: 143, pls. 44, 45 (*Trotteria*); ♀, pupa, larva. Holotype: ♀; Bahia Concha, Colombia; SMNS. Host: ex gall of an unidentified cecidomyiid on *Rivina humilis* (Phytolaccaceae). Distr.: Colombia.

salvadorensis Möhn 1963b: 3, figs. 9–17 (*Trotteria*); ♀, larva. Holotype: ♀; forest km 81, near Rio San Felipe, San Vicente, El Salvador; SMNS. Host: ex galls of *Asphondylia boerhaaviae* on *Boerhavia erecta* (Nyctaginaceae) and of *Asphondylia ayeniae* on *Ayenia pusilla* (Sterculiaceae). Distr.: Colombia, El Salvador. Ref.: Wünsch 1979.

SUPERTRIBE CECIDOMYIIDI

The supertribe Cecidomyiidi is unique for the shape of the male flagellomeres. Each is divided into two distinct nodes separated by an internode and has two or three separate, many-looped circumfila (fig. 141). These characters are not found elsewhere in the Cecidomyiidae. Male flagellomeres with one node and interconnected, flat circumfila are considered secondarily reduced (figs. 96, 124). This reduction occurs in several tribes and genera that otherwise have binodal flagellomeres, and I recently noticed the condition on one of a series of males of a European species, *Dicrodiplosis pseudococci* (Felt). The presence of 12 antennal flagellomeres and the evenly cylindrical shape of the female flagellomeres (figs. 97, 172) are good recognition characters for the supertribe, although these characters occur elsewhere in the Cecidomyiinae. A small budlike extension may be present at the apex of the 12th flagellomere in some genera, and it may even be articulated, but I consider this a secondary augmentation. Females of the Neotropical *Enallodiplosis discordis* have only 7 flagellomeres (males have the usual 12), and both sexes of some species of *Planetella*, a genus not yet known from the Neotropics, have as many as 25.

The Cecidomyiidi are divided here into nine tribes, the Anadiplosini, Aphidoletini, Asphondyliini, Cecidomyiini, Centrodiplosini, Clinodiplosini, Lestodiplosini, Lopesiini, and Mycodiplosini, and a group of "unplaced genera," that are not grouped further. Each of the tribes is presumably monophyletic but differs greatly in number of species and distribution. Relationships among the tribes and the unplaced genera are unclear.

To avoid repetition in the taxonomic diagnoses, I summarize many descriptive traits in Table 2. Character states in the grid are presented as morphoclines from 0 to 1, 2, or 3. The morphoclines do not necessarily imply polarity. The numbers to the left of the matrix correspond to the following characters:

1. Occipital process (the narrow, vertical extension of the occiput bearing two strong setae) present (0) or absent (1)

Table 2. Character matrix for the supertribe Cecidomyiidi. See text for explanation of characters. Some characters change within a taxon, but only the one closest to the 0 state is listed in the matrix. For example, although the Asphondyliini have one to four palpal segments, four (0) is noted in the character matrix. Any further changes within a taxon are noted in the treatment of the individual taxon. Numbers followed by an apostrophe indicate a separate morphocline from the same numbers without an apostrophe. A dash indicates that the state could not be determined. *Uleella*, a genus of Anadiplosini, is not included in the table because it is known from larvae only. *Andirodiplosis*, an unplaced genus, is too poorly known to include.

Tribe groupings of the genera (columns): Anadiplosini — *Alexomyia, Anadiplosis, Machaeriobia, Scopodiplosis*; Aphidoletini — *Aphidoletes, Asphondyliini (Table 3)*; Cecidomyiini — *Cecidomyia, Contarinia, Erosomyia, Prodiplosis, Sphaerodiplosis, Stenodiplosis, Taxodiomyia*; Centrodiplosini — *Centrodiplosis, Cystodiplosis, Jorgensenia*; Clinodiplosini — *Autodiplosis, Chauliodontomyia, Cleitodiplosis, Clinodiplosis, Houardodiplosis, Iatrophobia, Schismatodiplosis*; Lestodiplosini — *Arthrocnodax, Feltiella, Lestodiplosis, Pseudendaphis, Thripsobremia, Trisopsis, Dicrodiplosis*.

Character	Alexomyia	Anadiplosis	Machaeriobia	Scopodiplosis	Aphidoletes	Asphondyliini (Table 3)	Cecidomyia	Contarinia	Erosomyia	Prodiplosis	Sphaerodiplosis	Stenodiplosis	Taxodiomyia	Centrodiplosis	Cystodiplosis	Jorgensenia	Autodiplosis	Chauliodontomyia	Cleitodiplosis	Clinodiplosis	Houardodiplosis	Iatrophobia	Schismatodiplosis	Arthrocnodax	Feltiella	Lestodiplosis	Pseudendaphis	Thripsobremia	Trisopsis	Dicrodiplosis
1	1	1	1	1	0	1	1	0	0	0	–	1	0	–	–	–	0	0	0	0	0	0	0	0	0	0	0	0	0	0
2	0	0	0	0	0	0	0	0	2	0	0	2	2	–	–	–	0	0	0	0	0	0	0	2	2	0	1	–	2	0
3	3	2	2	3	0	0	0	0	0	0	0	0	0	3	1	2	0	0	0	0	0	0	0	0	0	0	0	0	0	0
4	–	1	1	–	0	1	0	0	0	0	0	0	0	0	0	0	0	1	0	0	0	0	0	0	0	0	0	0	0	0
5	–	0	0	–	0	2	0	1	1	0	1	1	0	0	0	0	0	2	0	0	0	0	0	0	0	0	0	1	0	0
6	0	0	0	0	0	1	0	0	1	0	–	1	0	0	0	0	0	0	0	0	0	0	0	2	1	0	2	0	1	0
7	0	0	0	0	1	2	1	1	1	1	–	1	1	2	2	2	1	1	1	1	1	1	1	2	1	1	1	2	1	1
8	0	0	0	0	1	1	1	1	1	1	–	1	1	1	1	1	1	1	1	1	1	1	1	1	1	1	1	1	1	1
9	0	0	0	0	1	0	0	0	0	0	–	0	1	–	–	–	0	0	0	0	0	0	0	1	1	1	1	1	1	0
10	0	0	0	0	1	1	1	1	1	1	–	1	1	–	–	–	1	1	1	1	1	1	1	1	1	1	1	1	1	1
11	1	1	1	1	0	1	1	1	1	1	–	1	1	–	–	–	0	0	0	0	0	0	0	1	1	0	1	0	1	0
12	1	1	1	1	0	1	0	0	0	0	–	0	0	–	–	–	0	0	0	0	0	0	0	0	0	0	0	0	0	0
13	1	1	1	1	0	1	1	1	0	1	1	1	1	1	1	1	1	1	1	0	1	1	1	1	0	1	1	1	1	0
14	1	1	1	1	1	1	2	2	2	2	2	–	2	2	2	2	2	0	0	0	0	0	0	2	2	1	2	2	2	1
15	1	1	1	1	1'	0	1'	0	0	0	–	0	1	1'	1'	1'	0	1	1	0	0	0	1	0	0	0	0	0	0	0
16	2	2	2	2	0	1	1	2	1	2	–	2	2	2	2	2	0	0	0	0	0	0	0	0	0	0	0	0	0	0
17	1	1	1	1	0	0	0	0	1	0	–	0	1	1	1	1	0	0	0	0	0	0	0	0	0	0	0	0	0	0
18	2	2	2	2	1	0	1	1	1	1	–	1	1	2	2	2	1	1	1	1	1	1	1	0	2	2	1	2	2	1
19	2	2	2	2	0	0	0	2	1	2	1	2	1	2	2	2	0	0	0	0	0	0	0	0	0	0	0	0	0	0
20	2	2	2	2	0	0	0	2	0	2	–	2	0	2	2	2	1	1	1	1	1	1	0	0	0	1	1	1	1	0
21	2	2	2	2	0	0	0	2	0	2	–	2	0	2	2	2	0	0	0	0	0	0	0	0	0	1	0	1	0	1

Lopesiini					Mycodiplosini		unplaced to tribe																									
Ctenodactylomyia	Liebeliola	Lopesia	Rochadiplosis	Tetradiplosis	Coquillettomyia	Mycodiplosis	Anasphodiplosis	Astrodiplosis	Austrodiplosis	Bremia	Charidiplosis	Compsodiplosis	Contodiplosis	Dactylodiplosis	Diadiplosis	Enallodiplosis	Epihormomyia	Frauenfeldiella	Gnesiodiplosis	Huradiplosis	Karshomyia	Megaulus	Moehniella	Neobaezomyia	Ouradiplosis	Plesiodiplosis	Resseliella	Schizodiplosis	Silvestrina	Styraxdiplosis	Xenasphondylia	Youngomyia
0	-	-	0	0	0	0	1	0	-	0	0	-	0	1	1	1	1	-	-	1	0	-	-	-	1	-	1	-	1	-	1	0
0	-	-	0	0	0	0	0	0	-	0	0	-	1	0	1	2		-	0	2	0	0	-	-	-	0	-	0	-	2	-	0
1	0	0	1	1	0	0	0	3	0	0	0	2	1	0	0	1	1	3	3	1	0	0	1	0	0	0	0	0	0	0	0	0
1	-	0	0	0	0	-	0	-	0	0	0	0	0	0	0	-	-	0	0	0	-	-	-	-	-	0	0	-	0	0	-	0
2	-	0	0	0	0	0	-	0	-	0	0	0	0	0	0	1	-	0	1	0	-	-	-	-	-	0	0	-	0	0	-	0
0	0	0	0	0	0	0	1	0	1	0	0	0	0	0	1	1	0	1	1	0	0	0	1	1	0	0	0	0	2	0	0	0
1	0	0	1	1	1	1	2	0	1	1	1	0	0	1	2	2	1	-	1	0	1	1	2	2	0	-	2	-	1	0	2	0
0	0	0	0	0	1	1	1	1	0	1	0	1	1	1	0	1	1	1	-	0	1	1	1	1	1	0	-	1	-	1	0	1
0	-	0	0	0		1	-	1	0	0	0	0	0	1	1	0	-	-	1	0	0	-	1	0	-	0	-	1	-	0	0	0
0	0	0	0	0	1	1	1	1	1	1	1	1	1	1	1	1	1	-	1	1	1	1	-	1	0	-	1	-	1	1	1	1
0	-	-	0	0	0	0	1	-	1	0	0	0	1	0	1	1	0	-	1	0	1	-	1	1	-	1	-	1	-	0	0	0
0	-	-	0	0	0	0	0	0	0	0	0	0	0	0	0	1	-	0	0	0	0	0	0	-	0	-	0	-	0	-	0	0
1'	1	-	1'	1'	0	0	1	1	1	0	1	1	1	0	0	1	1'	0	1	1	0	1	1	1	-	1	0	1	1	1	1	0
0	-	0	0	0	0	0	2	2	0	0	0	2	2	0	0	2	0	0	2	1	0	2	0	2	-	2	0	2	2	-	2	0
0	1	0	0	0	0	0	0	1	0	0	0	1	1	0	0	0	1	0	0	0	0	0	0	-	0	0	0	0	1	1'	0	
0	0	-	0	0	0	0	2	1	1	0	1	1	1	0	0	0	1	0	0	0	0	1	0	0	0	0	1	1	0	1	2	1
0	0	0	0	0	0	0	0	0	0	0	0	0	0	0	0	0	0	0	0	0	0	0	0	0	0	0	0	0	0	0	1	0
1	-	-	1	1	1	1	1	1	-	1	0	1	1	0	1	1	1	-	1	-	1	-	-	1	1	-	1	2	1	-	2	1
0	-	-	0	0	0	0	1	0	-	0	0	0	0	0	0	2	-	0	-	0	-	0	0	0	-	0	2	0	-	1	0	
1	-	-	1	1	1	0	2	0	-	1	0	-	1	1	0	1	1	-	1	1	1	-	0	1	0	-	1	2	1	-	2	0
1	1	-	1	1	0	0	0	0	0	1	0	0	0	0	0	0	0	-	0	0	0	-	0	0	0	0	0	2	0	-	2	0

2. Eye facets hexagonal (0), mostly hexagonal but circular laterally (1), or circular (2)
3. Palpus with four (0), three (1), two (2), or one segment (3)
4. Male flagellomeres with two nodes (0) or one (as in female) (1)
5. Male flagellomeres with three (0) or two separate circumfila (1), or with interconnected circumfila (as in female) (2)
6. R_5 longer than wing and joining C beyond wing apex (0), approximately as long as wing and joining C near wing apex (1), or shorter than wing and joining C before wing apex (2)
7. Rs partially as strong as R_1 but weak anteriorly (0), weaker than R_1 along its entire length but evident (1), or absent (2)
8. Base of M with slight curve (0) or straight (1)
9. M_3 fold present (0) or absent (1)
10. Rs stub beyond midlength of R_1 (0) or before (1)
11. CuP present (0) or absent (1)
12. First tarsomeres without (0) or with (1) spur
13. Claws toothed on at least forelegs (0), simple (1), or with more than one tooth (1′)
14. Claws bent near base (0), near midlength (1), or near two-thirds length (2)
15. Empodia as long as bend in claws (0), shorter (1), or longer (1′)
16. Ovipositor short, barely protrusible (0), segments 8–9 somewhat elongated (1), or segments 8–9 and cerci modified other than by simple elongation (2)
17. Female cerci separate (0) or fused (1)
18. Female eighth tergite completely sclerotized (0), sclerotized only anteriorly (1), or unsclerotized (2)
19. Female segment 9 with ventral setae (0), without (1), or otherwise greatly modified (2)
20. Female tergum 10 with setae (0), without (1), or greatly modified (2)
21. Female cerci with setae evenly distributed (0), more numerous and crowded together ventrally or ventroapically (1), or otherwise greatly modified (2)

Individual tribal and generic descriptions will make note of these character states only if they need further comment. Because the genera of this supertribe differ considerably in body size, wing length is given in the generic diagnoses. A wing length of 1 mm is considered small for cecidomyiids, 5 mm and over is considered large. Conspicuous markings and shapes are noted if peculiar to genera.

Key to Tribes and Genera of Neotropical Cecidomyiidi

This is mainly a key to the adult stage. Because some genera lack distinguishing adult features, they are separated in the key on differences of immature stages or the plant host or both. Ideally, one should have examples of all stages and know the plant host, if any, especially when keying Asphondyliini (couplets 2–22). *Andirodiplosis, Ouradiplosis, Plesiodiplosis,* and *Schizodiplosis* are not keyed here because the shape of their tarsal claws is unknown. Other genera can be keyed only so far, as indicated in footnotes, because certain body parts are lost or only one sex is known. It is nevertheless still possible to identify the missing genera by scanning the generic descriptions and illustrations.

1   Male flagellomeres cylindrical (fig. 96), rarely binodal (fig. 124); if flagellomeres binodal, the circumfila interconnected (fig. 124); gonostyli short, wide, situated dorsocaudally on gonocoxites (figs. 104, 123); female seventh abdominal sternite at least 1.5 times length of preceding sternite (fig. 95) . . . . Asphondyliini, 2

Male flagellomeres usually binodal (figs. 86, 141); when flagellomeres binodal, the circumfila separate; gonostyli usually elongate and narrow and situated caudally on gonocoxites (figs. 87, 143); female seventh abdominal sternite not longer than preceding sternite (fig. 142) 23

2(1)   Legs with ventroapical spine on first tarsomeres (figs. 100, 136) 3
Legs without ventroapical spine (fig. 137). 12

3(2)   Palpus four-segmented . . . . *Schizomyia* in part
Palpus with fewer than four segments 4

4(3)   Gonocoxites with setose mesobasal lobe (as in fig. 139) and/or pupa with acute ventral prominence at base of antennal horn (fig. 111) 5
Gonocoxites without setose mesobasal lobe (fig. 104) and pupa without acute ventral prominence at base of antennal horn (fig. 109) 6

5(4)   Tarsal claws robust, enlarged at base; denticles of gonostyli not completely fused, some apparent at least mesally (fig. 106) . . . . Zalepidota
Tarsal claws narrow throughout, not enlarged at base; denticles of gonostyli completely fused (as in fig. 104) . . . . Heterasphondylia

6(4)   Tarsal claws enlarged near midlength (fig. 102), dimorphic among legs and between sexes; larva with reduced, two-toothed spatula (fig. 119) and pair of large, corniform terminal papillae (fig. 120); in bud galls of Mimosoideae (Fabaceae) . . . . Hemiasphondylia
Tarsal claws not enlarged near midlength (fig. 101), but may be dimorphic; larva with large, three- or four-toothed spatula (fig. 116), or, if spatula two-toothed and reduced, terminal papillae without large, corniform pair (fig. 117). 7

7(6)   Tooth of gonostyli completely divided mesally (fig. 107); pupa without frontal horns (fig. 114) and with diminutive dorsal abdominal spines (fig. 113); area surrounding larval spatula not pigmented (fig. 118); in woody stem galls . . . . Bruggmanniella
Tooth of gonostyli entire (fig. 104); pupa with frontal horns (as in fig. 109) and robust dorsal abdominal spines (fig. 108); area surrounding larval spatula pigmented (fig. 116); usually in bud, flower, or fruit galls 8

8(7) Pupa without upper frontal horn; larval anal segment elongated; in spheroid galls of stem and leaves of some Asteraceae . . . . *Rhoasphondylia*
Pupa with upper frontal horn (as in fig. 108); larval anal segment not elongated; if on Asteraceae not in spheroid galls 9

9(8) Larva without spatula; in bud galls of *Salvia* (Lamiaceae). . . *Sciasphondylia*
Larva with spatula; on other hosts 10

10(9) Larva with the two lateral teeth of spatula smaller than the mesal teeth; in pod galls of *Mimosa* (Fabaceae) . . . . *Tavaresomyia*
Larva with the two lateral teeth of spatula larger than the mesal teeth (fig. 116); on other hosts 11

11(10) Female 9th through 11th flagellomeres subequal in length (fig. 99); the two mesal teeth of spatula tiny; in buds of *Chromolaena* and *Eupatorium* (Asteraceae) . . . . *Perasphondylia*
Female 9th through 11th flagellomeres each progressively shorter than preceding (fig. 98); the two mesal teeth of spatula variable with relation to lateral teeth; not in buds of *Chromolaena* and *Eupatorium* . . . . *Asphondylia*

12(2) Palpus one- or two-segmented; ovipositor short, unpigmented, with many long setae on 10th tergum (fig. 138); pupa without dorsal abdominal spines; larva without spiracles on eighth abdominal segment; in leaf galls of *Eugenia* and *Neomithranthes* (Myrtaceae) . . . . *Stephomyia*
Palpus three- or four-segmented; ovipositor, if short and unpigmented, without many long setae on 10th tergum; pupa with dorsal abdominal spines; larva with spiracles on eighth abdominal segment; on other hosts 13

13(12) Ovipositor greatly elongate, completely pigmented, the setae, if present, extremely tiny and rigid (fig. 134); aedeagus usually narrow and pointed (fig. 133) 14
Ovipositor not much longer than seventh sternite, not or only partly pigmented, with elongate, flexible setae (figs. 122, 129); aedeagus bulbous (figs. 128, 139) 19

14(13) Palpus three-segmented; larva without spatula 15
Palpus four-segmented; larva with spatula 16

15(14) In leaf galls of *Zexmenia* (Asteraceae) . . . . *Ameliomyia*
In leaf galls of *Pisonia* (Nyctaginaceae) . . . . *Pisoniamyia*

16(14) Ovipositor rigid, without setae; gonostyli bilobed apically, the tooth separated from a setose lobe; in flower galls of *Bursera* (Burseraceae) . . . .
. . . . *Burseramyia*
Ovipositor rigid or flexible (fig. 134), with distinct but short setae; gonostyli not indented apically (fig. 133); on other hosts 17

17(16) In hemispherical leaf gall on an unidentified Myrtaceae . . . . *Anasphondylia*
On other hosts. 18

18(17) In cylindrical galls on branch tips of an unidentified Malvaceae . . . .
. . . . *Metasphondylia*
On other hosts . . . . *Schizomyia*

19(13) Palpus four-segmented; spatula present 20
Palpus three-segmented; spatula absent 21

20(19) Ovipositor elongate (fig. 129); spatula clove-shaped (fig. 130); in leaf galls on
Quercus (Fagaceae) . . . . *Polystepha*
Ovipositor short, not protrusible; spatula absent; on other hosts . . . .
. . . . *Macroporpa*

21(19) Male antennal circumfila longitudinally sinuous; female unknown; on *Pisonia* (Nyctaginaceae) . . . . *Pisphondylia*
Antennal circumfila straight, although individual sections distinctly looped;
on *Pisonia* or other hosts 22

22(21) Male and usually female flagellomeres constricted near midlength to form
nearly separate nodes (figs. 124, 125); female with short, broad ovipositor
(fig. 121); on various hosts, including *Pisonia* . . . . *Bruggmannia*
Flagellomeres not constricted near midlength; ovipositor more attenuate (fig.
132); host unknown. . . . . *Proasphondylia*

23(1) First tarsomeres with spur (fig. 83) 24
First tarsomeres without spur (as in fig. 137) 27

24(23) Tarsal claws toothed, strongly bent near basal third (fig. 250); empodia
reaching bend in claw; host unknown . . . . *Epihormomyia*
Tarsal claws simple, bowed beyond midlength (fig. 84); empodia rudimen-
tary; in spherical leaf and stem galls on Fabaceae . . . .
. . . . Tribe Anadiplosini, 25

Uleella, known only from larvae, is not keyed beyond couplet 24.

25(24) Palpus two-segmented; ovipositor bulbous on basal third, tapered beyond
(fig. 88) . . . . *Anadiplosis* and *Machaeriobia*
Palpus one-segmented; ovipositor bulbous for most of its length, tapered only
at very apex (figs. 89, 90). 26

26(25) Female cerci elongate, with tiny setae (fig. 89) . . . . *Alexomyia*
Female cerci short, with long setae (fig. 90) . . . . *Scopodiplosis*

27(23) Tarsal claws bent or bowed near basal third (fig. 159) 28
Tarsal claws bowed at or beyond midlength (fig. 144) 57

28(27) Tarsal claws toothed at least on forelegs (fig. 212). 29
Tarsal claws simple (fig. 159) 45

29(28) Wing with Rs situated beyond midlength of R_1 (fig. 186); at least foretarsal
claws pectinate or with two teeth (figs. 187, 188) 30
Wing with Rs situated appreciably before midlength of R_1 (fig. 253); tarsal
claws not pectinate, usually with only one tooth (fig. 218). 35

30(29) Foreclaws with at least three teeth (fig. 187). 31
Foreclaws with at most one large tooth and one small tooth beneath (fig. 188)
. . . . 32

31(30) Male antenna binodal, tricircumfilar (as in fig. 199); from spherical, hairy
galls on stems of *Gourleia* (Fabaceae) . . . . *Allodiplosis*

Male antenna gynecoid (fig. 189); from blister galls on leaves of *Coccoloba* (Polygonaceae) .... *Ctenodactylomyia*

32(30) Palpus four-segmented; male flagellomeres with circumfilar loops all elongate and subequal (fig. 274)... 33

Palpus three-segmented; male flagellomeres with circumfilar loops dissimilar (fig. 190).. 34

33(32) Gonocoxites with spiny mesobasal lobe (fig. 273); female cerci elongate-cylindrical, without ventral concentration of short, sensory setae (fig. 278); host unknown .... *Youngomyia* in part

Gonocoxites without spiny mesobasal lobe (as in figs. 193, 194); female with concentration of short, sensory setae ventrally; from spherical, hairy galls on leaves of *Ossaea* (Melastomataceae)........................... *Lopesia*

34(32) Female flagellomere necks setulose; from spherical, hairy galls on leaves of *Tibouchina* (Melastomataceae)......................... *Rochadiplosis*

Female flagellomere necks bare; in stem swellings of *Prosopis* (Fabaceae)...

.... *Tetradiplosis*

35(29) Abdominal second through sixth tergites unsclerotized between the caudal and lateral groups of setae (fig. 254) .... *Karshomyia*

Abdominal tergites entire.. 36

36(35) Eyes divided laterally or the facets wider apart there than elsewhere (fig. 229); gonostyli completely setulose (as in fig. 242); predators of coccoids and whiteflies.. *Diadiplosis*

Eyes entire, the facets similar and closely approximated throughout; gonostyli with setulae only at base (fig. 273)............................. 37

37(36) Female flagellomeres with sinuous circumfila; palpus one-segmented; male unknown; from aerial roots of *Coussapoa* (Urticaceae)..... *Frauenfeldiella*

Female antenna with circumfila not conspicuously sinuous; palpus four-segmented.. 38

38(37) Tarsal claws usually toothed on all legs, the toothed claws with two teeth, the proximal tooth tiny (fig. 276) 39

Tarsal claws not toothed on hind legs, usually not toothed on midlegs, and fore claws with only a single tooth (as in fig. 212).................... 41

39(38) Rs joining R_1 about one-fourth distance from arculus to end of R_1; gonocoxites not splayed, without mesobasal lobe; male hypoproct bilobed.........

.... *Resseliella*

Rs joining R_1 about one-half distance from arculus to end of R_1; gonocoxites splayed, with mesobasal lobe; male hypoproct not subdivided into two lobes (fig. 273).. 40

40(39) Female cerci discoid, covered mesally with numerous, closely placed sensory setae (fig. 277), or elongate, cylindrical (fig. 279); male hypoproct much larger than the cerci, covered ventrally with dense, spiniform setulae (fig. 273); hosts unknown...................................... *Youngomyia*

Female cerci discoid, with only two apical sensory setae (fig. 227); male hypoproct not larger than the cerci, without dense, spiniform setulae (fig. 226); from spherical leaf galls of *Heisteria* (Olacaceae) .... *Dactylodiplosis*

41(38) Tarsal claws toothed on fore- and midlegs, simple on hind legs; male first and third circumfila with loops of uneven lengths, some longer than length of flagellomere (fig. 203); predators of aphids.................. *Aphidoletes*
Tarsal claws toothed only on forelegs; male circumfila with loops of similar length or second circumfilum bandlike (fig. 216)..................... 42

42(41) R$_5$ often darker than other major veins; male antenna with second circumfilum bandlike, unlooped, loops of other circumfila uneven in length (fig. 216); female cerci with posteroventral concentration of short, closely placed sensory setae (fig. 215)...................................... *Bremia*
R$_5$ not darker than other veins; male antenna with second circumfilum with at least short loops and other circumfila with loops of equal length; female cerci without posteroventral concentration of short, closely placed sensory setae 43

43(42) Aedeagus pigmented, usually lobed or variously divided (fig. 196).........
.... *Coquillettomyia*
Aedeagus unpigmented, unlobed 44

44(43) Male cerci convex apically (fig. 198); female flagellomeres with setulae on necks.. *Mycodiplosis*
Male cerci angular apically (as in fig. 168); female flagellomeres without setulae on necks .... *Clinodiplosis* in part

45(28) Tergites weakly sclerotized, unpigmented, with single, sparse row of setae posteriorly and no other chaetotaxy except for anterior pair of trichoid sensilla (fig. 265); predators of mites .... *Silvestrina*
Tergites mostly sclerotized, pigmented, extensively covered with setae and scales.. 46

46(45) Abdominal second through sixth tergites unsclerotized between the caudal and lateral groups of setae (fig. 254) .... *Karshomyia*
Abdominal tergites entire (as in fig. 142).......................... 47

47(46) Hypoproct elongate, approximately as wide and long as aedeagus, tapering to apex (fig. 220); female eighth tergite sclerotized, completely covered with setae (fig. 219)..................................... *Charidiplosis*
Hypoproct otherwise; female eighth tergite unsclerotized, with only weak setae posteriorly 48

48(47) Females.. 49
Males 51

49(48) Apical female flagellomeres longest, the circumfila sinuous; male unknown; from pods of *Mimosa* (Fabaceae)........................ *Moehniella*
Basal female flagellomeres longest, the circumfila not sinuous.......... 50

50(49) Cerci with apicoventral concentration of short, blunt setae; male unknown; from spherical twig gall on *Prosopis* (Fabaceae) .... *Liebeliola*
Cerci with setae equally distributed 51

Austrodiplosis, known only from the female, cannot be keyed beyond couplet 50.

51(48,50) Aedeagus pigmented, usually lobed or variously divided (fig. 196)........
.... Coquillettomyia
Aedeagus not pigmented, not lobed or divided (fig. 198) 52

52(51) Male cerci convex apically (fig. 198); female flagellomeres with setulae on necks.. Mycodiplosis
Male cerci angular or quadrate apically (figs. 166, 170); female flagellomeres

53(52) Rostrum greatly enlarged, labella narrow, strongly curved at midlength (fig. 163); adults associated with flowers of Pariana (Poaceae) ....
.... Chauliodontomyia
Rostrum not enlarged, labella ovoid (fig. 162) 54

54(53) Hypoproct with lobes recurved apically (fig. 166); from leaf galls of Manihot (Euphorbiaceae) .... Iatrophobia
Hypoproct with lobes not recurved 55

55(54) Gonostyli abruptly constricted just beyond base; from bud galls on Combretum (Combretaceae) .... Houardodiplosis
Gonostyli not or only slightly narrowed just beyond base 56

56(54) Gonostyli relatively short, robust; hypoproct with splayed lobes; associated with stem galls of Paspalum (Poaceae) .... Cleitodiplosis
Gonostyli elongate, narrow; hypoproct with parallel lobes; in various simple plant galls or mycophagous .... Clinodiplosis in part

57(27) Tarsal claws toothed at least on forelegs (fig. 152)................... 58
Tarsal claws simple on all legs (fig. 144) 60

58(57) Male flagellomeres bicircumfilar (as in fig. 141); female cerci fused (fig. 151); in bud galls of mango..................................... Erosomyia
Male flagellomeres tricircumfilar (as in fig. 222); female cerci separate .. 59

59(58) Gonocoxites with pigmented, glabrous mesobasal lobes (fig. 185); female cerci with posteroventral concentration of short sensory setae (as in fig. 177); predators of coccoids................................. Dicrodiplosis
Gonocoxites with unpigmented, setose or spiny mesobasal lobes; female cerci with only two posteroventral sensory setae; predators of tetranychid mites
.... Feltiella

60(57) Empodia reaching beyond bend in tarsal claws (fig. 271); if only slightly so, the ovipositor is aciculate and pigmented (fig. 270)................... 61
Empodia not reaching beyond bend in tarsal claws (fig. 144) 65

61(60) Ovipositor not needlelike; in resin on Pinus (Pinaceae)........ Cecidomyia
Ovipositor needlelike, modified for piercing plant tissue (figs. 153, 270)....
.... 62

62(61) Palpus four-segmented; host unknown................. Xenasphondylia
Palpus one- to three-segmented; from galls on Solanaceae 63

Neobaezomyia, known only from the female, cannot be keyed beyond couplet 62.

63(62) Palpus one-segmented; from leaf galls on *Grabowskia*....... *Cystodiplosis*
Palpus two- or three-segmented; from galls on *Lycium* 64

64(63) Palpus two-segmented; ovipositor recurved dorsally (fig. 154) ....
....*Jorgensenia*
Palpus three-segmented; ovipositor straight (fig. 153)....... *Centrodiplosis*

65(60) Neck and head greatly elongated, head three times as long as wide ....
.... *Contarinia* in part, *prolixa* only
Neck and head not conspicuously elongated......................... 66

66(65) Wing short with R_1 strongly bowed (fig. 199); ovipositor truncate apically, cerci very tiny (fig. 201); male unknown; from bud galls of *Aspidosperma*
.... *Anasphodiplosis*
Wing long with R_1 weakly bowed (fig. 140); if female cerci tiny, ovipositor gradually tapered, not truncate (fig. 142) 67

67(66) Distal half of protrusible part of ovipositor more than twice length of seventh tergite, attenuate to the tiny cerci, which are dorsoventrally flattened and closely appressed mesally (fig. 142) 68
Ovipositor otherwise, cerci either not closely appressed or completely fused; all males 70

68(67) Antennal circumfila looped .... *Megaulus*
Antennal circumfila appressed.................................... 69

69(68) Abdominal second through seventh tergites without lateral setae at midlength
.... *Stenodiplosis*
Abdominal tergites with several lateral setae at midlength...............
.... *Contarinia* and *Prodiplosis*

70(67) Distalmost male antennal flagellomeres gynecoid ....
.... *Prodiplosis* in part, *floricola* only
All male antennal flagellomeres binodal; all females 71

71(70) Empodia reaching only halfway to bend in tarsal claws; female cerci fused; gall makers on *Taxodium* (Taxodiaceae).................. *Taxodiomyia*
Empodia variable, usually reaching bend in tarsal claws (144); female cerci separate; on other hosts.. 72

72(71) Male flagellomeres bicircumfilar (fig. 241) 73
Male flagellomeres tricircumfilar (as in fig. 222) 78

73(72) Eyes divided laterally (fig. 237); gonostyli completely setulose (fig. 242); female antenna with seven pyriform flagellomeres (fig. 240); probably predaceous.. *Enallodiplosis*
Eyes not divided laterally; gonostyli setulose only at base (fig. 181); female antenna with 12 flagellomeres..................................... 74

74(73) Male circumfilar loops greatly unequal (fig. 182); gonocoxites with acute mesobasal lobe (fig. 181); female cerci with large number of short, closely set,

sensory setae (as in fig. 177)............................ *Thripsobremia*
Male circumfilar loops subequal (fig. 141); gonocoxites without mesobasal
lobe (fig. 143); female cerci with only two sensory setae (fig. 142)...... 75

75(74) Abdominal second to sixth tergites without lateral setae at midlength......
.... *Stenodiplosis*
Abdominal second to sixth tergites with lateral setae at midlength...... 76

76(75) Palpus three-segmented; tarsal claws strongly bent at midlength; from leaf
galls on *Hura* (Euphorbiaceae) .... *Huradiplosis*
Palpus four-segmented; tarsal claws bowed beyond midlength (fig. 144)....
.... 77

77(76) Empodia not reaching bend in tarsal claws; hypoproct weakly concave on
posterior margin; host unknown; female unknown........ *Sphaerodiplosis*
Empodia reaching bend in tarsal claws (fig. 144); hypoproct deeply bilobed;
hosts various .... *Contarinia*

78(72) Gonopods articulated dorsoventrally (fig. 148) .... *Prodiplosis*
Gonopods articulated lateromesally (fig. 143) 79

79(78) R$_5$ shorter than wing (fig. 175), or straight and joining C near end of wing (fig.
174); legs short, first and fifth tarsomeres nearly equal in length; predators
.... Lestodiplosini in part, 80
R$_5$ longer than wing, curved posteriorly to join C behind wing apex (fig. 206);
legs long, fifth tarsomere appreciably longer than first; plant or fungus feeders
.... 83

80(79) R$_5$ joining C before apex of wing (fig. 175) 81
R$_5$ joining C near apex of wing (fig. 174) 82

81(80) R$_5$ straight; scutum covered almost entirely with scales; internal parasitoids
of aphids .... *Pseudendaphis*
R$_5$ sigmoid (fig. 175); scutum without scale covering; predators of mites ...
.... *Arthrocnodax*

82(80) Eyes conspicuously divided laterally .... *Trisopsis*
Eyes not divided laterally................................. *Lestodiplosis*

83(79) Palpus one- to three-segmented.................................... 84
Palpus four-segmented 87

84(83) Empodia reaching bend in tarsal claws; from rosette bud gall on undetermined Rubiaceae .... *Gnesiodiplosis*
Empodia not reaching more than halfway to bend in tarsal claws (fig. 225)
.... 85

85(84) Palpus one-segmented; female postabdomen gradually tapered to cerci (fig.
204); aedeagus much narrower and no longer than hypoproct (fig. 208); from
galls of unknown vine .... *Astrodiplosis*
Palpus two- to three-segmented; female postabdomen abruptly tapered to
protrusible portion of ovipositor (fig. 224) 86

86(85) Empodia reaching halfway to bend in tarsal claws; male circumfilar loops
longer than width of node; cerci and hypoproct subequal in length; from leaf
galls of *Smilax* (Smilacaceae) .... *Compsodiplosis*

Empodia rudimentary (fig. 225); male circumfilar loops shorter than width of node; hypoproct much longer than cerci (fig. 223); from leaf galls of *Styrax* (Styracaceae).. *Contodiplosis*

87(83) Empodia rudimentary; from leaf galls of *Styrax* (Styracaceae) ....
.... *Styraxdiplosis*
Empodia reaching more than halfway to bend in tarsal claws.......... 88

88(87) Male cerci triangular (fig. 170) or broadly convex (fig.181) on posterior margin 89
Male cerci truncate or concave on posterior margin (fig. 168).......... 91

89(88) Intermediate circumfilum ringlike, not looped, remaining circumfila with loops of uneven length (fig. 182) .... *Thripsobremia*
All circumfila similar, the loops subequal in length 90

90(89) Empodia reaching bend in tarsal claws; from leaf roll of undetermined Fabaceae (fig. 160).. *Autodiplosis*
Empodia not reaching bend in tarsal claws; from aerial stem galls of Orchidaceae........................ *Clinodiplosis* in part, *cattleyae* only

91(88) Tarsal claws bowed (as in fig. 159); female 10th tergum with setae; in spherical leaf galls of *Lantana* (Verbenaceae) .... *Schismatodiplosis*
Tarsal claws subangular, usually expanded laterally beyond bend (fig. 158); female 10th tergum without setae; in various situations.................
.... *Clinodiplosis* in part

Tribe Anadiplosini

The tribe Anadiplosini includes eight species in five genera, all Neotropical and possibly all associated with spherical, hairy, leaf and stem galls on Fabaceae. Adults are distinctive for their generally large size (2.5–6.0 mm wing length), the placement of the Rs vein beyond the midlength of R_1 (fig. 85), the slightly curved base of M, the short tibiae relative to the femora and tarsi (figs. 81, 82), the presence of tibial spurs (figs. 83), the rudimentary empodia (fig. 84), and the uniquely modified ovipositor (figs. 88–90). Larvae are distinctive for their caudal lobes (fig. 92), reduced or lost spatula (fig. 93), and reduced papillae. Male antennal flagellomeres have regular, short-looped circumfila (fig. 86) and the distal node constricted between the the two circumfila. Female flagellomeres are long and are constricted proximad of the lower circumfilar ring (fig. 91).

Alexomyia and *Scopodiplosis*, associated with the Anadiplosini for the first time, are each known from a single female caught in flight. Both differ from the remaining genera of Anadiplosini by their shorter, more bulbous ovipositors and one-segmented palpi, and from each other in details of the ovipositor. The three remaining genera, *Anadiplosis*, *Machaeriobia*, and *Uleella*, all reared from Fabaceae, are separated on the basis of larval characters as outlined in the generic descriptions. The differences among these three genera seem superficial inasmuch as adults of *Anadiplosis* and *Machaeriobia* show no essential differences, and *Uleella* is still known only from larvae. I know of females of an undescribed species, reared from an unknown plant in

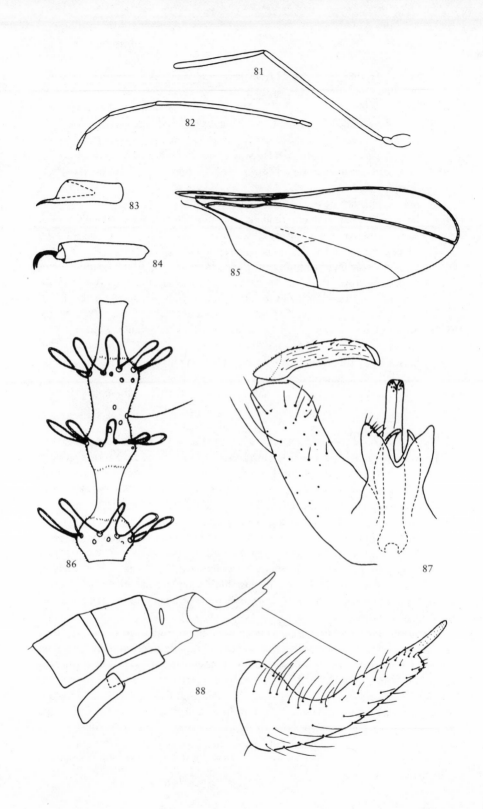

81

82

83

84

85

86

87

88

Colombia, with an ovipositor not referable to any of the known genera as presently delimited.

Alexomyia

Alexomyia Felt 1921a: 141. Type species, *ciliata* Felt (orig. des.).

This genus is known from one female caught in flight. It differs from other Ana-diplosini by its shorter, more bulbous ovipositor (fig. 89). The wing is somewhat narrower than in other genera of the tribe.

ciliata Felt 1921a: 141 (*Alexomyia*); ♀. Holotype: Prata, Pará, Brazil; Felt Coll. Distr.: Brazil.

Anadiplosis

Anadiplosis Tavares 1916: 41. Type species, *pulchra* Tavares (orig. des.).

The four described species included here form spherical leaf or stem galls on Fabaceae. One character Tavares (1920c) used to separate this genus from *Machaeriobia* and *Uleella* is the presence in the larva of mamelons beneath the lateral papillae and the absence of a spatula, but the larva of *A. pulchra*, the type species of *Anadiplosis*, is still unknown. Ref.: Tavares 1920c.

caetetensis Tavares 1920b: 110 (*Anadiplosis*); pupa. Syntypes: Caeteté, Bahia, Brazil; presumed lost. Host: unknown Fabaceae. Distr.: Brazil. Ref.: Tavares 1920c: 70 (formal description; was meant to precede Tavares 1920b).
procera Tavares 1920b: 112 (*Anadiplosis*); gall. Syntypes: near Itaparica, Bahia, Brazil; part presumed lost, but adults and pupae from type galls were sent to Felt by Tavares and are now in the Felt Collection. Host: unknown Mimosoideae (Fabaceae). Distr.: Brazil. Ref.: Tavares 1920c: 66 (formal description of ♂, ♀, pupa; was meant to precede Tavares 1920b).
pulchra Tavares 1916: 42, figs. 3–6 (*Anadiplosis*); ♂, ♀. Syntypes: Nova Friburgo, Rio de Janeiro, Brazil; part presumed lost, part sent to Felt by Tavares and now in Felt Collection. Host: *Machaerium* sp. (bico de pato) (Fabaceae). Distr.: Brazil. Ref.: Tavares 1920c.
venusta Tavares 1916: 46, figs. 7(A–E)–9 (*Anadiplosis*); ♂, ♀, larva. Syntypes: Nova

---

Figures 81–88. Anadiplosini, *Machaeriobia machaerii*: **81,** foreleg, coxa to tibia; **82,** foretibia; **83,** detail: first tarsomere; **84,** detail: fifth tarsomere; **85,** wing; **86,** male third flagellomere; **87,** male genitalia (dorsal); **88,** female postabdomen with detail of ninth segment and fused cerci (lateral).

Friburgo, Rio de Janeiro, Brazil; presumed lost. Host: *Machaerium* sp. (Fabaceae). Distr.: Brazil. Refs.: Tavares 1920b, 1920c.

Machaeriobia

Machaeriobia Rübsaamen 1916a: 448. Type species, *brasiliensis* Rübsaamen (mon.) = *Uleella machaerii* Kieffer.

Ruebsaamenodiplosis (as *Rübsaamenodiplosis*) Tavares 1920c: 61. Type species, *Uleella machaerii* Kieffer (orig. des.). New synonymy.

The one known species of this genus forms spherical leaf galls on *Machaerium*. The presence of a spatula (fig. 93) and the lack of mamelons beneath the lateral papillae separate this genus from *Anadiplosis* and *Uleella* (Tavares 1920c). Adult and larval characteristics are illustrated in figs. 81–88, 92–93.

machaerii Kieffer 1913a: 101 (*Uleella*; no description, cited Rübsaamen 1907: 120, fig. 2 (not 3, an error in Kieffer 1913a, and see Rübsaamen 1907: 155, #79); larva. Syntypes: Tubarão, Santa Catarina, Brazil; presumably in ZMHU. Distr.: Brazil. Host: *Machaerium* sp. (Fabaceae). Refs.: Rübsaamen 1907 (p. 120, fig. 2), Fernandes et al. 1987.
brasiliensis Rübsaamen 1916a: 448, figs. 18–23 (*Macheriobia*); ♀, pupa, larva. New synonym. Syntypes: Tubarão, Santa Catarina, Brazil; ZMHU. Host: *Machaerium* sp. (Fabaceae).

Scopodiplosis

Scopodiplosis Felt 1915c: 209. Type species, *speciosa* (orig. des.).

This genus is based on a single female caught in flight. It differs from the other Anadiplosini in its short ovipositor and long setae covering the cerci (fig. 90). The wing was reportedly dark-banded in life, but the coloration on the slide-mounted type specimen is now indistinct. A flagellomere is illustrated in fig. 91.

speciosa Felt 1915c: 210, fig. 15 (*Scopodiplosis*); ♀. Holotype: S. Bernardino, Paraguay; USNM. Host: unknown. Distr.: Paraguay.

Uleella

Uleella Rübsaamen 1907: 120. Type species, *dalbergiae* Rübsaamen (mon.).
Uleela Gagné 1968a: 26, error.

This genus is known only from the larval stage of a species taken from spherical leaf galls on *Dalbergia*. The lack of a spatula and mamelons beneath the lateral papillae separate this genus from *Anadiplosis* and *Machaeriobia* (Tavares 1920c). Other species previously referred here have been removed to *Bruggmannia* (Asphondyliini).

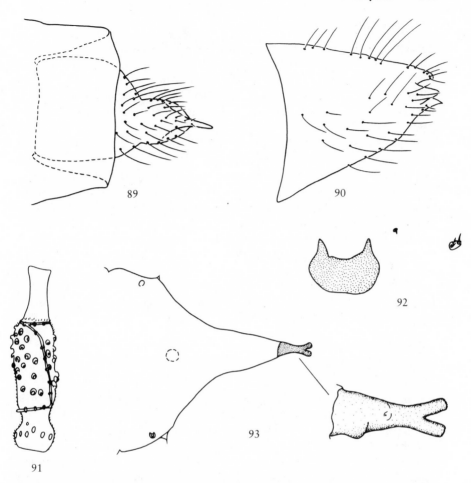

Figures 89–93. Anadiplosini. **89,** *Alexomyia ciliata,* female postabdomen (dorsal). **90, 91,** *Scopodiplosis speciosa*: **90,** female postabdomen (lateral); **91,** female third flagellomere. **92, 93,** *Machaeriobia machaerii*: **92,** larval spatula and associated papillae; **93,** larval eighth and terminal segments with detail of posterior process (dorsal).

dalbergiae Rübsaamen 1907: 121 (*Uleella*); larva. Syntypes: Jacarepagua, Rio de Janeiro, Brazil; ZMHU. Host: *Dalbergia* sp. (Fabaceae). Distr.: Brazil.

Tribe Aphidoletini

The tribe Aphidoletini is a small group of five aphidoid predators in two genera, the chiefly Holarctic *Aphidoletes* and the Palearctic *Monobremia*. The cosmopolitan *Aphidoletes aphidimyza* is the only species known to occur in the Neotropics. The unique character defining this tribe is the two-lobed terminal larval segment, each lobe with four equal-sized but short setae. Ref.: Harris 1966.

Aphidoletes

Aphidoletes Kieffer 1904: 385. Type species, *Bremia abietis* Kieffer (subseq. des. Felt 1911h: 53).

This is a Holarctic genus of four described species; one, *A. aphidimyza*, is widespread over the Holarctic Region and occurs also in Chile. Larvae are aphid predators.

The wing (fig. 202) is 2.0–2.5 mm long. The first and third circumfila of each male flagellomere (fig. 203) have very irregular loops, those on the dorsum exceeding the length of the flagellomere; associated setae are also greatly irregular in length. The tarsal claws are toothed on the fore- and midleg, simple on the hindleg. The gonocoxites are robust and unlobed, and the hypoproct is covered with thick, pliant setulae. The larva has elongate antennae, convex pseudopods on the venter, a clove-shaped spatula, and eight terminal papillae, four situated on each of two posterior extensions of the terminal segment. Ref.: Harris 1973.

aphidimyza Rondani 1847: 443 (*Cecidomya*); adult, pupa, larva. Syntypes: Italy; depos. unknown. Hosts: aphids. Distr.: Chile; Holarctic and elsewhere, possibly spread by commerce. Ref.: Harris 1973.

Tribe Asphondyliini

The tribe Asphondyliini is distributed throughout the world, but nowhere is it as diverse as in the Neotropics. All species live in plants and form galls. Möhn 1961b analyzed this group (as a supertribe) in great detail. He arranged the Holarctic and Neotropical genera into three tribes, and two of those tribes into four informal subtribes each. Asphondyliini that have been found since 1961, the discovery of two important discriminating characters (the paramere and the first tarsomere spur), and the reevaluation of other characters call for some changes in that scheme. Here, the Neotropical genera are divided into two subtribes, the Asphondyliina and the Schizomyiina.

The tribe Asphondyliini has generally been treated as a supertribe equivalent to the supertribe Cecidomyiidi, most recently in Skuhravá (1986). The reasons for this treatment are the distinctive cylindrical male flagellomeres and anastomozing male circumfila (fig. 96) found in most asphondyliines. But in some species of *Bruggmannia*, flagellomeres are binodal (fig. 124) and the male circumfila fairly simple, although not completely separate. I place the Asphondyliini in the supertribe Cecidomyiidi because the antenna have 12 regular flagellomeres that were possibly primitively binodal in the male. Larvae of Asphondyliini primitively have one enlarged pair of recurved, corniform terminal papillae (figs. 120, 131). This is a character shared with the tribe Cecidomyiini.

Asphondyliini form a well-circumscribed, monophyletic group, sharing unique characters of the adult postabdomen. These are the distinctive female seventh sternite (fig. 95) that is much longer than either the preceding sternites or the seventh tergite;

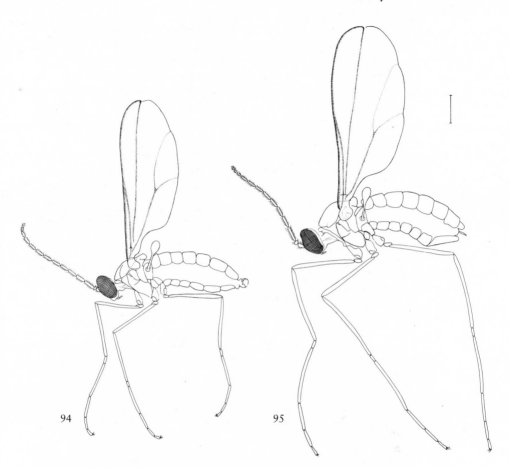

Figures 94, 95. *Asphondylia atriplicis* (scale line = 0.5 mm): **94,** male; **95,** female.

the strongly sclerotized, wide, and laterally notched female eighth tergite (fig. 103); the ventrally lengthened gonocoxites (figs. 104, 123); and the dorsally instead of apically disposed, short, flat, and broad gonostyli (fig. 104). Further characters distinguishing the tribe but shared with other tribes are the lack of a dorsal prominence on the postocciput; the lack of basal teeth on the tarsal claws (fig. 101); the barely evident Sc (fig. 94); and the loss of Rs (fig. 94).

Additional characters are peculiar to the Asphondyliini and of some use in separating genera. Most were discovered, surveyed, and polarized by Möhn (1961b). I list only the new or most useful here.

1. Relative sinuosity and the numbers of circumfilar strands (fig. 96). Both characters have developed separately many times within this tribe, even within genera.

2. The female 9th to 12th flagellomeres of most Asphondyliini are progressively and conspicuously shortened (fig. 98).

3. The palpus becomes reduced from four segments to as few as one, and the reduction has evidently occurred more than once within the tribe.

4. The first tarsomere has an apical spur (fig. 100) in the Asphondyliina but not in the Schizomyiina (fig. 137), except for a smaller spur in those *Schizomyia* species from *Vitis* (fig. 136). The apical spur occurs elsewhere in the Cecidomyiidi only in the Anadiplosini and *Epihormomyia*.

5. The empodia become secondarily reduced in some genera of Schizomyiina.

6. The gonocoxite of Schizomyiina has a small, smooth, apically setose paramere (fig. 123). This character is not present in the Asphondyliina with the exception of *Zalepidota*.

7. The separate denticles of the gonostyli are completely fused in the Asphondyliina (fig. 104), except only partially so in *Zalepidota* (fig. 106). They are not fused in the Schizomyiina (fig. 123).

8. The aedeagus is narrow and pointed (fig. 104) in genera with elongate ovipositors but wide and blunt (fig. 123) in those with short ovipositors, indicating that the wide aedeagus is primitive.

9. The ovipositor may be short, the distal half weakly sclerotized and with many long setae (fig. 121), to elongate and needlelike with short and few to no setae (fig. 123). I consider the short ovipositor with the long setae to be primitive because an ovipositor once strongly modified and with attendant reductions could not return to its former state. Relatively few genera of Asphondyliini with short, soft, long-setose ovipositors are known outside of the Neotropics. They are *Asphoxenomyia*, *Daphnephila*, and *Luzonomyia*, from Asia, *Eocincticornia* from Australia, and *Xenhormomyia* from southern Africa.

10. Species with elongate ovipositors may have a lobe, usually mesally divided, at the end of the eighth tergite (fig. 103). All Asphondyliina have this lobe.

11. The pupal frons may have large "horns," an upper set or a lower set or both (fig. 108). The upper may be centrally or laterally located. Horns help in cutting out of galls and are never present in species that do not also have antennal horns or that escape from the galls as larvae.

12. The larval anus is primitively ventral and slitlike, as seen in some Schizomyiina, to ventral and circular in other Schizomyiina, to dorsocaudal (fig. 117) or dorsal and ovoid or circular in Asphondyliina.

13. The spatula is primitively clove-shaped with six lateral papillae per side, as seen in some Schizomyiina, but both the spatula and papillae become reduced in that tribe. In the Asphondyliina the spatula is primitively enlarged and with only five or fewer lateral papillae per side (fig. 116), but then also becomes secondarily reduced in size and the associated papillae reduced in number.

14. The larval sternal and ventral papillae are primitively without setae (e.g., in *Schizomyia*) but they become setose in various genera.

Table 3. Character matrix for the tribe Asphondyliini. See text for explanation of characters. For all characters except 4 and 9, for which polarity is not implied, 0 indicates the plesiomorphic state, and 1 and 2 the apomorphic·states. A dash indicates that the state could not be determined from either specimens or descriptions.

	Asphondyliina									Schizomyiina											
Character	Asphondylia	Bruggmanniella	Hemiasphondylia	Heterasphondylia	Perasphondylia	Rhoasphondylia	Sciasphondylia	Tavaresomyia	Zalepidota	Ameliomyia	Anasphondylia	Bruggmannia	Burseramyia	Macroporpa	Metasphondylia	Pisoniamyia	Pisphondylia	Polystepha	Proasphondylia	Schizomyia	Stephomyia
1	1	1	1	–	1	1	1	1	1	1	1	0	1	0	1	1	–	0	–	1	1
2	0	0	0	–	1	2	0	0	2	0	2	0,1	0	0	0	0	–	0,2	0	0,2	2
3	1	1	1	1	1	1	1	1	1	1	1	0	1	0	0	0	1	1	0	0	1
4	1	1	1	–	1	1	–	1	1	–	–	0	–	–	0	–	–	0	0	0,1	0
5	0	0	0	0	0	0	0	0	0,1	1	1	0	0	1	0–	1	1	0	1	0,1	0
6	0,1	0	1	0	0	0	0	0	–	0	0	0	0	–	0	0	0	0	0	0	0
7	0,1	1	1	–	0	1	0	0	1	–	–	0	0	–	0	0	0	0	0	0	0
8	2	2	2	2	2	2	–	2	1	–	–	0	0	0	0	0	0	0	0	0	0
9	1	1	1	–	1	1	–	1	0	–	–	0	0	0	0	0	0	0	0	0	0
10	1	1	1	1	1	1	–	1	1	–	–	0	1	0	1	0	0	0	0	1	0
11	1	1	1	–	1	1	1	1	1	0	–	0	0	0	0	0	–	0	0	0	0
12	1	1	1	–	1	1	1	1	1	1	–	0	1	0	1	1	–	0	0	1	0
13	1	1	1	1	1	1	1	–	1	1	–	1	–	–	1	1	1	0	–	0	1
14	1	1	1	1	1	1	1	1	1	1	–	1			1	0,1	0	0	–	0	0

15. The terminal larval segment primitively has eight papillae (fig. 135), two of them corniform, but the segment becomes reduced in size and the number of papillae reduced (fig. 117), although in some Schizomyiina the segment becomes instead elongate-attenuate (fig. 122).

To avoid repetition in the subtribal and generic diagnoses that follow, I have summarized and mostly polarized the states of some of these and other characters in Table 3. All character states except 4 and 9 are polarized following Möhn 1961b, with 0 indicating the plesiomorphic state, 1 and 2 the apomorphic states. No polarization is implied for characters 4 and 9. The numbers to the left of the matrix correspond to the following characters:

1. Female 9th and 12th flagellomeres subequal in length (0) or progressively and conspicuously shortened (1)

2. Female circumfila straight (0), wavy (1), or reticulate (2)
3. Palpus with four segments (0) or fewer (1)
4. First tarsomeres with spur (0) or without (1)
5. Empodia as long as claws (0) or shorter (1)
6. Claws similar on all legs (0) or dissimilar (1)
7. Claws similar in both sexes (0) or dissimilar (1)
8. Teeth of gonostyli denticulate (0), partly denticulate (1), or solid (2)
9. Paramere present (0) or absent (1)
10. Aedeagus wide (0) or narrow (1)
11. Cercilike lobes not present just posterior to female eighth abdominal tergite (0) or present (1)
12. Female ninth abdominal segment short, pliable, with long setae (0) or elongate, rigid, with short setae (1)
13. Pupa pupates in ground (0) or in gall (1)
14. Pupal antennal horns absent (0) or present (1)

Subtribe Asphondyliina

The subtribe Asphondyliina is known in the Neotropics from 97 species in nine genera. The largest, *Asphondylia*, is cosmopolitan and contains 82 Neotropical species. Most species of *Asphondylia* are endemic to this region but a few extend into North America. The remaining seven genera are endemic and have a total of 10 species among them.

This subtribe is similar in concept to the treatment in Möhn 1961b (as a tribe), although *Zalepidota* has some characters in common with Schizomyiini. Asphondyliina have a reduced palpus of one to three segments, an apical spur on the first tarsomere of each leg, a large, bilobed structure at the posterior end of the female eighth tergite, a needlelike ovipositor, no parameres (except in *Zalepidota*), the denticles of the gonostyli at least partly fused into a solid tooth or teeth, the pupal head with antennal horns and usually frontal horns, most of pupal third through eighth abdominal segments covered with spines, the larva with five or fewer lateral papillae, four to six dorsal papillae per side, a reduced number of lateral larval papillae, a reduction in size of the anal segment, the anus more or less dorsal instead of ventral, and the ventral papillae setose.

Figures 96–107. Asphondyliina. 96–98, *Asphondylia websteri*: 96, male third flagellomere; 97, female third flagellomere; 98, female 8th through 12th flagellomeres. 99, *Perasphondylia reticulata*, female 10th through 12th flagellomeres. 100, 101, *Asphondylia websteri*: 100, first tarsomere; 101, fifth tarsomere. 102, *Hemiasphondylia enterolobii*, fifth tarsomere. 103, 104, *Asphondylia websteri*: 103, female postabdomen (lateral; arrow: lateral notch on eighth tergite); 104, male genitalia (dorsal). 105, 106, *Zalepidota piperis*: 105, wing; 106, gonostylus (dorsal). 107, *Bruggmanniella bumeliae*, gonostylus (dorsal).

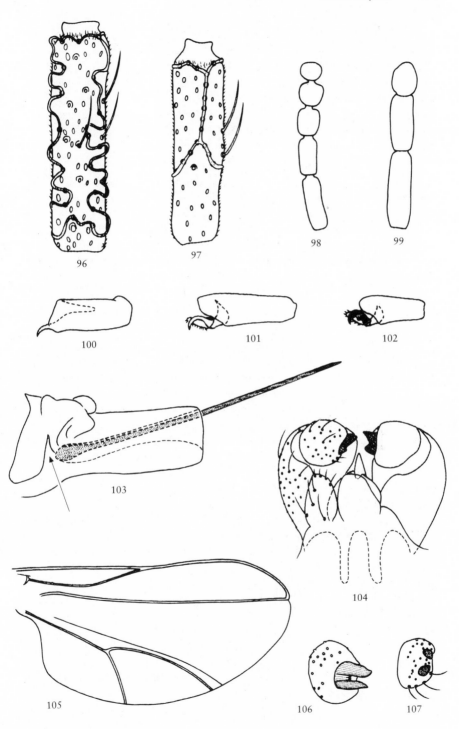

96   97   98   99
100   101   102
103   104
105   106   107

Asphondylia

Asphondylia Loew 1850: 21, 37. Type species, *Cecidomyia sarothamni* Loew (subseq. des. Karsch 1877: 15).
Gisonobasis Rübsaamen 1916a: 432. Type species, *tournefortiae* Rübsaamen (subseq. des. Rübsaamen 1916b: 12).
Eoasphondylia (as subgenus of *Asphondylia*) Möhn 1960a: 214. Type species, *Asphondylia convolvuli* Möhn (orig. des.).

This genus is cosmopolitan, with some 260 described species, of which 82 occur in the Neotropics. Fourteen more are listed in chapter 6 as *Asphondylia* sp. only. *Asphondylia* species form galls in flowers and fruit of a great many families of plants. This genus can be defined by the subtribal definition after the removal of the genera separated from it, some on slight but not necessarily insignificant differences. Figures 94–98, 100, 101, and 103–104 illustrate characters of this genus. *Asphondylia* will doubtless someday be further divided, with shared host groups as an important part of the distinction.

achatocarpi Möhn 1959: 304, pl. 37, figs. 19, 20 (*Asphondylia*); ♀, larva. Holotype: ♀; San Diego, E La Libertad, La Libertad, El Salvador; SMNS. Host: *Achatocarpus nigricans* (Phytolaccaceae). Distr.: El Salvador.
ajallai Möhn 1959: 366, pl. 41, figs. 120–122 (*Asphondylia*); ♀, larva. Holotype: ♀; Los Chorros, La Libertad, El Salvador; SMNS. Host: *Vernonia canescens, V. patens* (Asteraceae). Distr.: El Salvador. Ref.: Möhn 1960a: 213.
altani Felt 1915c: 206 (*Asphondylia*); ♀. Holotype: San Marcos, Nicaragua; USNM. Host: unknown. Distr.: Nicaragua.
ambrosiae Gagné 1975c: 51, figs. 13–16 (*Asphondylia*); ♂, ♀, pupa, larva. Holotype: pupa; Miami, Florida, USA; USNM. Hosts: *Ambrosia artemisiifolia** and *Ambrosia* spp. Distr.: USA (Florida; California and Texas).
anaceliae Möhn 1959: 355, pl. 40, figs. 103–105 (*Asphondylia*); ♂, larva. Holotype: ♂; Hacienda Chilata, Cumbre, La Libertad, El Savador; SMNS. Host: *Bidens refracta, B. squarrosa* (Asteraceae). Distr.: El Salvador. Ref.: Möhn 1960a: 213.
andirae Möhn 1959: 314, pl. 38, figs. 35–37 (*Asphondylia*); pupa, larva. Holotype: larva; San Diego, E La Libertad, La Libertad, El Salvador; SMNS. Host: *Andira inermis* (Fabaceae). Distr.: El Salvador.
annonae Möhn 1959: 309, pl. 38, figs. 26–28 (*Asphondylia*); ♀, larva. Holotype: ♀; N El Roblarón, Santa Ana, El Salvador; SMNS. Host: *Annona cherimola* (Annonaceae). Distr.: El Salvador.
ardisiae Möhn 1959: 333, pl. 39, figs. 67, 68 (*Asphondylia*); ♀, larva. Holotype: ♀; N El Cimarrón, La Libertad, El Salvador; SMNS. Host: *Ardisia paschalis* (Myrsinaceae). Distr.: El Salvador.
attenuatata Felt 1909: 300 (*Asphondylia*). Holotype: West Indies; USNM. Host: *Clerodendrum aculeatum* (Verbenaceae). Distr.: West Indies.
ayeniae Möhn 1960a: 216, pl. 24, figs. 49–65 (*Asphondylia*); ♂, ♀, pupa, larva.

Holotype: ♂; NE Santa Ana, road to Metapán, Santa Ana, El Salvador; SMNS. Host: *Ayenia pusilla* (Sterculiaceae). Distr.: El Salvador.

baccharis Kieffer and Herbst 1905: 65 (*Asphondylia*); gall. Syntypes: San Vicente, Chile; presumed lost. Host: *Baccharis* sp. (Asteraceae). Distr.: Chile. Ref.: Kieffer and Herbst 1906: 330.

bahiensis Tavares 1917b: 156, fig. 4 (A–C) (*Asphondylia*); ♂, ♀, pupa. Syntypes: near Roça da Madre de Deus, between Bahia and Rio Vermelho, Bahia, Brazil; part presumed lost; adults in Felt Collection. Host: Undetermined species of Rubiaceae (corredeira). Distr.: Brazil.

betheli Cockerell 1907: 324 (*Asphondylia*); ♂, pupa. Syntypes: Boulder, Colorado, USA; USNM. Hosts: *Opuntia* spp. (Cactaceae). Distr.: Mexico; USA (southwestern). Refs.: Felt 1916 (as *opuntiae*), Mann 1969.

    arizonensis Felt 1907c: 13 (*Asphondylia*); ♂, ♀. Syntypes: Fort Grant, Arizona, USA; USNM. Host: *Opuntia* sp.

    opuntiae Felt 1908b: 377 (*Asphondylia*); adult. Syntypes: Sinton, Texas, USA; USNM. Host: *Opuntia* sp.

blechi Wünsch 1979: 31, pl. 8 (*Asphondylia*); ♂, pupa. Holotype: ♂; Rio Buritaca, Magdalena, Colombia; SMNS. Host: *Blechum pyramidatum* (Acanthaceae). Distr.: Colombia.

boerhaaviae Möhn 1959: 302, pl. 37, figs. 16–18 (*Asphondylia*); ♂, ♀, larva. Holotype: ♂; near Rio San Felipe, San Vicente, El Salvador; SMNS. Hosts: *Boerhavia coccinea*, *B. diffusa*, *B. erecta*, and *Commicarpus scandens* (Nyctaginaceae). Distr.: Colombia, El Salvador, Jamaica. Refs.: Wünsch 1979, Freeman and Geoghagen 1987, 1989.

borreriae Rübsaamen 1905: 74 (*Asphondylia*); pupa. Syntypes: Harpoador, Rio de Janeiro, Brazil; ZMHU? Hosts: *Borreria* sp. and *Diodia* sp. (Rubiaceae). Ref.: Rübsaamen 1907. Distr.: Brazil.

borrichiae Rossi and Strong 1990: 732, figs. 1–6 (*Asphondylia*); ♂, ♀, pupa, larva. Holotype: pupa; St. Mark's Wildlife Refuge, Wakulla Co., Florida, USA; USNM. Host: *Borrichia frutescens* (Asteraceae). Distr.: Mexico, USA (Florida, South Carolina). Ref.: Stiling et al. 1992.

buddleia Felt 1935: 2 (*Asphondylia*); ♂, ♀. Syntypes: Austin, Texas, USA; Felt Coll. Hosts: *Buddleja americana* and *B. racemosa* (Loganiaceae). Distr.: El Salvador; USA (Texas). Ref.: Möhn 1960a.

cabezasae Möhn 1959: 312, pl. 38, figs. 32–34 (*Asphondylia*); ♂, ♀, larva. Holotype: ♂; Los Chorros, La Libertad, El Salvador; SMNS. Hosts: *Crotalaria mucronata*, *Desmodium nicaraguense*, and *Gliricidia sepium* (Fabaceae). Distr.: El Salvador. Ref.: Möhn 1960a.

caleae Möhn 1959: 357, pl. 41, figs. 106–108 (*Asphondylia*); ♂, larva. Holotype: ♂; lower half of the remainder of the cloud forest of Picacho, San Salvador, El Salvador; SMNS. Host: *Calea integrifolia* (Asteraceae). Distr.: El Salvador.

camarae Möhn 1959: 340, pl. 40, figs. 78–81 (*Asphondylia*); ♂, ♀, larva. Holotype: ♂; km 27, E San Martin, Cuscatlán, El Salvador; SMNS. Hosts: *Lantana camara* and *L. urticifolia* (Verbenaceae). Distr.: El Salvador, Mexico, Panama.

canavaliae Wünsch 1979: 33, pls. 9, 10 (*Asphondylia*); ♀, pupa, larva. Holotype: ♀; Cañaverales, Magdalena, Colombia; SMNS. Host: *Canavalia maritima* (Fabaceae). Distr.: Colombia.

caprariae Wünsch 1979: 53, pl. 16 (*Asphondylia*); ♂, ♀, pupa, larva. Holotype: ♂; Bahia Nenguange, Magdalena, Colombia; SMNS. Host: *Capraria biflora* (Scrophulariaceae). Distr.: Colombia.

cestri Möhn 1959: 344, pl. 40, figs. 86, 87 (*Asphondylia*); ♂, ♀, larva. Holotype: ♂; Los Chorros, La Libertad, El Salvador; SMNS. Host: *Cestrum nocturnum* (Solanaceae). Distr.: El Salvador.

cissi Möhn 1959: 324, pl. 39, figs. 51, 52 (*Asphondylia*); ♂, larva. Holotype: ♂; ca. 8 km before La Herradura, La Paz, El Salvador; SMNS. Host: *Cissus sicyoides* (Vitaceae). Distr.: El Salvador.

clematidis Felt 1935: 3 (*Asphondylia*); ♂, ♀. Syntypes: Austin, Texas, USA; Felt Coll. Host: *Clematis dioica, C. drummondii**, and *C. grossa* (Ranunculaceae). Distr.: El Salvador, USA (Texas). Ref.: Möhn 1960a.

convolvuli Möhn 1960a: 214, pl. 23, figs. 33–46, pl. 24, figs. 47, 48 (*Asphondylia*); ♂, ♀, pupa, larva. Holotype: ♂; NE Sitio del Niño, W lava fields, La Libertad, El Salvador; SMNS. Host: *Ipomoea arborescens* (Convolvulaceae). Distr.: El Salvador.

corbulae Möhn 1960a: 197, pl. 22, figs. 1–3 (*Asphondylia*); ♂, ♀, larva. Holotype: ♂; NW Las Chinamas, Ahuachapán, El Salvador; SMNS. Hosts: *Chromolaena odorata, Eupatorium* sp., and *Fleischmannia microstemon* (Asteraceae). Distr.: El Salvador, Trinidad. Ref.: Gagné 1977b: 117, fig. 22.

cordiae Möhn 1959: 338, pl. 40, figs. 75–77 (*Asphondylia*); ♂, ♀, larva. Holotype: ♂; W Ateos, La Libertad, El Salvador; SMNS. Hosts: *Cordia alba* and *C. dentata* (Boraginaceae). Distr.: El Salvador.

curatellae Möhn 1959: 330, pl. 39, figs. 62, 63 (*Asphondylia*); ♂, ♀, larva. Holotype: ♂; SE La Palma, Chalatenango, El Salvador; SMNS. Host: *Curatella americana* (Dilleniaceae). Distr.: El Salvador.

duplicornis Wünsch 1979: 51, pl. 15 (*Asphondylia*); ♀, pupa. Holotype: ♀; Cañaverales, Magdalena, Colombia; SMNS. Host: *Melanthera aspera* (Asteraceae). Distr.: Colombia.

erythroxylis Möhn 1959: 315, pl. 38, figs. 38, 39 (*Asphondylia*); ♀, larva. Holotype: ♀; NE Sitio del Niño, W lava fields, La Libertad, El Salvador; SMNS. Host: *Erythroxylum mexicanum* (Erythroxylaceae). Distr.: El Salvador.

evae Wünsch 1979: 39, pls. 11, 12 (*Asphondylia*); ♂, ♀, pupa, larva. Holotype: ♂; Bahia Gairaca, Magdalena, Colombia; SMNS. Host: *Chamissoa* sp. (Amaranthaceae). Distr.: Colombia.

godmaniae Möhn 1959: 346, pl. 40, figs. 89–91 (*Asphondylia*); pupa, larva. Holotype: larva; W Chalatenango, Chalatenango, El Salvador; SMNS. Host: *Godmania aesculifolia* (Bignoniaceae). Distr.: El Salvador.

guareae Möhn 1959: 316, pl. 38, figs. 40–42 (*Asphondylia*); ♂, ♀, larva. Holotype: ♂; forest of San Marcos Lempa, Usulután, El Salvador; SMNS. Host: *Guarea* sp. (Meliaceae). Distr.: El Salvador.

helicteris Möhn 1959: 327, pl. 39, figs. 56–58 (*Asphondylia*); ♀, larva. Holotype: ♀;

NW Olocuilta, La Paz, El Salvador; SMNS. Host: *Helicteres mexicana* (Sterculiaceae). Distr.: El Salvador.

herculesi Möhn 1959: 363, pl. 41, figs. 117–119 (*Asphondylia*); ♂, ♀, larva. Holotype: ♂; SE La Palma, Chalatenango, El Salvador; SMNS. Host: *Vernonia canescens* (Asteraceae). Distr.: El Salvador.

hyptis Möhn 1960a: 206, pl. 22, fig. 15 (*Asphondylia*); pupa. Holotype: pupa; ravine near ITIC, San Salvador, San Salvador, El Salvador; SMNS. Host: *Hyptis verticillata* (Lamiaceae). Distr.: El Salvador.

indigoferae Möhn 1960a: 204, pl. 22, figs. 15, 16 (*Asphondylia*); ♂, larva. Holotype: ♂; ravine near ITIC, San Salvador, San Salvador, El Salvador; SMNS. Host: *Indigofera suffruticosa* (Fabaceae). Distr.: El Salvador.

jatrophae Möhn 1959: 319, pl. 38, figs. 46, 47 (*Asphondylia*); ♀, larva. Holotype: ♀; ca. 8 km before La Herradura, La Paz, El Salvador; SMNS. Host: *Jatropha curcas* (Euphorbiaceae). Distr.: El Salvador.

lippiae Möhn 1959: 342, pl. 40, figs. 82–84 (*Asphondylia*); ♂, ♀, larva. Holotype: ♂; E San Marcos, San Salvador, El Salvador; SMNS. Host: *Lippia cardiostegia* (Verbenaceae). Distr.: El Salvador.

lonchocarpi Möhn 1960a: 205, pl. 22, figs. 17, 18 (*Asphondylia*); ♀, larva. Holotype: ♀; N Los Cóbanos, Sonsonate, El Salvador; SMNS. Host: *Lonchocarpus minimiflorus* (Fabaceae). Distr.: El Salvador.

lopezae Wünsch 1979: 43, pls. 12, 13 (*Asphondylia*); ♂, ♀, pupa, larva. Holotype: ♂; vic. Santa Marta, Magdalena, Colombia; SMNS. Hosts: *Amaranthus dubius*, *A. spinosus*, and *Iresine angustifolia* (Amaranthaceae). Distr.: Colombia.

malvavisci Möhn 1959: 325, pl. 39, figs. 53–55 (*Asphondylia*); ♂, ♀, larva. Holotype: ♂; ravine near ITIC, San Salvador, San Salvador, El Salvador; SMNS. Host: *Malvaviscus arboreus* (Malvaceae). Distr.: El Salvador.

melantherae Möhn 1959: 360, pl. 41, figs. 112–114 (*Asphondylia*); ♂, ♀, larva. Holotype: ♂; N El Delirio, San Miguel, El Salvador; SMNS. Host: *Melanthera nivea* (Asteraceae). Distr.: El Salvador.

moehni Skuhravá 1989: 203 (*Asphondylia*), new name for *tavaresi* Möhn. Host: *Mikania guaco* (Asteraceae). Distr.: Brazil.

   tavaresi Möhn 1973: 4, pl. 1, figs. 7–13 (*Asphondylia*); ♂, pupa, larva. Preoccupied by Rübsaamen 1916b. Holotype: ♂; São Leopoldo, Rio Grande do Sul, Brazil; Tavares Coll.

ocimi Möhn 1959: 343, pl. 40, fig. 85 (*Asphondylia*); ♂. Holotype: ♂; SW Jocoro, Morazán, El Salvador; SMNS. Host: *Ocimum micranthum* (Lamiaceae). Distr.: El Salvador.

palaciosi Möhn 1959: 310, pl. 38, figs. 29–31 (*Asphondylia*); ♂, ♀, larva. Holotype: ♂; Ciudad Arce, Santa Ana, El Salvador; SMNS. Host: *Crotalaria longirostrata* (Fabaceae). Distr.: El Salvador.

parasiticola Möhn 1960a: 202, pl. 22, figs. 10–13 (*Asphondylia*); ♀, pupa, larva. Holotype: ♀; near Ahuachapán, Ahuachapán, El Salvador; SMNS. Host: *Struthanthus marginatus* (Loranthaceae). Distr.: El Salvador.

parva Tavares 1917b: 159 (*Asphondylia*); ♂, ♀, pupa. Syntypes: near Roça de

Madre de Deus, Bahia, Bahia, Brazil; presumed lost. Host: undetermined species of Rubiaceae (carqueja). Distr.: Brazil.

pattersoni Felt 1911d: 301 (*Asphondylia*); ♂, ♀. Syntypes: St. Vincent; USNM. Host: *Citharexylum quadrangulare* (Verbenaceae). Distr.: St. Vincent.

portulacae Möhn 1959: 308, pl. 37, figs. 24, 25 (*Asphondylia*); pupa, larva. Holotype: pupa; San Diego, E La Libertad, La Libertad, El Salvador; SMNS. Host: *Portulaca oleracea* (Portulacaceae). Distr.: Argentina, Bolivia, Colombia, El Salvador, Jamaica, Montserrat, Nevis, St. Kitts, Trinidad; USA (Florida). Ref.: Wünsch 1979.

psychotriae Möhn 1959: 349, pl. 40, figs. 95, 96 (*Asphondylia*); ♂, ♀, larva. Holotype: ♂; ravine near ITIC, San Salvador, San Salvador, El Salvador; SMNS. Host: *Psychotria pubescens* (Rubiaceae). Distr.: El Salvador.

randiae Möhn 1959: 350, pl. 40, figs. 97–99 (*Asphondylia*); ♂, ♀, larva. Holotype: ♂; SE Ciudad Arce, La Libertad, El Salvador; SMNS. Host: *Randia aculeata* (Rubiaceae). Distr.: El Salvador.

rochae Tavares 1918b: 686, pl. III, 9–11 (*Asphondylia*); ♂, ♀, pupa. Syntypes: Fortaleza, Ceará, Brazil; presumed lost. Host: *Jussiaea* sp. (Onagraceae). Distr.: Brazil.

rondeletiae Möhn 1959: 352, pl. 40, figs. 100–102 (*Asphondylia*); ♂, ♀, larva. Holotype: ♂; lower half of remaining rain forest of Picacho, San Salvador, El Salvador; SMNS. Host: *Rondeletia strigosa* (Rubiaceae). Distr.: El Salvador.

ruelliae Möhn 1959: 347, pl. 40, figs. 92–94 (*Asphondylia*); ♂, ♀, larva. Holotype: ♂; Lago de Coatepeque, Santa Ana, El Salvador; SMNS. Host: *Ruellia albicaulis* (Acanthaceae). Distr.: El Salvador.

salvadorensis Möhn 1959: 358, pl. 41, figs. 109–111 (*Asphondylia*); ♂, ♀, larva. Holotype: ♂; ravine near ITIC, San Salvador, San Salvador, El Salvador; SMNS. Host: *Hymenostephium cordatum* (Asteraceae). Distr.: El Salvador.

scopariae Möhn 1959: 345, pl. 40, fig. 88 (*Asphondylia*); ♂, ♀. Holotype: ♂; E Santa Barbara, Chalatenango, El Salvador; SMNS. Host: *Scoparia dulcis* (Scrophulariaceae). Distr.: El Salvador.

serjaneae Möhn 1959: 322, pl. 38, figs. 48–50 (*Asphondylia*); ♂, larva. Holotype: ♂; ravine near ITIC, San Salvador, San Salvador, El Salvador; SMNS. Host: *Serjania goniocarpa* (Sapindaceae). Distr.: El Salvador.

siccae Felt 1908b: 376 (*Asphondylia*); adult. Syntypes: Jamaica; Felt Coll. Host: *Phyllanthus distichus* (Euphorbiaceae). Distr.: Jamaica. Ref.: Felt 1916.

sidae Wünsch 1979: 36, pls. 10, 11 (*Asphondylia*); ♂, ♀, pupa, larva. Holotype: ♂; Cañaverales, Magdalena, Colombia; SMNS. Host: *Sida acuta* and *S. rhombifolia* (Malvaceae). Distr.: Colombia.

stachytarpheta Barnes 1932: 482 (*Asphondylia*); ♂, ♀. Syntypes: Trinidad; BMNH. Host: *Stachytarpheta cayennensis* (Verbenaceae). Distr.: Trinidad.

struthanthi Rübsaamen 1916a: 432, fig. 1 (*Gisonobasis*); ♂, larva. Syntypes: Serra do Baturité, Ceará, Brazil; ZMHU? Host: *Struthanthus* sp. (Loranthaceae). Distr.: Brazil. Refs.: Möhn 1960a, 1973.

sulphurea Tavares 1909: 14, pl. I, fig. 1 (*Asphondylia*); ♂. Syntypes: Rio Grande do

Sul, Brazil; presumed lost. Host: possibly *Smilax* (Smilacaceae). Distr.: Brazil. Ref.: Möhn 1973.

swaedicola Kieffer and Jörgensen 1910: 435 (*Asphondylia*); ♂, ♀, pupa. Syntypes: San Ignacio, Potrevillos, and La Paz, Mendoza, and Cancete, San Juan, Argentina; presumed lost. Host: *Suaeda divaricata* (Chenopodiaceae). Distr.: Argentina.

tabernaemontanae Möhn 1959: 333, pl. 39, figs. 69–71 (*Asphondylia*); ♂, ♀, larva. Holotype: ♂; N Los Cóbanos, Sonsonate, El Salvador; SMNS. Host: *Tabernaemontana amygdalifolia* (Apocynaceae). Distr.: El Salvador.

talini Möhn 1959: 305, pl. 37, figs. 21–23 (*Asphondylia*); ♂, larva. Holotype: ♂; Hacienda Chilata, Cumbre, La Libertad, El Salvador; SMNS. Host: *Talinum paniculatum* (Portulacaceae). Distr.: El Salvador.

thevetiae Möhn 1959: 337, pl. 39, figs. 72–74 (*Asphondylia*); ♂, ♀, larva. Holotype: ♂; SE Santo Domingo, km 54, San Vicente, El Salvador; SMNS. Host: *Thevetia plumeriaefolia* (Apocynaceae). Distr.: El Salvador.

tithoniae Möhn 1960a: 200, pl. 22, figs. 6, 7 (*Asphondylia*); ♂, ♀, pupa. Holotype: ♂; E Sitio del Niño, La Libertad, El Salvador; SMNS. Host: *Tithonia rotundifolia* (Asteraceae). Distr.: El Salvador.

tournefortiae Rübsaamen 1916a: 433, figs. 2a–c, 3b (*Gisonobasis*); ♀, pupa, larva. Syntypes: Auristella and S. Francisco, Acre R., Amazonas, Brazil; ZMHU? Hosts: *Tournefortia angustiflora* and *T. volubilis* (Boraginaceae). Distr.: Brazil, El Salvador. Ref.: Möhn 1960a.

trichiliae Möhn 1959: 318, pl. 38, figs. 43–45 (*Asphondylia*); ♂, ♀, larva. Holotype: ♂; NE Sitio del Niño, W lava fields, La Libertad, El Salvador; SMNS. Host: *Trichilia* sp. (Meliaceae). Distr.: El Salvador.

trixidis Möhn 1959: 362, pl. 41, figs. 115, 116 (*Asphondylia*); ♂, larva. Holotype: ♂; NE Sitio del Niño, W lava fields, La Libertad, El Salvador; SMNS. Host: *Trixis radialis* (Asteraceae). Distr.: El Salvador.

ulei Rübsaamen 1907: 172 (*Asphondylia*); pupa, larva. Syntypes: Palmeiras, Rio de Janeiro, Brazil; ZMHU? Host: *Mikania* sp. (Asteraceae). Distr.: Brazil. Ref.: Möhn 1973.

vincenti Felt 1911b: 109 (*Asphondylia*); ♂, ♀, pupa, larva. Syntypes: St. Vincent; Felt Coll. Hosts: *Jussiaea angustifolia, J. linifolia,* and *J. suffruticosa* (Onagraceae). Distr.: El Salvador, Puerto Rico, St. Vincent. Refs.: Needham 1941, Möhn 1960a.

waltheriae Möhn 1959: 328, pl. 39, figs. 59–61 (*Asphondylia*); ♂, ♀, larva. Holotype: ♂; E Tablón del Coco, Santa Ana, El Salvador; SMNS. Host: *Waltheria americana* (Sterculiaceae). Distr.: El Salvador.

websteri Felt 1917b: 562 (*Asphondylia*). Lectotype: ♀ (designated by Gagné and Wuensche 1986); Tempe, Arizona, USA; Felt Coll. Hosts: *Cyamopsis tetragonoloba* and *Phaseolus* sp.; also *Medicago sativa**, and *Parkinsonia aculeata* (Fabaceae). Distr.: Dominican Republic, Honduras; Mexico (northern), USA (southwestern). Ref.: Gagné and Woods 1988.

ximeniae Möhn 1959: 301, pl. 37, figs. 1–15 (*Asphondylia*); ♂, ♀, larva. Holotype: ♂; NE Sitio del Niño, W lava fields, La Libertad, El Salvador; SMNS. Host: *Ximenia americana* (Olacaceae). Distr.: Brazil?, El Salvador.

xylosmatis Möhn 1959: 331, pl. 39, figs. 64–66 (*Asphondylia*); ♂, ♀, larva. Holotype: ♂; SE Candelaria, Hacienda Cortez, Santa Ana, El Salvador; SMNS. Host: *Xylosma flexuosa* (Flacourtiaceae).

yukawai Wünsch 1979: 47, pls. 14, 15 (*Asphondylia*); ♂, ♀, pupa, larva. Holotype: ♂; vic. Santa Marta, Magdalena, Colombia; SMNS. Hosts: *Arenaria* sp. (Caryophyllaceae) and *Melochia caracasana*, *M. lupulina*, and *M. nodiflora* (Sterculiaceae). Distr.: Colombia.

zacatechichi Möhn 1960a: 199, pl. 22, figs, 4, 5 (*Asphondylia*); larva. Holotype: SE Santo Domingo, km 54, San Vicente, El Salvador; SMNS. Host: *Calea zacatechichi* (Asteraceae). Distr.: El Salvador.

zexmeniae Möhn 1960a: 201, pl. 22, figs. 8, 9 (*Asphondylia*); pupa, larva. Holotype: pupa; ravine near ITIC, San Salvador, San Salvador, El Salvador; SMNS. Host: *Zexmenia frutescens* (Asteraceae). Distr.: El Salvador.

Bruggmanniella

Bruggmanniella Tavares 1909: 19. Type species, *braziliensis* Tavares (mon.).
Hemibruggmanniella Möhn 1961a: 6. New synonymy. Type species, *Bruggmanniella oblita* Tavares (orig. des.).

This genus consists of three described species, the two listed below and *Bruggmanniella bumeliae* (Felt), new combination, from Texas, USA. All three species live in woody stem swellings but in different host families. The Texas species is from Sapotaceae.

Hemibruggmanniella was differentiated from *Bruggmanniella* on the basis of a difference in the number of teeth on the larval spatula, three in the former, four in the latter (Möhn 1961a). The male of the one species of *Hemibruggmanniella*, *H. oblita*, was unknown, but the more recent discovery of the larva of *B. bumeliae* with a spatula (fig. 118) shaped as for *H. oblita* and male genitalia similar to *Bruggmanniella braziliensis* indicate that the reduced spatula of *H. oblita* is of too little importance as the basis for a separate genus.

Adults resemble *Asphondylia* except that the tooth of the gonostylus of *Bruggmanniella* is completely divided mesally, resulting in two separate teeth (fig. 107). A similar secondary development is found in the Palearctic and Australasian genus *Pseudasphondylia*, but pupae of that genus resemble those of *Asphondylia* (Yukawa 1974, Coutin 1980). Pupae of *Bruggmanniella* have well-developed antennal horns but no frontal horns (fig. 114), and the dorsal abdominal spines are tiny, numerous,

---

Figures 108–114. Asphondyliina pupae. 108, *Asphondylia atriplicis* (lateral; scale line = 0.5 mm). 109, 110, *Asphondylia websteri*: 109, head (ventral); 110, terminal segments (dorsal). 111, 112, *Zalepidota piperis*: 111, pupal head (ventral); 112, pupal abdomen, terminal segments (dorsal). 113, 114, *Bruggmanniella bumeliae*: 113, terminal segments (dorsal); 114, head (ventral).

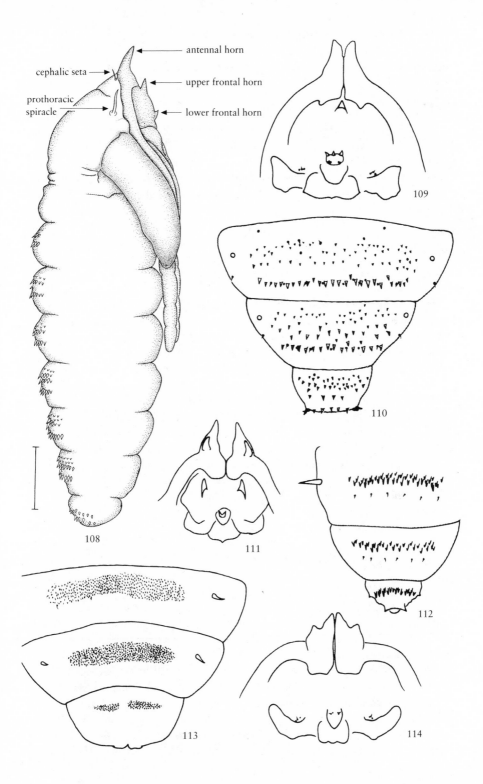

antennal horn

cephalic seta

upper frontal horn

prothoracic
spiracle

lower frontal horn

109

110

108

111

112

113

114

and continuous from the anterior margin of the tergum to the row of dorsal papillae (fig. 113).

braziliensis Tavares 1909: 20 (*Bruggmanniella*); ♂, ♀, pupa. Lectotype (Möhn 1963a: 5): ♀; São Leopoldo, Rio Grande do Sul, Brazil; Tavares Coll. Host: possibly *Sorocea ilicifolia* (Moraceae). Distr.: Brazil. Ref.: Möhn 1963a.

oblita Tavares 1920a: 34 (*Bruggmanniella*); ♀, pupa. Restored combination. Lectotype (Möhn 1961a: 7): ♀; Nova Friburgo, Rio de Janeiro, Brazil; Tavares Coll. Host: *Schinus* sp. (Anacardiaceae). Distr.: Brazil. Ref.: Möhn 1961a: 7, redescr. ♀, pupa, larva from orig. Tavares galls.

Hemiasphondylia

Hemiasphondylia Möhn 1960a: 229. Type species, *mimosae* Möhn (orig. des.).

This genus is known from two Neotropical and one Nearctic species, all from bud galls of Mimosoideae (Fabaceae). Adults are separated from other Asphondyliina by the modified tarsal claws that widen near midlength (fig. 102). Larvae differ from other Asphondyliini by the short-shafted spatula in combination with the large pair of corniform terminal papillae (figs. 119, 120). Refs.: Gagné 1978a, Wünsch 1979: 57.

enterolobii Gagné 1978a: 514, figs. 1, 2, 4–7 (*Asphondylia*): ♂, ♀, pupa, larva. New combination. Holotype: pupa; Santa Rosa National Park, Guanacaste, Costa Rica; USNM. Host: *Enterolobium cyclocarpum* (Fabaceae). Distr.: Costa Rica.

mimosae Möhn 1960a: 230, pl. 27, figs. 121–137 (*Hemiasphondylia*): ♂, ♀, pupa, larva. Restored name and combination. Holotype: ♂; N El Cimmarón, La Libertad, El Salvador; SMNS. Host: *Mimosa albida*, *M. leiocarpa*, and *Prosopis juliflora*. Distr.: Colombia, El Salvador. Ref.: Wünsch 1979.

mimosicola Gagné 1978a: 514. New name for *mimosae* Möhn 1960a, as secondary homonym of *Asphondylia mimosae* Felt 1934.

Heterasphondylia

Heterasphondylia Möhn 1960a: 227. Type species, *salvadorensis* Möhn (orig. des.).

This genus is known from one species in stem galls of *Piper* (Piperaceae). I have not seen adults so do not know whether they have a spur on the first tarsomeres and whether the male genitalia have a paramere. The pupa, however, is similar to that of *Zalepidota* with the ventral lobe on each of the antennal horns (see fig. 111), in combination with widely separated upper frontal horns and elongated spiracles on the first through seventh abdominal segments (see fig. 112). Also, the dorsal abdominal spines, as in most Schizomyiina, consist of one to two rows that do not attain the dorsal row of papillae. The pupa of *Zalepidota* has no spiracles on the eighth segment, but whether these occur on *Heterasphondylia* is unknown. The larva of

Heterasphondylia has no spatula, differing from that of *Zalepidota*, which has a short one.

salvadorensis Möhn 1960a: 228, pl. 26, figs. 111–114, pl. 27, figs. 115–120 (*Heterasphondylia*); ♂, pupa, larva. Holotype: ♂; ravine near ITIC, San Salvador, San Salvador, El Salvador; SMNS. Host: *Piper* sp. (Piperaceae). Distr.: El Salvador.

Perasphondylia

Perasphondylia Möhn 1960a: 218. Type species, *reticulata* Möhn (orig. des.).

This Neotropical genus is known from one species on bud galls of Eupatorieae (Asteraceae). Antennal circumfila anastomose more than in most Asphondyliini, and the apical female flagellomeres (fig. 99) are not so short as in other genera of this tribe. The larva has the mesal teeth of the spatula much shortened but otherwise resembles *Asphondylia*.

reticulata Möhn 1960a: 219, pl. 24, figs. 66–70, pl. 25, figs. 71–82 (*Perasphondylia*): ♂, ♀, pupa, larva. Holotype: ♂; W Chalatenango, Chalatenango, El Salvador; SMNS. Hosts: *Chromolaena odorata* and *Eupatorium* sp. (Asteraceae). Distr.: Bolivia, Brazil, El Salvador, Mexico, and Trinidad. Ref.: Gagné 1977b.

Rhoasphondylia

Rhoasphondylia Möhn 1960a: 224. Type species, *Oxasphondylia friburgensis* Tavares (orig. des.).

This genus contains three described and at least one undescribed Neotropical species that live in complex galls on leaves and stems of some Asteraceae. Adults have two-segmented palpi, and the antennal circumfila are more reticulate, especially in females, than in other Asphondyliini. The pupa has no upper frontal horn and only a small single-pointed lower frontal horn. The genus is most striking for the elongate anal segment of the larva. The spatula is short with two narrow teeth. Only two lateral papillae are present on each side of the spatula.

crassipalpis Kieffer and Jörgensen 1910: 365 (*Asphondylia*); ♂, ♀, pupa, larva. New combination. Syntypes: Chacras de Coria and Pedregal, Mendoza, and Cancete, San Juan, Argentina; presumed lost. Host: *Baccharis salicifolia* (Asteraceae). Distr.: Argentina.

friburgensis Tavares 1917b: 126, Fig. A (1–6) (*Oxasphondylia*); ♀, pupa. Syntypes: Nova Friburgo, Rio de Janeiro, Brazil; presumed lost. Hosts: *Baccharis dracunculifolia*, *B. schultzii*, and *B. trinervis* (Asteraceae). Distr.: Brazil, El Salvador. Ref.: Möhn 1960a.

sanpedri Wünsch 1979: 63, pls. 19, 20 (*Rhoasphondylia*); ♂, pupa, larva. Holotype:

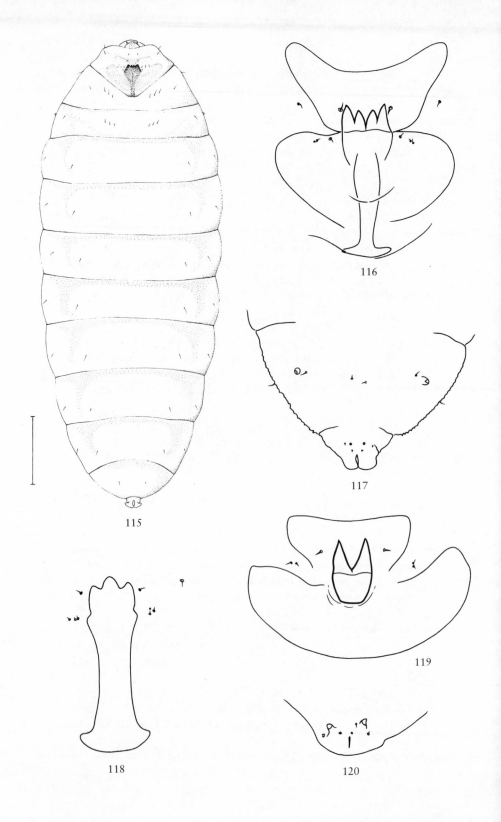

115

116

117

118

119

120

♂; vic. Santa Marta, Magdalena, Colombia; SMNS. Host: *Chaetospira funcki* (Asteraceae). Distr.: Colombia.

Sciasphondylia

Sciasphondylia Möhn 1960a: 221. Type species, *salviae* Möhn (orig. des.).

This genus is known from one Neotropical species, noteworthy only because the larva lacks a spatula. The male is thus far unknown and could have characters that would set the genus off better.

salviae Möhn 1960a: 222, pl. 25, figs. 83–90, pl. 26, figs. 91, 92 (*Sciasphondylia*); ♀, pupa, larva. Holotype: ♀; road to La Palma, ca. 15 km before La Palma, Chalatenango, El Salvador; SMNS. Host: *Salvia* sp. (Lamiaceae). Distr.: El Salvador.

Tavaresomyia

Tavaresomyia Möhn 1961a: 9. Type species, *Schizomyia mimosae* Tavares (orig. des.).

This genus is known from one species reared from fruit of Mimosoideae (Fabaceae). It is distinguished from other Asphondyliina by the shape of the spatula, which is four-pronged, the two middle prongs longest (Möhn 1961a). The pupa is unknown, but that stage could have additional distinguishing characters to further set off this genus.

mimosae Tavares 1925: 28, fig. 9 (*Schizomyia*); ♂, ♀. Syntypes: Ceará, Brazil; presumed lost. Host: *Mimosa caesalpinifolia* (Fabaceae). Ref.: Möhn 1961a.

Zalepidota

Zalepidota Rübsaamen 1907: 38. Type species, *piperis* Rübsaamen (mon.).
Ozobia Kieffer 1913a: 97. Type species, *tavaresi* Kieffer (orig. des.).
Oxasphondylia Felt 1915c: 204. Type species, *reticulata* Felt (orig. des.).

This genus is known from four species, two from stem galls of *Piper* (Piperaceae), one caught in flight, and the other only doubtfully placed here. This genus does not fit cleanly with the Asphondyliina because it has parameres, unlike all other genera of

---

Figures 115–120. Asphondyliina larvae. **115**, *Asphondylia atriplicis* (ventral; scale line = 0.5 mm). **116, 117**, *Asphondylia websteri*: **116**, spatula and associated papillae; **117**, terminal segments (dorsal). **118**, *Bruggmanniella bumeliae*, spatula and associated papillae. **119, 120**, *Hemiasphondylia enterolobii*: **119**, spatula and associated papillae; **120**, terminal segment (dorsal).

the subtribe, and the teeth of the gonostyli have some separate denticles (fig. 106) instead of being completely solid. *Zalepidota* is also distinct for its dense network of circumfila, one- to two-segmented palpus, the wide wing (fig. 105), the swollen basal part of the tarsal claws, and the single lobe at the base of the ovipositor. The pupa has a ventral lobe on each of the antennal horns (fig. 111), widely separated upper frontal horns, elongated spiracles on the first through seventh abdominal segments (fig. 112), no spiracles on the eighth segment, and dorsal abdominal spines as in most Schizomyiina, consisting of one to two rows that do not attain the dorsal row of papillae. The larva has a short spatula. The pupa is similar to that of *Heterasphondylia*, which also occurs on *Piper*, but males of the two genera differ at least in the shape of the teeth on the gonostyli. Refs.: Rübsaamen 1908, Tavares 1925, Möhn 1963a.

ituensis Tavares 1917b: 139 (*Oxasphondylia*); ♀, pupa. Syntypes: Itu, São Paulo, Brazil; presumed lost. Host: *Porophyllum* sp. (Asteraceae). Note: Automatically transferred here when Möhn (1963a) combined *Oxasphondylia* with *Zalepidota*, this species must be found again before it can be placed with certainty. The narrow claws and long empodia, as described, indicate that this species fits elsewhere.

piperis Rübsaamen 1907: 121 (*Zalepidota*); ♂, ♀, pupa. Syntypes: Tijuca, Rio de Janeiro, Brazil (Rübsaamen 1908: 38); ZMHU? Host: *Piper* sp. (Piperaceae). Refs.: Rübsaamen 1908, Möhn 1963a.

reticulata Felt 1915c: 205, 3 figs. (*Proasphondylia*); ♀. Holotype: Trece Aguas, Alta V Paz, Guatemala; USNM. Host: original description listed "cacao," but the specimen could merely have been swept from cacao. Ref.: Möhn 1963a.

tavaresi Kieffer 1913a: 98 (no descr.; cited Tavares 1909: 24 ["*Zalepidota* sp."]) (*Ozobia*); pupa. Lectotype (Möhn 1963a: 15): pupa; São Leopoldo, Rio Grande do Sul, Brazil; Tavares Coll. Host: *Piper* sp. (Piperaceae). Ref.: Möhn 1963a.

Subtribe Schizomyiina

The subtribe Schizomyiina is known in the Neotropics from 35 species distributed among 12 genera. Only two of the genera that occur in the Neotropics extend outside the region: *Schizomyia* is cosmopolitan and attacks many different groups of plants, and *Polystepha* occurs exclusively on oaks and is chiefly Holarctic but extends south into El Salvador. Except for *Bruggmannia*, with 15 described species, the remaining genera in this region contain one to four species, most of which occur on either Nyctaginaceae or Myrtaceae.

Schizomyiina have a palpus of one to four segments and lack an apical spur on the first tarsomere of each leg (fig. 137) (except in one *Schizomyia* sp.; fig. 136). There is usually no lobe (occasionally a very small one) at the posterior end of the female eighth tergite. The ovipositor is soft or needlelike. Male genitalia have a paramere, and the denticles of the gonostyli are not fused. The pupa may or may not have antennal horns and usually has reduced frontal horns, and spines cover only the anterior part of the third to eighth abdominal segments. The larva has six or fewer lateral papillae, a reduced number of lateral larval papillae, and a ventral anus.

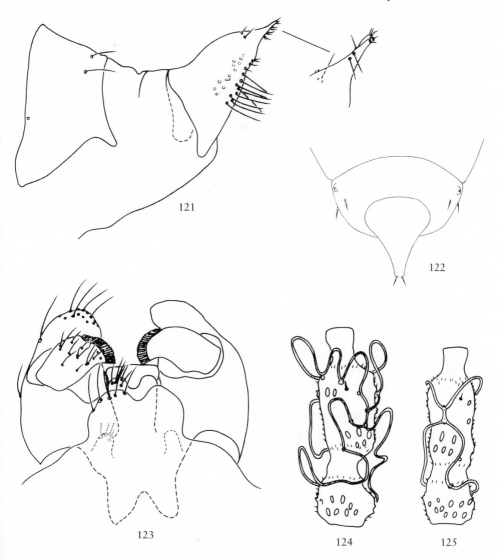

Figures 121–125. Schizomyiina. **121,** *Bruggmannia pisonifolia*, female postabdomen with detail of cerci (lateral). **122,** *Bruggmannia* sp., larval terminal segments (dorsal). **123–125,** *Bruggmannia pisonifolia*: **123,** male genitalia (dorsal); **124,** male third flagellomere; **125,** female third flagellomere.

Ameliomyia

Ameliomyia Möhn 1960a: 237. Type species, *zexmeniae* Möhn (orig. des.).

This genus is known from one species that forms leaf galls on *Zexmenia* (Asteraceae). The male is unknown. The female has a three-segmented palpus and a needlelike ovipositor without a lobe at its base. Pupation occurs in the gall and, accordingly, the pupal antennal horns are well developed. Two tiny but well-separated upper frontal horns are present on the pupa. The larva lacks a spatula and its terminal segment is convex.

zexmeniae Möhn 1960a: 237, pl. 28, figs. 163–172 (*Ameliomyia*); ♀, pupa, larva. Holotype: ♀; S Candelaria, Santa Ana, El Salvador; SMNS. Host: *Zexmenia frutescens* (Asteraceae). Distr.: El Salvador.

Anasphondylia

Anasphondylia Tavares 1920a: 38. Type species, *Asphondylia myrtacea* Tavares (orig. des.).

This genus is known from the female and pupa of a species that forms a hemispherical leaf gall on an unidentified Myrtaceae. The female has a needlelike ovipositor but with no lobe at its base. The adult emerged from the gall. Although the specific host is unknown, the gall may be located someday and the genus found. Ref.: Tavares 1922.

myrtacea Tavares 1920b: 125, pl. III, figs. 10–13 (*Anasphondylia*); ♀. Syntypes: Nova Friburgo, Rio de Janeiro, Brazil; presumed lost. Host: undetermined species of Myrtaceae. Distr.: Brazil. Ref.: Tavares 1922.

Bruggmannia

Bruggmannia Tavares 1906: 81. Type species, *braziliensis* Tavares (orig. des.).
Feltomyia Kieffer 1913a: 100. Type species, *Uleella pisonifolia* Felt (orig. des.).

This genus contains 15 described species. It includes all except one of the species previously carried in *Uleella*, which is now restricted to one species of Anadiplosini. These species form complex galls on leaves of various plants, mainly Nyctaginaceae, but also Myrsinaceae and Rubiaceae. Male flagellomeres and sometimes also female flagellomeres are constricted near the middle of the nodes (figs. 124, 125), the necks are longer and the circumfila are less appressed to the flagellomeres than in other Asphondyliini; the ovipositor (fig. 121) is short, unpigmented, with elongate ventral setae, sparse dorsal setae and tiny, separate cerci; the palpus is three-segmented; and the empodia are much shorter than the claws. Pupae have only weakly developed or no antennal and frontal horns and only one or two anterior rows of dorsal spines on the second to eighth abdominal segments. Larvae lack a spatula, have a ventral, circular anus and a convex or elongate and tapered terminal segment (fig. 122). This genus differs from *Pisphondylia* in that the male circumfila are more irregular and the flagellomeres are constricted near midlength.

braziliensis Tavares 1906: 82 (*Bruggmannia*); ♂, ♀, pupa. Lectotype (Möhn 1962: 212): ♂; São Leopoldo, Rio Grande do Sul, Brazil; Tavares Coll. Host: *Myrsine* sp. (Myrsinaceae). Ref.: Möhn 1962.

depressa Kieffer 1913a: 101 (no description, cites Rübsaamen 1908: 26) (*Uleella*); larva. New combination. Syntypes: Tubarão, Santa Catarina, Brazil; ZMHU? Host: *Neea* sp. (Nyctaginaceae). Distr.: Brazil.

globulifex Kieffer 1913a: 101 (no description, cites Rübsaamen 1908: 25) (*Uleella*); larva. New combination. Syntypes: Marary Juruá, Brazil; ZMHU? Host: *Neea* sp. (Nyctaginaceae). Distr.: Brazil.

lignicola Kieffer 1913a: 101 (no description, cites Rübsaamen 1905: 133) (*Uleella*); larva. New combination. Syntypes: Serra do Macahé, Rio de Janeiro, Brazil; ZMHU? Host: *Neea* sp. (orig. given as *Dalbergia* sp.) (Nyctaginaceae). Distr.: Brazil.

longicauda Kieffer 1913a: 101 (no description, cites Rübsaamen 1908: 26) (*Uleella*); larva. New combination. Syntypes: Rio de Janeiro, Rio de Janeiro, Brazil; ZMHU? Host: *Neea* sp. (Nyctaginaceae). Distr.: Brazil.

longiseta Kieffer 1913a: 101 (no description, cites Rübsaamen 1908: 23) (*Uleella*); larva. New combination. Syntypes: Marary Juruá, Brazil; ZMHU? Host: *Neea* sp. (Nyctaginaceae). Distr.: Brazil.

micrura Kieffer 1913a: 101 (no description, cites Rübsaamen 1908: 26) (*Uleella*); larva. New combination. Syntypes: Marary Juruá, Brazil; ZMHU? Host: *Neea* sp. (Nyctaginaceae). Distr.: Brazil.

neeana Kieffer 1913a: 101 (no description, cites Rübsaamen 1907: 121, fig. 3) (*Uleella*); larva. New combination. Syntypes: Rio de Janeiro, Rio de Janeiro, Brazil; ZMHU? Host: *Neea* sp. (Nyctaginaceae). Distr.: Brazil.

pisoniae Cook 1909: 144 (*Cecidomyia*); gall. Syntypes: Cuba; depos. unknown. New combination. Host: *Pisonia* sp. (Nyctaginaceae). Distr.: Cuba.

pisonioides Gagné. New name for *Bruggmanniella pisoniae* Felt, secondary homonym. Host: *Pisonia nigricans* (Nyctaginaceae). Distr.: St. Vincent.

  pisoniae Felt 1912a: 174 (*Bruggmanniella*); ♂, ♀, larva, pupa. Syntypes: St. Vincent; Felt Coll.

pisonifolia Felt 1912b: 353 (*Uleella*); ♂, ♀, pupa, larva. Syntypes: St. Vincent; Felt Coll. Host: *Pisonia nigricans* (Nyctaginaceae). Distr.: St. Vincent. Ref.: Möhn 1962: 217 (redescribes all stages from original material). Note: Gagné (1989) listed this species from Florida, but the specimens belong to another species.

psychotriae Möhn 1960b: 352, pl. 51, figs. 71–78, pl. 52, figs. 79–91 (*Bruggmannia*): ♂, ♀, pupa, larva. Holotype: ♂; San Diego, E La Libertad, La Libertad, El Salvador, SMNS. Host: *Psychotria carthaginensis* (Rubiaceae). Distr.: El Salvador.

pustulans Möhn 1960b: 348, pl. 50, figs. 49–52, pl. 51, figs. 53–70 (*Bruggmannia*);

♂, ♀, pupa, larva. Holotype: ♂; SE Ciudad Arce, La Libertad, El Salvador; SMNS. Host: *Pisonia micranthocarpa* (Nyctaginaceae). Distr.: El Salvador.

randiae Möhn 1960b: 354, pl. 52, figs. 92–100, pl. 53, figs. 101–109 (*Bruggmannia*); ♂, ♀, pupa, larva. Holotype: ♂; ravine near ITIC, San Salvador, San Salvador, El Salvador; SMNS. Host: *Randia spinosa* (Rubiaceae). Distr.: El Salvador.

ruebsaameni (as *rübsaameni*) Kieffer 1913a: 101 (no description, cites Rübsaamen 1905: 133) (*Uleella*); larva. New combination. Syntypes: Pedras Grandes, Santa Catarina, Brazil; ZMHU? Host: *Neea* sp. (orig. given as *Dalbergia* sp.) (Nyctaginaceae). Distr.: Brazil.

Burseramyia

Burseramyia Möhn 1960a: 232. Type species, *burserae* Möhn (orig. des.).

This genus, known from one species from flower galls of *Bursera* (Burseraceae), differs from *Schizomyia* in that the apex of the gonostylus is equally subdivided into toothed and setose portions and the ovipositor has lost all of its setae. Pupation occurs in the soil.

burserae Möhn 1960a: 233, pl. 28, figs. 138–153 (*Burseramyia*); ♂, ♀, larva. Holotype: ♂; Sitio del Niño, W lava fields, La Libertad, El Salvador; SMNS. Host: *Bursera simaruba* (Burseraceae). Distr.: El Salvador.

Macroporpa

Macroporpa Rübsaamen 1916a: 437. Type species, *peruviana* Rübsaamen (subseq. des. Gagné 1968a: 17).

Two species are placed in this genus. Their male genitalia, as illustrated by Möhn (1962), show great differences, and the larva of one and the female of the other are unknown. It is possible these species are not closely related. Both were reared from leaf galls but on different host families. Flagellomeres of both species are similar to those of *Bruggmannia* except that the nodes are not constricted near midlength. Unlike *Bruggmannia*, the palpus is four-segmented. The female of the type species is distinctive for its short, soft ovipositor with short, scattered setae, and separate cerci (Möhn 1962). The larva of *M. ulei* has an elongate anal segment as do most *Bruggmannia*, but has a distinct, although reduced, pointed spatula. Ref.: Möhn 1962.

peruviana Rübsaamen 1916a: 437, figs. 4a–c, 5 (*Macroporpa*); ♂, ♀, larva. Syntypes: Auristella, Acre R., Amazonas, Brazil; ZMHU? Host: unidentified sp. of Malpighiaceae. Distr.: Brazil. Ref.: Möhn 1962.

ulei Rübsaamen 1916a: 441, figs. 9, 10a–b, 11a–c (*Macroporpa*); ♂, larva. Syntypes: S. Francisco, Acre R., Amazonas, Brazil; ZMHU? Host: Unidentified species of Lauraceae. Distr.: Peru. Ref.: Möhn 1962.

Metasphondylia

Metasphondylia Tavares 1918a: 41. Type species, *squamosa* Tavares (orig. des.).

This genus is known from one species that forms cylindrical galls on the branch tips of an unknown Malvaceae. Adults could fit into *Schizomyia*. The female has an extremely long, attenuate, sclerotized ovipositor with short setae. The pupa has strongly developed antennal horns but no frontal horns. The larva has a short, wide, two-toothed spatula, but other characters of that stage were not described.

squamosa Tavares 1918a: 42, pl. II, figs. 6, 7, 9, 10 (*Metasphondylia*); ♂, ♀, pupa, larva. Syntypes: Retior, near Bahia, Bahia, Brazil; part presumed lost; adults and pupa sent to Felt by Tavares are deposited in the Felt Collection. Host: undetermined species of Malvaceae. Distr.: Brazil.

Pisoniamyia

Pisoniamyia Möhn 1960b: 338. Type species, *armeniae* Möhn (orig. des.).

This genus is known from three species from the same host genus. The type species is from an inflorescence gall on *Pisonia*; the remaining two, from stem galls, can be placed here only tentatively because females are unknown and the male genitalia look very different from one another and from those of the type species. All three species have a three-segmented palpus, very short empodia, and a needlelike ovipositor without a dorsal lobe at the base. Pupae may or may not have well-developed antennal horns and small frontal horns, and have spines only anteriorly on abdominal segments 2–8. Larvae lack a spatula and have a rounded anal segment with a ventral anus.

armeniae Möhn 1960b: 339, pl. 48, figs. 13–17, pl. 49, figs. 18–30 (*Pisoniamyia*); ♂, ♀, pupa, larva. Holotype: ♂; W Ateos, La Libertad, El Salvador; SMNS. Host: *Pisonia micranthocarpa* (Nyctaginaceae). Distr.: El Salvador.
mexicana Felt 1911e: 465 (*Bruggmanniella*); gall. Syntypes: Tampico, Veracruz, Mexico (Felt 1911f); lost. Host: *Pisonia aculeata* (Nyctaginaceae). Distr.: Mexico. Refs.: Felt 1911f, Möhn 1962.
similis Möhn 1960b: 342, pl. 49, figs. 31–33, pl. 50, figs. 34–38 (*Pisoniamyia*); ♂, pupa, larva. Holotype: ♂; W Ateos, La Libertad, El Salvador; SMNS. Host: *Pisonia micranthocarpa* (Nyctaginaceae). Distr.: El Salvador.

Pisphondylia

Pisphondylia Möhn 1960b: 344. Type species, *salvadorensis* Möhn (orig. des.).

This genus is known from one species reared from hairy leaf galls on *Pisonia*. Larvae resemble those of *Bruggmannia* in lacking a spatula and having an elongated anal segment. Male circumfila are more sinuous than in *Bruggmannia*, and the flagellomeres are not constricted near midlength. Otherwise, males could pass for those of *Bruggmannia*. The female of *Pisphondylia* is unknown.

salvadorensis Möhn 1960b: 345, pl. 50, figs. 39–48 (*Pisoniamyia*); ♂, pupa, larva. Holotype: ♂; W Ateos, La Libertad, El Salvador; SMNS. Host: *Pisonia micrantho-carpa* (Nyctaginaceae). Distr.: El Salvador.

Polystepha

Polystepha Kieffer 1897: 11. Type species, *quercus* Kieffer (orig. des.).

This genus, restricted to leaf galls on oaks, is mainly Holarctic, but extends into Central America. The apical antennal segments of the female are not progressively shortened as in most other Asphondyliini. Male genitalia (fig. 128) have the largest parameres of the tribe, the teeth of the gonostyli are short but very wide, and the cerci, hypoproct, and aedeagus are also broad. The ovipositor (fig. 129) is elongate, the distal half sclerotized ventrally, and the cerci, while separate and distinct, are closely appressed mesally. The long ventral setae of the ovipositor, arranged in two separate rows, are reminiscent of those in *Bruggmannia*, but *Polystepha* has many more dorsal setae than does *Bruggmannia*. The many transverse rows of circumfila (figs. 126, 127) seen in both *Polystepha* and the Palearctic genus *Kochiomyia* (part formerly known as *Turkmenomyia* and all found exclusively on Chenopodiaceae) caused the two genera to be grouped together into the tribe Polystephini (Möhn 1961b). But some species of *Polystepha* have simple circumfila (see Felt 1916: 158), so the multiple circumfila of some species most likely developed independently of those in *Kochiomyia*. There is no other reason to consider these two genera as especially closely related. The spatula and terminal segment of the larva of a *Polystepha* are illustrated in figures 130, 131.

lapalmae Möhn 1960b: 335, pl. 48, figs. 11, 12 (*Polystepha*); larva. Holotype: road to La Palma, ca. 15 km before La Palma, Chalatenango, El Salvador; SMNS. Host: *Quercus hondurensis* (Fagaceae). Distr.: El Salvador.
salvadorensis Möhn 1960b: 333, pl. 48, figs. 1–10 (*Polystepha*); ♂, pupa, larva. Holotype: ♂; road to La Palma, ca. 15 km before La Palma, Chalatenango, El Salvador; SMNS. Host: *Quercus eugeniaefolia* (Fagaceae). Distr.: El Salvador.

Proasphondylia

Proasphondylia Felt 1915c: 203. Type species, *brasiliensis* (orig. des.).

The host and larva of the one included species are unknown, and the pupa, originally only cursorily described, is lost. Adults are similar to *Bruggmannia* in many ways. They have a three-segmented palpus. The flagellomeres have long necks and circumfila that are not closely appressed to the nodes. The ovipositor has distinct cerci

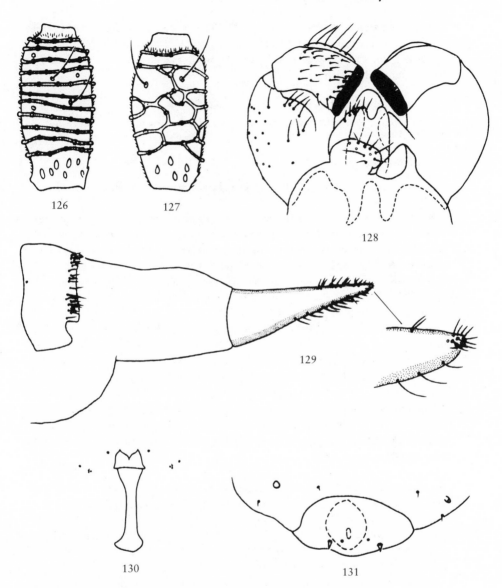

Figures 126–131. *Polystepha pilulae*: **126,** male third flagellomere; **127,** female third flagello-mere; **128,** male genitalia (dorsal); **129,** female postabdomen with detail of cerci (lateral); **130,** larval spatula and associated papillae; **131,** larval terminal segments (dorsal).

and two longitudinal rows of setae (fig. 132). The only apparent difference between the two genera is that the distal half of the ovipositor of *Proasphondylia* is narrow, attenuate, and strongly sclerotized ventrally, while that of *Bruggmannia* is short, broad, and weakly sclerotized.

brasiliensis Felt 1915c: 203, fig. 6 (*Proasphondylia*); ♂, ♀, pupa. Syntypes: Bonito, Pernambuco, Brazil; USNM. Host: unknown. Distr.: Brazil. Ref.: Möhn 1962.

Schizomyia

Schizomyia Kieffer 1889: 183. Type species, *galiorum* Kieffer (mon.).

Schizomyia serves as a catchall genus for those species with needlelike ovipositors, four-segmented palpi, and larvae with all four pairs of terminal papillae present. The four described Neotropical species are listed below. I know also of several undescribed species, including one from Jamaica on *Pimenta dioica* (Myrtaceae) (Parnell and Chapman 1966) and one from Costa Rica on *Vitis* sp. (Gagné unpubl.). The species from *Vitis* is particularly noteworthy as the only schizomyiine with a ventroapical spur on the first tarsomeres.

Möhn (1960a) originally placed *S. serjaneae* in *Placochela* because of the resemblance between the simple circumfila of his species and that of *Placochela*, a monotypic European genus. But *Placochela* was described with distinct cerci at the tip of the ovipositor, which *S. serjaneae* does not appear to have (fig. 134). Rather than tentatively retain *S. serjaneae* in a European genus and leave that arrangement open to conjecture concerning intercontinental relationships, I prefer for the present to place this species under the broader concept of *Schizomyia*. Representative male genitalia and larval terminal segments of *Schizomyia* are illustrated in figures 133, 135, respectively.

ipomoeae Felt 1910a: 160 (*Schizomyia*); ♂, ♀, pupa. Syntypes: St. Vincent; Felt Coll. Host: *Ipomoea* sp. (Convolvulaceae). Distr.: St. Vincent. Ref.: Felt 1910b.
manihoti Tavares 1925: 22, figs. 3–8 (*Schizomyia*); ♂, ♀, pupa, larva. Syntypes: Ceará, Brazil; presumed lost. Host: *Manihot utilissima* (Euphorbiaceae). Distr.: Brazil, Colombia. Ref.: Wünsch 1979.
serjaneae Möhn 1960a: 235, pl. 28, figs. 154–162 (*Placochela*); ♀, larva. Holotype: ♂; ravine near ITIC, San Salvador, San Salvador, El Salvador; SMNS. Host: *Serjania gonicarpa* (Sapindaceae). Distr.: El Salvador.

Figures 132–139. Schizomyiina. **132,** *Proasphondylia brasiliensis*, female postabdomen with detail of cerci (lateral). **133, 134,** *Schizomyia ipomoeae*: **133,** male genitalia (dorsal); **134,** female postabdomen with detail of cerci (dorsolateral). **135, 136,** *Schizomyia vitispomum*: **135,** larval terminal segments (dorsal); **136,** first tarsomere. **137,** *Schizomyia ipomoeae*, first tarsomere. **138, 139,** *Stephomyia* sp.: **138,** female postabdomen with detail of cerci (lateral); **139,** male genitalia (dorsal; arrow: paramere).

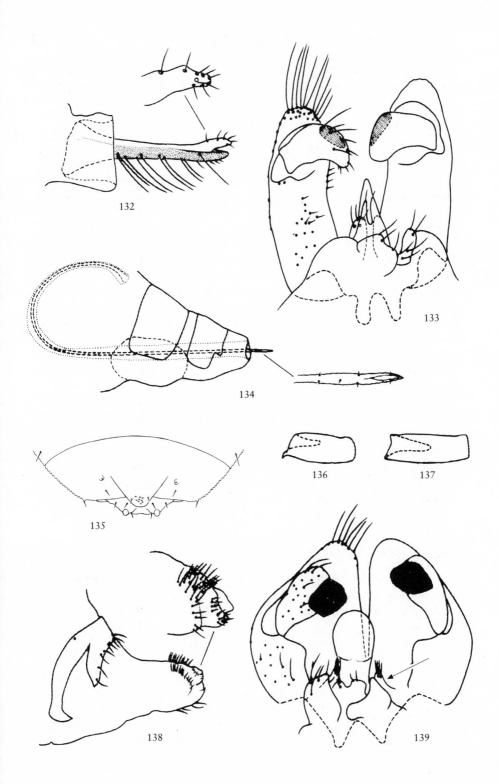

132

133

134

135

136   137

138   139

stachytarphetae Barnes 1932: 483 (*Schizomyia*); ♂, ♀, larva. Syntypes: ♂, ♀; Trinidad; BMNH. Host: *Stachytarpheta cayennensis* (Verbenaceae). Distr.: Trinidad.

Stephomyia

Stephomyia Tavares 1916: 54. Type species, *eugeniae* Tavares (mon.).

This Neotropical genus is known from three described and many undescribed species from *Eugenia* (Myrtaceae). All form complex leaf galls. Adults have a one- to two-segmented palpus, simple to anastomozing circumfila, and short-necked flagellomeres. The ovipositor (fig. 138) is very short and soft with many subapical dorsal setae and sparse ventral setae; the cerci are separate and distinct, at least apically. Representative male genitalia are illustrated in figure 139. Pupae lack cephalic horns and dorsal abdominal spines. Larvae have a well-developed spatula but lack spiracles on the eighth abdominal segment. Pupation occurs in the galls.

clavata Tavares 1920b: 125, pl. III, fig. 14 (*Oxasphondylia*); ♂. New combination. Syntypes: Roça da Madre de Deus, vic. Bahia, Bahia, Brazil; presumed lost. Host: undetermined species of Myrtaceae. Distr.: Brazil. Ref.: Tavares 1921: 93. Note: This species is placed here mainly because it forms galls similar to those made by other *Stephomyia* spp. on Myrtaceae.

eugeniae Felt 1913c: 175 (*Cystodiplosis*); ♂, ♀, pupa. Syntypes: Key West, Florida, USA; Felt Coll. Host: *Eugenia buxifolia* (Myrtaceae). Distr.: USA (Florida).

epeugenia Gagné, New name for *eugeniae* Tavares. Host: *Eugenia* sp. (Myrtaceae). Distr.: Brazil. Ref.: Möhn 1962.

eugeniae Tavares 1916: 54, figs. 12–14 (*Stephomyia*); ♂, ♀. Secondary homonym of *eugeniae* Felt. Lectotype (Möhn 1962: 234): ♂; Botanical Garden, Rio de Janeiro, Rio de Janeiro, Brazil; lectotype in Tavares Coll.; part of original series sent to Felt by Tavares and now in Felt Collection.

Tribe Cecidomyiini

The tribe Cecidomyiini appears to be much more diverse in the rest of the world than in the Neotropical Region. Only 11 species in 7 genera are known from the Neotropics. Two of the genera are each known from one immigrant species.

Cecidomyiini are all plant feeders and occur chiefly in buds and flowers. Some form galls, but many are free-living and gregarious. The main distinguishing character is

---

Figures 140–147. Cecidomyiini. **140–144**, *Contarinia gossypii*: **140**, wing; **141**, male third flagellomere; **142**, female abdomen, third segment to end, with detail of cerci (lateral); **143**, male genitalia (dorsal); **144**, tarsal claw and empodium. **145**, *Contarinia* sp. near *perfoliata*, larval terminal segments (dorsal; arrow: one of modified, recurved pair of papillae). **146**, *Cecidomyia reburrata*, larval terminal segments (dorsal). **147**, *Contarinia schulzi*, larva (ventral; scale line = 0.5 mm).

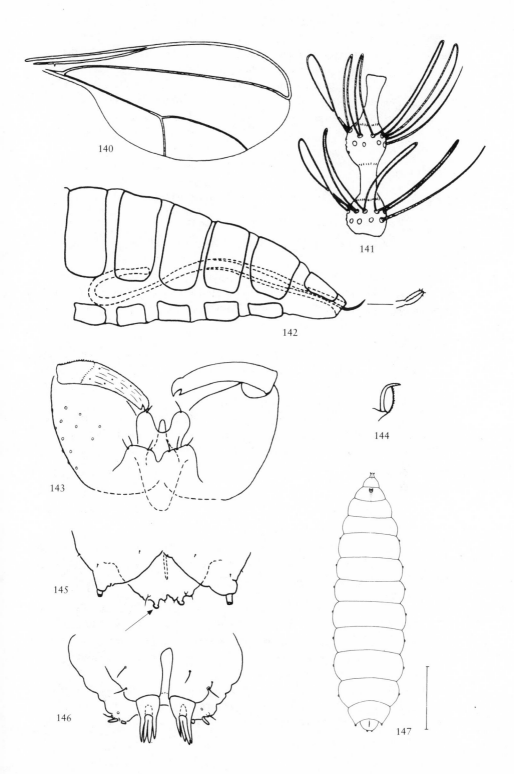

the makeup of the terminal papillae of the larvae (fig. 145); one pair of these is corniform and recurved (unless secondarily reduced); the remaining three pairs are short and setose and generally similar to one another, unless secondarily modified. Adults are generally small, with a wing length of 1–3 mm. All genera except the immigrant *Erosomyia* have simple tarsal claws. The gonocoxites are robust and without parameres.

Cecidomyia

<u>Itonida</u> Meigen 1800: 19. Type species, *Tipula pini* DeGeer (subseq. desig. Coquillett 1910: 556). Suppressed by International Commission on Zoological Nomenclature 1963.
Cecidomyia Meigen 1803: 261. Type species, *Tipula pini* DeGeer (mon.).

This is a chiefly Holarctic genus of 19 species, one of which extends into the Neotropics. Larvae live in resin, mainly on pines, but also on spruce and fir. They have greatly enlarged and caudally directed hind spiracles (fig. 146) that allow them to breathe when immersed in resin.

The wing is about 3 mm long. The ovipositor is the least modified in the tribe, only slightly protrusible, its posterior half approximately as long as the seventh tergite; the cerci are ovoid, separate, have two apical sensory setae and short, regular setae elsewhere. Ref.: Gagné 1978c.

reburrata Gagné 1978c: 6, figs. 6–9 (*Cecidomyia*); ♂, ♀, pupa, larva. Holotype: ♂; Cedarville State Park, Charles Co., Maryland, USA; USNM. Host: *Pinus caribaea* and *P. taeda* (Pinaceae). Distr.: Cuba, USA (Florida; north to Maryland).

Contarinia

Contarinia Rondani 1860: 289. Type species, *Tipula loti* DeGeer (orig. des.).

This cosmopolitan, catchall genus has almost 300 species assigned to it. Although only four species have been described from the Neotropics, I know of several others still unnamed in the USNM collections, including a species near *Contarinia perfoliata* Felt reared from flowers of *Chromolaena odorata* in Trinidad (Gagné 1977b). Two undescribed fossil species are known, one from Mexican amber (Gagné 1973a), the other from Dominican amber, both of Upper Oligocene to Lower Miocene age. Larvae of most species are gregarious and free-ranging in flowers and buds. *Stenodiplosis* (including the sorghum midge, *Stenodiplosis sorghicola*, new combination) is being removed from *Contarinia*.

The wing of *Contarinia* (fig. 140) is 1–2 mm long. The head and neck of one species, *C. prolixa*, are greatly elongated and otherwise modified, and I know also of several other undescribed Neotropical species with elongate mouthparts. The ovipositor (fig. 142) of *Contarinia* species can be extremely long and attenuate, occasionally longer than the abdomen so that it is curved when retracted. The tiny, dorsoventrally flattened cerci are closely appressed to one another mesally but not fused, are mostly glabrous and striated, and have two sensory setae apicolaterally and several short setae elsewhere. Hind spiracles of many flower-inhabiting larvae are situated at the caudolateral margins of the eighth segment (figs. 145, 147). A male third antennal flagellomere, male genitalia, and tarsal claw are illustrated in figures 141, 143, and 144, respectively.

gossypii Felt 1908a: 210 (*Contarinia*); ♂, ♀. Syntypes: Antigua; Felt Coll. and USNM. Host: *Gossypium* sp. (Malvaceae). Distr.: Antigua, Barbados, Colombia, Montserrat, St. Croix, Tortola. Ref.: Callan 1940.

lycopersici Felt 1911d: 303 (*Contarinia*); ♂, ♀, larva. Syntypes: St. Vincent; USNM. Host: *Lycopersicon esculentum* (Solanaceae). Distr.: Barbados, Belize, Dominica, Grenada, Guyana, St. Lucia, St. Vincent, Trinidad. Refs.: Barnes 1932, Callan 1940, 1941.

partheniicola Cockerell 1900: 201 (*Diplosis*); ♂, ♀, pupa, larva. Lectotype (desig. Gagné 1975c): ♂; Picacho Mt., Mesilla Valley, New Mexico, USA. Host: *Parthenium incanum**, *Ambrosia artemisiifolia*, and *A.* spp. (Asteraceae). Distr.: USA (Florida; California to Texas).

prolixa Gagné and Byers 1985: 736 (*Contarinia*); ♂. Holotype: Zulia, Venezuela; USNM. Host: unknown. Distr.: Venezuela.

Erosomyia

Erosomyia Felt 1911h: 49. Type species, *mangiferae* Felt (orig. des.).
Mangodiplosis Tavares 1918a: 47. Type species, *mangiferae* Tavares (orig. des.).

This genus is known from a single species introduced with mango from India. The larva, generally similar to that of some *Contarinia* species, causes swellings on growing tips and inflorescences. Adults, however, are unusual for Cecidomyiini because the tarsal claws are toothed.

The wing is 1.0–1.5 mm long. Eye facets are circular, but farther apart laterally than in other genera. The tarsal claws (fig. 152) are toothed. The ovipositor (fig. 151) is not greatly attenuate; the protrusible part, or ninth segment, is slightly more than twice the length of the seventh tergite. The cerci are fused to form a single lobe. Male gonocoxites (fig. 150) are elongate and, unlike those of all other Cecidomyiini, have a short mesobasal lobe.

mangiferae Felt 1911h: 49 (*Erosomyia*); ♂, ♀, pupa, larva. Syntypes: St. Vincent; Felt Coll. Host: *Mangifera indica* (Anacardiaceae). Distr.: Brazil, St. Lucia, St. Vincent, and Trinidad; India. Refs.: Callan 1940, Prasad 1971 (as *Erosomyia indica* Grover).

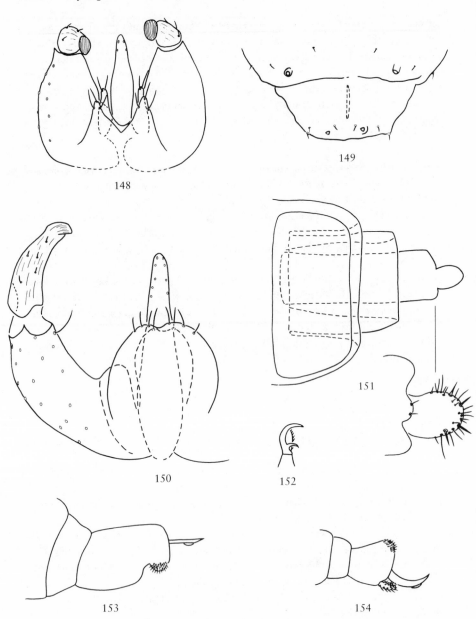

Figures 148–152. Cecidomyiini. **148, 149,** *Prodiplosis longifila*: **148,** male genitalia (dorsal); **149,** larval terminal segments (dorsal). **150–152,** *Erosomyia mangiferae*: **150,** male genitalia (dorsal); **151,** female postabdomen with detail of fused cerci (ventral); **152,** tarsal claw and empodium. **Figures 153, 154.** Centrodiplosini. **153,** *Centrodiplosis crassipes*, female postabdomen (lateral; redrawn from Kieffer 1913a). **154,** *Jorgensenia falcigera*, female postabdomen (lateral; redrawn from Kieffer 1913a).

mangiferae Tavares 1918a: 48, pl. II, figs. 1–5 (*Mangodiplosis*); ♂, ♀, pupa, larva. Syntypes: Roça da Madre de Dios, near Bahia, Bahia, Brazil; part presumed lost, part sent to Felt by Tavares and now deposited in the Felt Collection.

Prodiplosis

Prodiplosis Felt 1908b: 403. Type species, *Cecidomyia floricola* Felt (orig. des.).

This genus is known from one European and nine North American species, of which two occur also in the Neotropics (Gagné 1986). I know of other, undescribed species, two from flowers of *Crescentia* (Bignoniaceae) in Nicaragua (Maes 1990), and one associated with cacao flowers in Bolivia. Larvae of *Prodiplosis* live mainly in flowers and buds. Some species, including the two named Neotropical species, are widespread and have a broad host range. The genus is diverse, the distinguishing character being the robust, dorsoventrally articulated gonopods (fig. 148).

The wing is 1–2 mm long. Male flagellomeres are variable, binodal to cylindrical, sometimes in the same specimen, with two or three circumfila, which may have loops of greatly uneven length or may be shaped like those of the female. The aedeagus is variable, sometimes greatly modified. The ovipositor is elongate-attenuate, the protrusible part 4 to 11 times the length of seventh tergite, with the longer one curving inside the abdomen when retracted; the cerci are tiny, dorsoventrally flattened, and closely appressed to one another mesally like those of many *Contarinia* (fig. 142), mostly devoid of setulae and striated, with two sensory setae apicolaterally and several short setae elsewhere. Larvae (fig. 149) are similar to those of some *Contarinia*. Ref.: Gagné 1986.

floricola Felt 1907a: 21 (*Cecidomyia*); ♂, ♀. Syntypes: Albany, New York, USA; Felt Coll. Hosts: *Caryocar brasiliense* (Caryocaraceae); *Clematis* sp.* (Ranunculaceae), and *Spiraea* sp.* (Rosaceae). Distr.: Brazil; USA (New York). Ref.: Gagné 1986.
longifila Gagné 1986: 240, figs. 3, 14 (*Prodiplosis*); ♂, ♀, larva. Holotype: ♂; Marco I., Monroe Co., Florida, USA; USNM. Hosts: buds of plants from many families, including alfalfa, castorbean, citrus, green beans, potato, tomato, wild cotton, and wormseed. Distr.: Colombia, Ecuador, Peru (Lambayeque), and USA (Florida). Refs.: Diaz B. 1981, Peña et al. 1989. Note: Reports in the literature of *Contarinia medicaginis* Kieffer from Colombia and Peru on *Medicago sativa* refer to *P. longifila*.

Sphaerodiplosis

Sphaerodiplosis Rübsaamen 1916a: 461. Type species, *dubia* Rübsaamen (mon.).

This genus is known from one species based on a male reared from an unknown gall and plant. The bifilar and binodal male flagellomeres drawn in the original description pass for those of a *Contarinia*. The detailed original drawing of the male genitalia should allow this species to be recognized again.

The wing is about 3 mm long. Rübsaamen (1916a) reported that the empodia were shorter than the tarsal claws, but we do not know how much shorter or whether they only appeared that way from other than lateral view.

dubia Rübsaamen 1916a: 461, figs. 35–37 (*Sphaerodiplosis*); ♂. Syntypes: type locality unstated, Peru or Brazil; ZMHU. Host: unidentified shrub.

Stenodiplosis

Stenodiplosis Reuter 1895: 9. Type species, *geniculati* Reuter (orig. des.). Restored status.
Allocontarinia Solinas 1986: 23. Type species, *Diplosis sorghicola* Coquillett (orig. des.). New synonym.

This genus is cosmopolitan, with about 20 known species, and includes most Cecidomyiini that live in aborted grass seeds. *Stenodiplosis* is superficially similar to many *Contarinia* but differs from them chiefly in the lack of lateral setae on the adult abdominal tergites and the reduced size and number of larval papillae. The second-instar skin of the sorghum midge, *S. sorghicola*, forms a puparium for the third instar. This immigrant species, the only one of the genus known to occur in the Neotropics, probably originated in Africa (Harris 1976) and is now known wherever sorghum is grown.

The wing is about 2 mm long. Eye facets are circular and usually farther apart laterally than elsewhere on the eye. The larva lacks a spatula and has very short setae and only one pair of terminal papillae.

sorghicola Coquillett 1899: 82 (*Diplosis*); ♂, ♀. New combination. Syntypes: College Station, Texas, USA; USNM. Hosts: *Sorghum bicolor* and *Sorghum* spp. (Poaceae). Distr.: Argentina, Brazil, Colombia, Curaçao, Mexico, Peru, Puerto Rico, St. Vincent, Trinidad, U.S. Virgin Is., and Venezuela; originally Africa, now on all continents wherever sorghum is grown. Refs.: Harris 1964, 1976; Cermeño et al. 1984; Solinas 1986.
palposa Blanchard 1958: 423 (*Contarinia*); ♂, ♀. Syntypes: Argentina; INTAC.

Taxodiomyia

Taxodiomyia Gagné 1969a: 271. Type species, *Cecidomyia cupressiananassa* Osten Sacken (orig. des.).

This genus is presently known only from southeastern North America, including one from southern Florida. Four species are known, all from complex leaf and twig galls of *Taxodium* (Taxodiaceae).

The wing is about 2 mm long. The empodia reach only half way to the bend of the claws. Male flagellomeres have either two or three circumfila. Female cerci are fused to form a single, elongate lobe. Larval terminal papillae are reduced to two pairs, one corniform, the other setose. Refs.: Gagné 1969a, 1989.

cupressiananassa Osten Sacken 1878: 3 (Cecidomyia); validated in catalog with reference to Riley's (1870) description and invalid name. Lectotype (desig. Gagné 1969a: 273): ♂; Savannah, Tennessee, USA: USNM. Host: Taxodium distichum (Taxodiaceae). Distr.: USA (Florida; southeastern USA). Refs.: Chen and Appleby 1984a,b.
    cupressi ananassa Riley 1870: 244 (Cecidomyia); ♂, ♀, pupa, larva. Lectotype, see under cupressiananassa. This name is unavailable because it is a polynomial.

Centrodiplosini, new tribe

The tribe Centrodiplosini is erected to include three genera: Centrodiplosis, Cystodiplosis, and Jorgensenia. Each is known from a single species reared from galls on Solanaceae. They share the unique character of a strongly sclerotized, aciculate, more-or-less curved ovipositor with fused cerci (figs. 153, 154).

Centrodiplosis

Centrodiplosis Kieffer and Jörgensen 1910: 405. Type species, crassipes Kieffer and Jörgensen (subseq. des. Gagné 1968a: 31).

This genus is known from a species taken from swellings on stems of Lycium. Kieffer's (1913a) illustration of the female postabdomen is redrawn here (fig. 153). The pupa has no dorsal spines on the abdomen.

crassipes Kieffer and Jörgensen 1910: 406, figs. 34, 35 (Centrodiplosis): ♂, ♀, pupa, egg. Syntypes: Chacras de Coria, Mendoza, Argentina; presumed lost. Host: Lycium chilense (Solanaceae). Distr.: Argentina. Refs.: Jörgensen 1917, Kieffer 1913a.

Cystodiplosis

Cystodiplosis Kieffer and Jörgensen 1910: 395. Type species, longipennis Kieffer and Jörgensen (mon.).

This genus is known from a species that forms raised blister galls on the leaves of Grabowskia. The ovipositor is apparently short and curved dorsally. The male genitalia were not described in any detail. The larval spatula is broad and anteriorly convex, and the terminal segment has eight short setae.

longipennis Kieffer and Jörgensen 1910: 395, fig. 25 (*Cystodiplosis*); ♂, ♀, pupa, larva, egg. Syntypes: Chacras de Coria, Mendoza, Argentina; presumed lost. Host: *Grabowskia obtusa* (Solanaceae). Distr.: Argentina. Refs.: Jörgensen 1917, Kieffer 1913a.

Jorgensenia

Jorgensenia Kieffer 1913a: 148. Type species, *Centrodiplosis falcigera* Kieffer and Jörgensen (orig. des.).

This genus is known from the female of one species reared from bud galls on *Lycium*. The ovipositor (fig. 154), redrawn here from Kieffer 1913a, is stiff and curved, and the eighth abdominal segment shows an unusual ventral lobe.

falcigera Kieffer and Jörgensen 1910: 411, figs. 37–39 (*Centrodiplosis*); ♀. Syntypes: Mendoza, Argentina; presumed lost. Host: *Lycium gracile* (Solanaceae). Distr.: Argentina. Refs.: Jörgensen 1917, Kieffer 1913a.

Tribe Clinodiplosini

This large tribe is cosmopolitan, but feeding habits of Neotropical Clinodiplosini are generally different from the mainly fungivorous Holarctic species. Many Neotropical species are not only primary plant feeders, but cause simple or even complex galls. Seven genera are represented here, six of them endemic and represented by one or two species. The seventh genus, *Clinodiplosis*, is cosmopolitan but is a catchall repository for species that do not belong to other named genera. The 13 Neotropical species assigned to this genus are endemic.

The main distinguishing characters of this tribe are the distinctive terminal segment of the larva, with its three pairs of corniform terminal papillae, one pair smaller than the other two, and one setiform pair (figs. 155, 156). Additional characters are the obtuse mesobasal lobe of the gonocoxite (fig. 166), the usually quadrate or secondarily bilobed male cerci, and the generalized female postabdomen with its slightly protrusible ovipositor (fig. 164). The diversity in the shape of the tarsal claws is remarkable (figs. 158–161), from bowed to strongly bent at basal third, to bowed at or beyond midlength, and with or without teeth. The great variety in the claws makes

---

Figures 155–165. Clinodiplosini. 155, *Clinodiplosis* sp., larva (scale line = 0.5 mm). 156, 157, *Iatrophobia brasiliensis*: 156, larval terminal segments; 157, wing. 158, *Clinodiplosis americana*, tarsal claw and empodium. 159, *Clinodiplosis marcetiae*, tarsal claw and empodium. 160, *Autodiplosis parva*, tarsal claw and empodium. 161, *Chauliodontomyia egregia*, tarsal claw and empodium. 162, *Iatrophobia brasiliensis*, head (antennae removed; arrow: occipital process). 163, *Chauliodontomyia egregia*, mouthparts (arrow: modified labellum). 164, *Iatrophobia brasiliensis*, female postabdomen (lateral). 165, *Clinodiplosis cattleyae*, female cerci and 10th abdominal sternite (lateral).

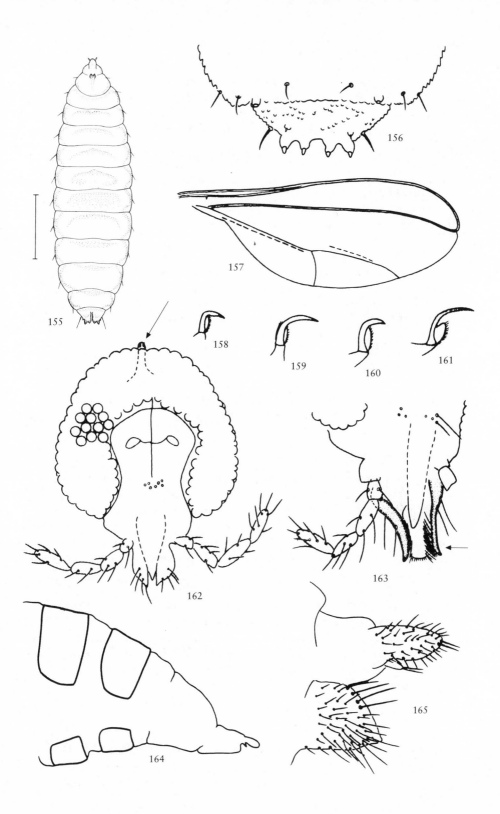

155

156

157

158

159

160

161

162

163

164

165

it difficult to assess their importance in delimiting genera. Until a better way to group species becomes apparent, I have retained all of the previously used genera and defined all of them narrowly except *Clinodiplosis*, which remains a catchall genus. All the genera here except *Clinodiplosis* are endemic to this region.

Autodiplosis

Eudiplosis Tavares 1916: 51. Type species, *parva* Tavares (orig. des.). Preocc. Kieffer 1894a.

Autodiplosis Tavares 1920b: 106, new name for *Eudiplosis* Tavares. Type species, *Eudiplosis parva* Tavares (aut.).

Tavares listed seven Brazilian species under this genus but I have kept only the type species here. It differs from the remainder, all removed to *Clinodiplosis*, by the tarsal claws (fig. 160), which are bowed at midlength instead of at the basal third, and the simply rounded instead of quadrate posterior margins of the male cerci. The wing is 1.5 mm long.

parva Tavares 1916: 51, figs. 10 (A–D), 11 (*Eudiplosis*); ♂, ♀. Syntypes: between Bahia and Rio Vermelho, Bahia, Brazil; part presumed lost, but ♀♀ sent to Felt by Tavares are deposited in the Felt Collection. Host: Faboideae (flor de S. Foão) (Fabaceae). Distr.: Brazil. Refs.: Tavares 1917b, 1920b.

Chauliodontomyia

Chauliodontomyia Gagné 1969c: 108. Type species, *parianae* Gagné (orig. des.).

Two species are known of this genus, both associated as adults with flowers of a grass in Venezuela. They are distinctive among Clinodiplosini for their modified mouthparts (fig. 163), with the swollen rostrum and elongate, narrow, and strongly curved labella, and for the gynecoid male antennae. The wing is 2 mm long and has a very strongly curved R_5.

egregia Gagné 1969c: 111 (*Chauliodontomyia*); ♂, ♀. Holotype: ♂; Rancho Grande, near Maracay, Venezuela; USNM. Host: *Pariana stenolemma* (Poaceae). Distr.: Venezuela.

parianae Gagné 1969c: 110 (*Chauliodontomyia*); ♂, ♀. Holotype: ♂; Rancho Grande, near Maracay, Venezuela; USNM. Host: *Pariana stenolemma* (Poaceae). Distr.: Brazil, Venezuela.

Cleitodiplosis

Cleitodiplosis Tavares 1921: 108. Type species, *Necrophlebia graminis* Tavares (mon.).

This genus is known from one Brazilian species associated with foreshortened grass stems with bunched leaves. The larva associated with this genus has a four-toothed spatula (Tavares 1921), so it may belong to another group of gall midges. In North America, similar galls on grasses are made by *Chilophaga* (Oligotrophidi: Alycaulini). The distinguishing characters of *Cleitodiplosis* are the splayed lobes of the male hypoproct (fig. 171) and the relatively short gonostyli. The wing is 2.5–3.0 mm long.

graminis Tavares 1916: 37, figs. 1A–C, 2 (*Necrophlebia*); ♂, ♀. Syntypes: Bahia and Rio de Janeiro, Brazil; part presumed lost, but ♀♀ sent to Felt by Tavares are deposited in the Felt Collection. Host: *Paspalum conjugatum* (Poaceae). Refs.: Tavares 1921, 1922.

Clinodiplosis

Clinodiplosis Kieffer 1895a: cclxxx. Type species, *Diplosis cilicrus* Kieffer (orig. des.).

This is a worldwide, catchall genus with 75 described species. In the Holarctic Region, most species are associated with fungi, although some form simple bud galls, but many Neotropical species are primary plant feeders and some form complex galls. I include very diverse Neotropical species here rather than erect a series of small genera to contain them. As more species are described and those already described are better known, it should be possible to classify them in a more satisfactory way. The USNM collection has a great number of undescribed species, most caught in flight. Those with host data are listed in the host section as *Clinodiplosis* species. Several other species are known from Chiapas and Dominican amber of Upper Oligocene to Lower Miocene age; a species near the Nearctic *Clinodiplosis terrestris* (Felt) was reported from Chiapas amber (Gagné 1973a).

Wings are 1–3 mm long. In a few species the wings are conspicuously marked. The tarsal claws are variable, toothed (only foreclaws in Neotropical species) or un-toothed, bent or curved near basal third (fig. 159) or beyond midlength (fig. 158); empodia usually reach to the curve or bend of the claws, but are sometimes shorter. The male cerci may be quadrate or secondarily lobed (fig. 169), or acute as in *C. cattleyae* (fig. 170). The aedeagus is large and bulbous in some species (fig. 169). A representative female 10th abdominal segment is shown in figure 165.

americana Felt 1911i: 192 (*Hyperdiplosis*); ♂, ♀, larva. Syntypes: Paraiso, Panama; USNM. Host: rotting fig branch. Distr.: Panama.
bahiensis Tavares 1917b: 145, pl. IX, figs. 1–4, 9–11 (*Eudiplosis*): ♀, pupa, larva. New combination. Syntypes: Roça da Madre de Deus, between Bahia and Rio Vermelho, Bahia, Brazil; part presumed lost, but ♀♀ and a larva sent to Felt by Tavares are deposited in the Felt Collection. Host: an unknown Asteraceae, but gall shown in Tavares 1917b. Distr.: Brazil.

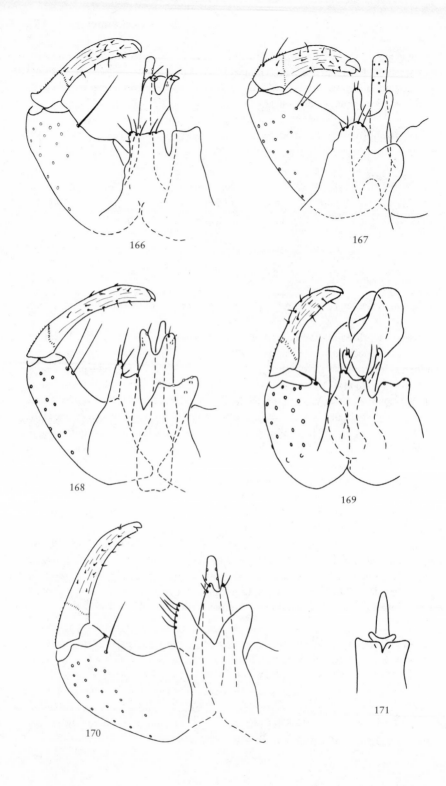

166

167

168

169

170

171

cattleyae Molliard 1903: 165, pl. 2, figs. 1–18 (*Cecidomyia*); ♂, ♀. Syntypes: Brazil (through the orchid trade via England to France); depos. unknown. Hosts: *Cattleya* spp., *Epidendrum* spp., *Laelia* spp., and orchids (Orchidaceae). Distr.: Mexico to Ecuador and Brazil, and Jamaica.

<u>*cattleyae*</u> Felt 1908b: 412 (*Clinodiplosis*); ♂, ♀. Syntypes: Neotropical Region (through the orchid trade to Natick, Massachusetts, USA); USNM.

cearensis Tavares 1917b: 153, pl. IX, figs. 5–7 (*Eudiplosis*); ♂, ♀. New combination. Syntypes: Fortaleza, Ceará, Brazil; part presumed lost, but a ♀ sent to Felt by Tavares is deposited in the Felt Collection. Host: unknown plant. Distr.: Brazil.

chlorophorae Rübsaamen 1905: 82 (*Clinodiplosis*); larva. Syntypes: Fabrica, Rio de Janeiro, Brazil; ZMHU. Host: *Chlorophora tinctoria* (Moraceae). Distr.: Brazil.

coffeae Felt 1911d: 305 (*Hyperdiplosis*); ♂, ♀. Syntypes: St. Vincent; USNM. Host: *Coffea liberica* (Rubiaceae). Distr.: Dominica, St. Vincent.

denotata Gagné 1984: 127, figs. 18–23 (*Clinodiplosis*); ♂, ♀. Holotype: ♂; Finca La Lola, near Siquirres, Limon, Costa Rica; USNM. Host: unknown. Distr.: Costa Rica.

eupatorii Felt 1911b: 110 (*Hyperdiplosis*); ♂, ♀, pupa. Syntypes: St. Vincent; Felt Coll. Hosts: *Chromolaena ivaefolia*, *C. odorata*, and *Eupatorium* sp. (Asteraceae). Distr.: Brazil (Pará), Costa Rica, St. Vincent, and Trinidad. Ref.: Gagné 1977b.

iheringi Tavares 1925: 45, figs. 13–14 (*Autodiplosis*); ♀, pupa, larva. New combination. Syntypes: Joinville, Santa Catarina, Brazil; presumed lost. Host: *Aegiphila arborescens* (Verbenaceae). Distr.: Brazil.

marcetiae Tavares 1917a: 22, figs. 1–5 (*Eudiplosis*); ♂, ♀, pupa. New combination. Syntypes: Nova Friburgo, Rio de Janeiro, Brazil; part presumed lost, but ♂♂ and ♀♀ sent to Felt by Tavares are deposited in the Felt Collection. Host: *Marcetia* sp. (Melastomataceae). Distr.: Brazil. Ref.: Tavares 1917b.

picturata Felt 1915a: 156 (*Lestodiplosis*); ♀. New combination. Syntypes: Bartica, British Guiana; Felt Coll. Host: unknown. Distr.: Guyana.

pulchra Tavares 1917b: pl. X, figs. 3–7, pl. XI, fig. 6 (*Eudiplosis*); ♂, larva. New combination. Syntypes: Roça da Madre de Deus, near Bahia, Bahia, Brazil; presumed lost. Host: *Lantana* sp. (Verbenaceae). Distr.: Brazil. Ref.: Tavares 1918a.

rubiae Tavares 1918a: 70, pl. III, fig. 12 (*Eudiplosis*); ♂, ♀. New combination. Syntypes: Nova Friburgo, Rio de Janeiro, Brazil; presumed lost. Host: *Rubia* sp. (Rubiaceae). Distr.: Brazil.

Houardodiplosis

Houardodiplosis Tavares 1925: 13. Type species, *rochae* Tavares (orig. des.).

Figures 166–171. Clinodiplosini. Male genitalia (dorsal): **166**, *Iatrophobia brasiliensis*; **167**, *Clinodiplosis eupatorii*; **168**, *Schismatodiplosis lantanae*; **169**, *Clinodiplosis coffeae*; **170**, *Clinodiplosis cattleyae*; **171**, *Cleitodiplosis graminis* (cerci, hypoproct, and aedeagus only).

This genus is known from one species that causes bud swellings on *Combretum*. The distinguishing character of *Houardodiplosis* is the basally constricted gonostyli; otherwise the original illustrations show genitalia that are generally similar to Schismatodiplosis. The wing is 2.0–2.5 mm long.

rochae Tavares 1925: 14, figs. 1, 2, Pl. I, fig. 4 (*Houardodiplosis*): ♂, ♀, pupa. Syntypes: type locality not specifically stated, but collected by Dias da Rocha, who regularly sent Tavares galls collected in vicinity of Ceará, Brazil; presumed lost. Host: *Combretum leprosum* (Combretaceae). Distr.: Brazil.

Iatrophobia

Iatrophobia Rübsaamen 1916a: 469. Type species, *Clinodiplosis brasiliensis* Rübsaamen (mon.).
Jatrophobia Barnes 1946: 32. Emend.

This genus is known from one species that causes cylindrical leaf galls on cassava. The distinguishing generic character, the recurved apical lobes of the male hypoproct (fig. 166), is uniform if seemingly superficial. The wing (fig. 157) is 2.0–2.5 mm long. A head, female abdomen, and larval terminal segments are illustrated in figures 162, 164, and 156, respectively.

brasiliensis Rübsaamen 1907: 156 (*Clinodiplosis*); pupa, larva. Syntypes: Tubarão, Rio de Janeiro, Brazil; ZMHU? Hosts: *Manihot dichotoma*, *M. palmata*, and *M. utilissima* (Euphorbiaceae). Distr.: Brazil, Colombia, Costa Rica, Guyana, Peru, St. Vincent, Tobago, and Trinidad. Refs.: Rübsaamen 1916a, Tavares 1918a, Korytkowski and Sarmiento 1967 (as *Hyperdiplosis* sp.).
    manihot Felt 1910b: 268 (*Cecidomyia*); ♂, ♀. Syntypes: St. Vincent; Felt Coll.

Schismatodiplosis

Schismatoplosis Rübsaamen 1916a: 467. Type species, *Clinodiplosis lantanae* Rübsaamen (mon.).

This genus is known from one species that forms spherical leaf galls on *Lantana*. The distinguishing characters are the secondarily lobed male cerci (fig. 168) and the simple, bowed tarsal claws. The female is unusual among the Clinodiplosini for its setose 10th tergum. The wing is 2.0–2.5 mm long. The empodia are short, not quite attaining the bend of the claws.

lantanae Rübsaamen 1907: 151 (*Clinodiplosis*); ♂, ♀, pupa, larva. Syntypes: Cabo Frio and Palmeiras, Rio de Janeiro, and Tubarão, Santa Catarina, Brazil; ZMHU. Hosts: *Lantana camara*, *L. hispida*, *L. urticifolia*, and *Lantana* spp. (Verbenaceae).

Distr.: Brazil, Mexico (Vera Cruz, Tabasco, and Quintana Roo), Trinidad. Refs.: Rübsaamen 1916a, Tavares 1917a.

Tribe Lestodiplosini

The tribe Lestodiplosini is numerous, diverse, and cosmopolitan. Adults are among the tiniest cecidomyiids. All larvae are predators of insects or mites. Larvae have long antennae and are free-ranging on the substrate, some aided in their locomotion by pseudopods. Lestodiplosini are represented in the Neotropics by 15 species in 7 genera. All the genera except *Thripsobremia* are cosmopolitan, and a few of the species are also widespread.

Wings are 1–3 mm long. Male flagellomeres are tricircumfilar or, less commonly, bicircumfilar. The palpus usually has four segments, rarely three. R_5 is usually straight, reaching C before or at the apex of the wing (figs. 174, 175), but in some genera it is curved posteriorly to join C posterior to the wing apex; Rs is usually absent, M+rm is straight, and the M_3 fold is usually not visible. Tarsal claws are bowed near the basal third or at or beyond midlength, and they are toothed only in *Dicrodiplosis* and *Feltiella*; the empodia reach the bow in the claw. The female eighth abdominal tergite is usually unsclerotized and has posterior setae; the ovipositor is short, the 10th tergum usually bare, the cerci ovoid, with either a pair or a group of closed placed, ventroapical, sensory setae (fig. 177). Pupae have short dorsal spines. The larva usually has a smooth cuticle, often with ventral pseudopods. The spatula is reduced or absent and the anus is dorsal.

Arthrocnodax

Arthrocnodax Rübsaamen 1895a: 189. Type species, *vitis* Rübsaamen (subseq. des. Coquillett 1910: 510).

This cosmopolitan genus of 29 species is known in the Neotropics from two described and several undescribed species known to me. Larvae are mite predators distinguished by their extremely long cephalic apodemes. Adults are tiny, with a wing length R_5 of about 1 mm. The wing is distinct for its slightly shortened, sigmoid R_5 (fig. 175) that ends before the wing apex. The cubital vein may be simple or forked. Eye facets are circular and not closely appressed. Female flagellomeres have short necks. Female cerci are elongate-ovoid, each with two long, sensory setae and sparse, scattered setae no longer than sensory setae. The male genitalia have a short, unlobed gonocoxite, short, entire hypoproct, and a short aedeagus.

abdominalis Felt 1911c: 128 (*Endaphis*); ♂, ♀. Syntypes: Piura, Peru; USNM. Host: mites (Acari). Distr.: Peru.

meridionalis Felt 1912a: 176 (*Arthrocnodax*): ♂, ♀, larva. Syntypes: St. Vincent; Felt Coll. Hosts: *Eriophyes* (Acari: Eriophyidae) galls on various plant hosts. Distr.: St. Vincent, Trinidad, USA (Florida). Ref.: Gagné 1977b.

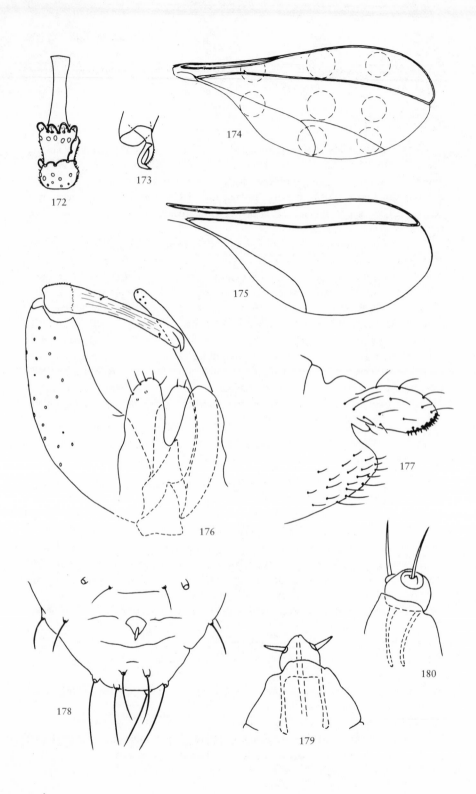

172

173

174

175

176

177

178

179

180

Dicrodiplosis

Dicrodiplosis Kieffer 1895c: cxciv. Type species, *fasciata* Kieffer (orig. des.).

This is a widespread genus of 10 described species with one described and at least one other undescribed species from this region. Larvae are predators of various insects, but of mealybugs in the Neotropics. *Dicrodiplosis* species are larger than other Lestodiplosini, with a wing length of 2–3 mm. Male circumfilar loops are short. Legs and antennae may be conspicuously banded. The tarsal claws are toothed on all legs and are strongly curved just before midlength, as are those of *Feltiella* species (fig. 184); the empodia attain the bend in the claws. Female cerci have a large field of short, blunt setae ventrally as do those of *Lestodiplosis*. Male genitalia (fig. 185) are very large with respect to the rest of the abdomen; gonocoxites have prominent, sclerotized, acute, setose mesobasal lobes. Larvae have elongate, narrow antennae, pseudopods, a dorsoventral anus, and eight terminal papillae (four long, four short). Refs.: Harris 1968, 1981b.

guatemalensis Felt 1938: 43 (*Dicrodiplosis*); ♂, ♀. Syntypes: Guatemala; USNM and BBM. Host: mealy bugs (Homoptera: Pseudococcidae). Ref.: Harris 1968.

Feltiella

Feltiella Rübsaamen 1910: 285. Type species, *tetranychi* Rübsaamen (orig. des.).

This cosmopolitan genus of 21 described species is known in the Neotropics from three species. Larvae have a characteristically rounded head capsule (fig. 180) and are predators of tetranychid mites (Acari: Tetranychidae). Wings are only about 1 mm long. Tarsal claws (fig. 184) are curved near midlength and toothed at least on forelegs, the teeth large. The male genitalia could pass for those of some *Lestodiplosis* species. Females have short flagellomere necks and lack setae on the eighth segment except for the basal pair of sensory setae.

curtistylus Gagné 1984: 128, figs. 24–30 (*Feltiella*); ♂, ♀, larva. Holotype: Petrolina, Pernambuco, Brazil; MZSP. Host: *Tetranychus evansi* (Acari: Tetranychidae). *insularis* Felt 1913a: 305 (*Mycodiplosis*); ♂, ♀, larva. New combination. Syntypes: Rio Piedras, Puerto Rico; Felt Coll. Hosts: *Tetranychus bimaculatus*, *T. desertorum*, and *Tetranychus* spp. (Acari: Tetranychidae) on castorbean, tomato, soy

Figures 172–180. Lestodiplosini. **172–174, 176–179,** *Lestodiplosis callipus*: **172,** female third flagellomere; **173,** tarsal claw and empodium; **174,** wing (circles approximating the areas covered with darkened scales); **176,** male genitalia (dorsal); **177,** female terminal abdominal segments (lateral); **178,** larval posterior segments (dorsal); **179,** larval head (dorsal). **175.** *Arthrocnodax* sp., wing. **180.** *Feltiella curtistylus*, larval head (dorsolateral).

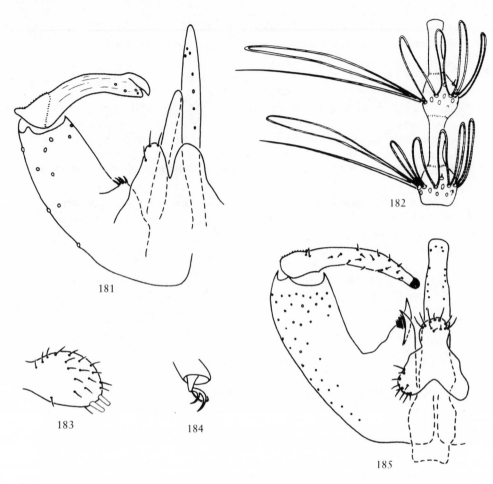

Figures 181–185. Lestodiplosini. **181, 182,** *Thripsobremia liothripis*: **181,** male genitalia (dorsal); **182,** male third flagellomere. **183,** *Pseudendaphis* sp., female cercus (lateral). **184,** *Feltiella curtistylus*, foretarsal claw and empodium. **185,** *Dicrodiplosis guatemalensis*, male genitalia (dorsal, the cerci directed anteriorly).

bean, and hibiscus. Distr.: Colombia, Jamaica, Puerto Rico, Trinidad, USA (Florida).

constricta Felt 1914b: 481 (*Arthrocnodax*); ♂, ♀. New combination. New synonym. Syntypes: Rio Piedras, Puerto Rico; Felt Coll.

pini Felt 1907a: 31 (*Mycodiplosis*); ♂. Holotype: Albany, New York, USA; Felt Coll. Hosts: originally unknown, now known from *Tetranychus urticae, Tetranychus* sp. on beans, and corn-infesting mites (Acari: Tetranychidae). Distr.: Bahamas, Nicaragua, USA (Florida); widespread North America.

Lestodiplosis

Leptodiplosis Kieffer 1894a: xxviii. Invalidated original spelling of Lestodiplosis (International Commission on Zoological Nomenclature 1958).
Lestodiplosis Kieffer 1894b: 84. Type species, Cecidomyia pictipennis Perris (subseq. des. Kieffer 1895a: cclxxx).
Coprodiplosis Kieffer 1894b: 84. Type species, cryphali Kieffer (mon.).

This is a large, cosmopolitan genus of 160 described species, five of them from the Neotropics. Many more undescribed Neotropical species are represented in the USNM collection, and two fossil species have been reported from Chiapas amber (Gagné 1973a). Larvae are predators of various arthropods, including Acarina, Coccoidea, and other gall midges. The wings are 1–2 mm long, and on many species they have subtle to distinct dark spots (fig. 174). Legs and antennae may be banded with light and dark areas. Male flagellomeres usually have three, occasionally two (although not in the Neotropics, but see Thripsobremia) circumfila, the loops of different lengths and occasionally very long in some species; flagellomeres of females (fig. 172) have long necks and their circumfila may be appressed or looped. Tarsal claws (fig. 173) are simple, sometimes widened after the bend. Females have a field of numerous coniform, closely adjacent sensory setae on the venter of the cerci (fig. 177). Male genitalia usually have prominent, acute, mesobasal gonocoxite lobes (as in fig. 181), but not on L. callipus (fig. 176).

braziliensis Tavares 1920b: 113 (Coprodiplosis); ♀. Syntypes: Nova Friburgo, Rio de Janeiro, Brazil; presumed lost. Host: ex gall of Anadiplosis pulchra (Diptera: Cecidomyiidae). Distr.: Brazil.
callipus Gagné 1977b: 120, figs. 30–34 (Lestodiplosis); ♂, ♀, larva. Holotype: male; Simla, Trinidad; USNM. Hosts: larvae of various arthropods in flowers of Chromolaena odorata (Asteraceae). Distr.: Trinidad.
gagnei Baylac 1987: 125, figs. 1–18 (Lestodiplosis); ♂, ♀, larva. Holotype: male; Colombia; MNHNP. Host: Elaeidobius subvittatus (Coleoptera: Curculionidae). Distr.: Colombia.
grassator Fyles 1883: 237, fig. 25 (Diplosis); adults, pupa, larva. Syntypes: Cowansville, Quebec, Canada; depos. unknown. Hosts: Phylloxera vastatrix* (Homoptera: Phylloxeridae) and flower heads and bud galls of Chromolaena odorata (Asteraceae). Distr.: Trinidad; North America. Ref.: Gagné 1977b.
peruviana Felt 1911a: 10 (Lestodiplosis); ♂, larva. Syntypes: Piura, Peru; USNM. Host: Hemichionaspis minor (Homoptera: Diaspididae). Distr.: Peru. Ref.: Harris 1968.

Pseudendaphis

Pseudendaphis Barnes 1954: 772. Type species, maculans Barnes (mon.).

This genus is known from a species from Trinidad, an endoparasite of aphids. I know of an undescribed Neotropical species associated with aphids on cacao in Costa Rica.

Wings are 1.0 mm long. Eye facets are hexagonal except at midheight, where they are circular and farther apart than elsewhere on the eye. The scutum is covered extensively with scales. The two sensory setae at the apex of the female cerci are enlarged (fig. 183). The gonocoxite has no mesobasal lobe; the hypoproct is approximately the same length and breadth as the cerci, and the aedeagus is longer than the hypoproct.

maculans Barnes 1954: 772 (*Pseudendaphis*); ♂, ♀, larva. Holotype: ♂; Trinidad; BMNH. Host: *Toxoptera aurantii* (Homoptera: Aphididae). Distr.: Trinidad. Ref.: Kirkpatrick 1954.

Thripsobremia

Thripsobremia Barnes 1930: 331. Type species, *liothripis* Barnes (mon.).

This genus contains one species, a predator of thrips. It does not essentially differ from *Lestodiplosis*. The male antennae of *Thripsobremia* have three circumfila on the first five flagellomeres but only two beyond (fig. 182), on which the intermediate circumfilum is lost. On the first five flagellomeres, that circumfilum is ringlike and difficult to see. The antennal setae and loops of the first and third circumfila are very uneven. The USNM has larvae of an undetermined lestodiplosine from *Liothrips* in Brazil that may be this species. Because Lestodiplosini of the Neotropics are poorly known, the genus might serve some future service in grouping species, so it is retained for the present. Wings are 1.5 mm long. Female flagellomeres have long necks, and the circumfilar sections are conspicuously bowed. Male genitalia are illustrated in figure 181; those illustrated for this species in the original description show a lobe, the mirror image of the hypoproct, that is not present on the type specimen.

liothripis Barnes 1930: 331 (*Thripsobremia*); ♂, ♀. Syntypes: Trinidad; BMNH. Host: *Liothrips urichi* (Thysanoptera: Phlaeothripidae). Distr.: Trinidad.

Trisopsis

Trisopsis Kieffer 1912b: 171. Type species, *oleae* Kieffer (orig. des.).

This cosmopolitan genus of 23 described species is known in the Neotropics from two described and several undescribed species in the USNM collection. Larvae are predators of various arthropods, including Acarina and Coccoidea. In addition to the two species listed below, I have recently seen adults of an unknown species reared from larvae associated with the coffee leaf miner in Ecuador. Adults roost on spider webs. The laterally divided eyes resulting in "three eyes" characteristic of these species could have arisen more than once in the Lestodiplosini, so the species placed here are not necessarily monophyletic. Wings are 1.0–1.5 mm long. Cu may be simple or forked. Female cerci are small, ovoid or elongate-ovoid, with two or three sensory setae at the apex and otherwise covered with sparse setae. Male genitalia are generally as for *Lestodiplosis* but without mesobasal gonocoxite lobes.

incisa Felt 1907a: 43 (*Cecidomyia*); ♂. Holotype: Albany, New York, USA: Felt Coll. Hosts: eriophyid mites (Acari). Distr.: Trinidad; Europe, USA. Ref.: Gagné 1977b.
oleae Kieffer 1912b: 171 (*Trisopsis*); ♂, ♀. Syntypes: Wellington, South Africa; presumed lost. Hosts: unknown arthropod in fruit of *Olea verrucosa** and mealy-bug (Homoptera: Pseudococcidae) on cacao. Distr.: Colombia; Republic of South Africa. Ref.: Harris 1968.

Lopesiini, new tribe

The tribe Lopesiini is known from seven Neotropical species in six genera and one species from Africa. All the genera except *Lopesia* are endemic to the Neotropics. At least two species of *Lopesia* occur in Africa, the first example of a phytophagous genus known only from South America and Africa. These gall midges occur on Fabaceae, Melastomataceae, Rosaceae, and Boraginaceae (an undescribed species from *Cordia*), and in very different kinds of galls, from hairy, spherical leaf galls to swollen stems. Among the tribe's distinguishing characters are the bend in R_5 at its juncture with Rs (fig. 186), the Rs situated beyond midlength of R_1, the extra teeth on the tarsal claws (figs. 187, 188) (except for *Liebeliola prosopidis*, which apparently has simple claws), and the short female postabdomen (figs. 191, 192), its cerci with many short setae concentrated ventrally or ventroapically. Some Lopesiini show various reductions in the male antennae, from reduced circumfilar loops (fig. 190) to gynecoidy (fig. 189). Gynecoidy may be common in this tribe; besides its occurrence in *Ctenodactylomyia*, I know it in two undescribed species not yet assigned to genus. A few of the several undescribed species I know do not fit well in the genera below, which will undergo some conceptual changes in due course. Pupae vary considerably in the shape of the head, and abdominal spines occur on abdominal segments 1–7 on most species I have seen, except for *Tetradiplosis* species, which have none. The larval spatula is also quite variable in this tribe. Larvae have a full complement of papillae; those of the terminal segment are short but corniform, each on a terminal projection (fig. 195).

Allodiplosis

Allodiplosis Kieffer and Jörgensen 1910: 389. Type species, *crassa* Kieffer and Jörgensen (mon.).

This genus is known from one species that forms spherical, hairy stem galls on *Geoffraea* in Argentina. The wing as originally illustrated resembles that of *Tetradiplosis*, and the claws were illustrated with three teeth. These characters place this genus with the Lopesiini, but the male genitalia are unlike those of other lopesiines for

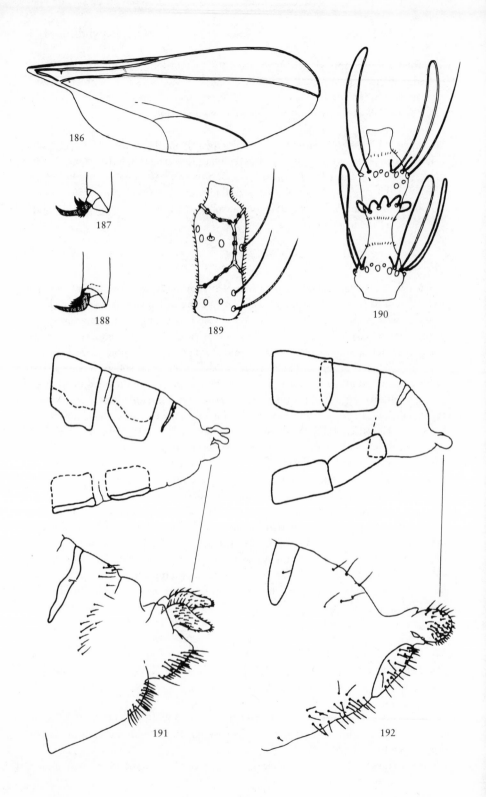

186

187

188

189

190

191

192

their short, although splayed, gonopods. Lagrange's (1980) illustration of the genitalia agrees with the sketchy original description. The pupal antennal horns are strongly developed, and the larval spatula has two long anterior teeth but lacks the shaft.

crassa Kieffer and Jörgensen 1910: 389 (*Allodiplosis*); ♂, ♀, pupa, larva, egg. Syntypes: Mendoza, Argentina; presumed lost. Host: *Geoffraea decorticans* (Fabaceae). Distr.: Argentina. Ref.: Lagrange 1980.

Ctenodactylomyia

Ctenodactylomyia Felt 1915c: 199. Type species, *watsoni* Felt (orig. des.).

This genus is known from two, possibly synonymous, Neotropical species that form leaf blister galls on *Coccoloba*. Adults are large, with a wing length of 3–4 mm. The male flagellomeres are gynecoid (fig. 189), and the tarsal claws are pectinate (fig. 188). The female postabdomen and male genitalia are illustrated in figures 192 and 194, respectively.

cocolobae Cook 1909: 145 (*Cecidomyia*); gall. New combination. Syntypes: Cuba; depos. unknown. Host: *Coccoloba uvifera* (Polygonaceae).
watsoni Felt 1915c: 199 (*Ctenodactylomyia*); ♂, ♀, pupa, larva. Syntypes: Palm Beach, Florida, USA; USNM. Host: *Coccoloba uvifera* (Polygonaceae). Distr.: Florida. Note: Wolcott (1936) recorded this species from Puerto Rico on *Coccoloba uvifera*, but it could as well have been C. *cocolobae*, assuming the species are not synonymous. The USNM has a series of adults from St. Croix, Virgin Is. and pupae from Puerto Rico and Mexico that generally fit C. *watsoni*. Ref.: Mead 1970.

Liebeliola

Lieboliola Kieffer and Jörgensen 1910: 428. Type species, *prosopidis* Kieffer and Jörgensen (mon.).

This genus is known from the female of one species reared from a large, spherical twig gall on *Prosopis*. One other species from Western Samoa has been described in

---

Figures 186–192. Lopesiini. 186, 187, *Tetradiplosis* sp. on *Prosopis*: 186, wing; 187, tarsal claw and empodium. 188, 189, *Ctenodactylomyia watsoni*: 188, tarsal claw and empodium; 189, male third flagellomere. 190, 191, *Tetradiplosis* sp. on *Prosopis*: 190, male third flagellomere (not all circumfilar loops shown); 191, female postabdomen with detail of posterior segments (dorsolateral). 192, *Ctenodactylomyia watsoni*, female postabdomen with detail of posterior segments (lateral).

this genus by Barnes (1928), but, because it was based on only a female, its generic placement cannot be verified. Kieffer's (1913a) description of a wing and cerci of *Liebeliola* resemble those seen in Lopesiini, but the claws of the two genera are reportedly different, those of *Liebeliola* simple, those of *Tetradiplosis* many-toothed. For purposes of the key, I have placed this genus with those that have the claws bent near the base; Kieffer did not specify that the claws had that shape, but they should if *Liebeliola* is a lopesiine.

prosopidis Kieffer and Jörgensen 1910: 428 (*Liebeliola*); ♀, egg. Syntypes: Mendoza and San Juan, Argentina; presumed lost. Host: *Prosopis strombulifera* (Fabaceae). Distr.: Argentina.

Lopesia

Lopesia Rübsaamen 1908: 29. Type species, *brasiliensis* Rübsaamen (mon.).

Lopesia is known from one Brazilian and one African species. The Brazilian species forms a spherical leaf gall on *Ossaea* (Melastomataceae), the African a pediceled lenticular leaf gall on *Parinaria* (Rosaceae) (Tavares 1908). Adults are large, with a wing length of 3.0–4.5 mm. Antennal flagellomeres are binodal and tricircumfilar; the tarsal claws possibly have no more than one tooth. As originally illustrated, the male genitalia generally resemble those of *Rochadiplosis*. Ref.: Tavares 1908.

brasiliensis Rübsaamen 1908: 30, figs. 11, 12 (*Lopesia*); ♂, ♀, larva. Syntypes: Rio de Janeiro, RJ, and Tubarão, Santa Catarina, Brazil; ZMHU. Host: *Ossaea* sp. (Melastomataceae). Distr.: Brazil.

Rochadiplosis

Rochadiplosis Tavares 1917a: 33. Type species, *tibouchinae* Tavares (orig. des.).

This Neotropical genus is known from one described species that forms spherical leaf galls on *Tibouchina* (Melastomataceae). The wing is 3–4 mm long. Flagellomeres of males are similar to those of *Tetradiplosis*, with the first and third circumfila much longer than those of the second. Tarsal claws have two teeth, the lower tiny. The female and male abdomens are also like those of *Tetradiplosis*. I know of an undescribed species from similar galls on *Leandra* (Melastomataceae) that would fit here except that the antennae of the male are gynecoid.

tibouchinae Tavares 1917a: 34, figs. 5, 6, pl. V, figs. 1–10 (*Rochadiplosis*); ♂, ♀, pupa, larva. Syntypes: Nova Friburgo, Rio de Janeiro, Brazil; part presumed lost, but ♂♂ sent to Felt by Tavares are deposited in the Felt Collection. Host: *Tibouchina* sp. (Melastomataceae). Distr.: Brazil.

Tetradiplosis

Tetradiplosis Kieffer and Jörgensen 1910: 421. Type species, *sexdentatus* Kieffer and Jörgensen (mon.).

This genus is known from one species forming stem galls on *Prosopis*. The wing (fig. 186) is 3–4 mm long. Male flagellomeres (fig. 190) have the first and third circumfila with long loops, the second with much shorter loops. Tarsal claws (fig. 187) have two teeth, the lower tiny. Female cerci (fig. 191) are elongate-ovoid, tapering apically, with closely set setae apicoventrally. The pupa has long, pointed antennal horns and a strongly developed conical prothorax that protrudes a considerable distance beyond the antennal horns. The larval spatula is robust, with teeth along the mesal surface of the anterior lobes. I have reared adults from *Prosopis nigra* and adults, pupae, and larvae from *Prosopis flexuosa* but cannot determine whether they belong to *T. sexdentatus*.

sexdentatus Kieffer and Jörgensen 1910: 421, figs. 46–49 (*Tetradiplosis*); ♀, pupa, larva. Syntypes: Mendoza, Argentina; presumed lost. Hosts: *Prosopis alpataco* and *P. campestris* (Fabaceae). Distr.: Argentina.

Tribe Mycodiplosini

The two cosmopolitan genera of the tribe Mycodiplosini placed here, *Coquillettomyia* and the broadly defined *Mycodiplosis*, have so far only six Neotropical species assigned to them. Larvae are fungus-feeding, with those of *Mycodiplosis* usually associated with rust fungi. All of the known species are endemic, except for one, possibly an emigrant, recently discovered in Hawaii.

Coquillettomyia

Coquillettomyia Felt 1908b: 398. Type species, *Mycodiplosis lobata* Felt (orig. des.).

This is a worldwide genus of 22 described species, two of them from the Neotropics. I know of many additional but undescribed species from this region. Larvae are free-living in decaying plant matter or among fungus (Möhn 1955a,b). Female flagellomere necks are setulose in some species. The wing is 2.0–2.5 mm long. The aedeagus is strongly sclerotized, pigmented, often divided longitudinally (fig. 196), the gonocoxites are short and robust, usually with a conspicuous mesobasal lobe, and the gonostyli are occasionally bilaterally flattened or flanged.

obliqua Gagné 1984: 125, figs. 10–17 (*Coquillettomyia*); ♂, ♀. Holotype: ♂; Finca La Lola, near Siquirres, Limon, Costa Rica; USNM. Host: adults found associated with cacao flowers.

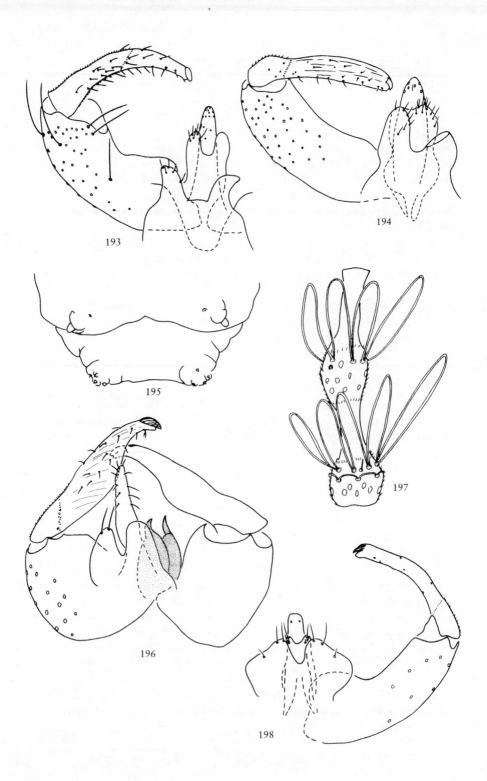

193

194

195

197

196

198

townsendi Felt 1912d: 155 (*Karschomyia*); ♂, ♀. New combination. Syntypes: Jaén, Cajamarca, Peru; Felt Coll. Host: unknown.

Mycodiplosis

Mycodiplosis Rübsaamen 1895b: 186. Type species, *Cecidomyia coniophaga* Winnertz (orig. des.).
Toxomyia Felt 1911d: 302. Type species, *fungicola* Felt (orig. des.). New synonym.

Four described species of *Mycodiplosis* are known from the Neotropics, and I know of several more in the USNM collection. Larvae feed on the spores of rust fungi, although in Europe they are known to live on hyphae of other fungi growing on plants (Holz 1970). Wings are 1–2 mm long. Male flagellomeres are binodal or, in *M. cylindrica*, gynecoid; when binodal they have either two or three circumfila. In *M. fungicola* (fig. 197), two circumfila occur on the basal node and only one on the distal. This character is unique for Cecidomyiidae. Female flagellomere necks are setulose. Tarsal claws are bowed or strongly bent, toothed or simple, although toothed on forelegs in this region. The gonocoxites (fig. 198) have an obtuse mesobasal lobe, cerci are triangular or quadrate, and the hypoproct is elongate, usually narrow and divided only apically. The larval spatula is clove-shaped, and the terminal papillae characteristically have two pairs of papillae with long setae and two with corniform setae. Refs.: Parnell 1969, Holz 1970.

australis Brèthes 1914: 152 (*Neobaeomyza*); ♀. New combination. Holotype: General Urquiza, Buenos Aires, Argentina; MACN. Host: unknown. Distr.: Argentina.
cylindrica Gagné *in* Gagné and Rios de Saluso 1987: 4, figs. 1–6 (*Mycodiplosis*); ♂, ♀, larva. Holotype: male; Paraná, Entre Rios, Argentina; UNLP. Host: *Puccinia coronata*. Distr.: Argentina.
fungicola Felt 1911d: 302 (*Toxomyia*); ♂, ♀, larva. New combination. Syntypes: St. Vincent; USNM. Hosts: *Puccinia emiliae, P. melanpodii, P. psidii, P. substriata, Puccinia* sp., *Ravenelia humphreyana, Uromyces bidenticola, U. euphorbiae,* and *U. pisi*; also *Coleosporium domingensis* in Hawaii. Distr.: Brazil, Jamaica, St. Vincent; Hawaii. Ref.: Parnell 1969.
    rubida Felt 1911g: 194 (*Toxomyia*); ♂, ♀. Syntypes: St. Vincent; USNM.
ligulata Gagné 1984: 124, figs. 1–9 (*Mycodiplosis*); ♂, ♀. Holotype: ♂; Finca La Lola, near Siquirres, Limon, Costa Rica; USNM. Host: adults associated with cacao flowers. Distr.: Costa Rica.

Figures 193–195. Lopesiini. **193,** *Tetradiplosis* sp. on *Prosopis*, male genitalia (dorsal). **194,** *Ctenodactylomyia watsoni*, male genitalia (dorsal). **195,** Lopesiine on *Cordia* sp., larval terminal segments (dorsal). **Figures 196–198.** Mycodiplosini. **196,** *Coquillettomyia townsendi*, male genitalia (dorsal, the pigmented, two-pronged aedeagus slightly askew). **197, 198,** *Mycodiplosis fungicola*: **197,** male third flagellomere; **198,** male genitalia (dorsal).

Unplaced Genera of Cecidomyiidi

Twenty-seven genera of Cecidomyiidi found in the Neotropics show no close relationship to one another or to the formal tribes. They include phytophagous, predaceous, and fungivorous species. Twenty-two of the genera are each known from a single species, one genus has two, two genera have three, and one genus has nine species. The remaining genus, *Resseliella*, is recorded here on the basis of two undescribed species. All except five of the 27 genera and all of the species are endemic. *Bremia*, *Karshomyia*, *Resseliella*, and *Youngomyia* are cosmopolitan, and *Diadiplosis* is pantropical. The last, with its nine Neotropical species that feed mostly on Coccoidea, is the best known of these ungrouped genera.

Anasphodiplosis, new name

Anasphondylia Blanchard 1939: 173. Type species, *aspidospermae* Blanchard (orig. des.). Preocc. Tavares 1920a.

Anasphodiplosis Gagné, new name for *Anasphondylia* Blanchard. Type species, *Anasphondylia aspidospermae* Blanchard (aut.).

This genus is known from one species reared from bud galls on *Aspidosperma*. Only the female is known. The ovipositor (figs. 200, 201), although much shorter than that in *Contarinia*, is reminiscent of that genus for the dorsoventrally flattened and diminutive cerci. The wing (fig. 199) is broad, short, 2 mm long. The antennae and legs are foreshortened. The antennal circumfilar loops are uneven, some much longer than their flagellomeres. The abdomen has short, straplike sclerites; cerci (fig. 201) are tiny, with two prominent sensory setae and scattered short setae.

aspidospermae Blanchard 1939: 173, figs. a–e (*Anasphondylia*); ♀. Syntypes: San Cristobal, Santa Fe, Argentina; INTAC. Host: *Aspidosperma quebracho-blanco* (Apocynaceae). Distr.: Argentina.

Andirodiplosis

Andirodiplosis Tavares 1920b: 93. Type species, *bahiensis* Tavares (mon.).

This genus is known from one species that forms tight, vermiform leaf folds on *Andira*. Tavares's original description did not adequately differentiate this genus, for which additional reared material will be necessary, but his figures of the anterior parts of a pupa and the four- or five-toothed larval spatula should help in determining this genus.

Figures 199–201. *Anasphodiplosis aspidospermae*: 199, wing; 200, female postabdomen; 201, detail of terminal segments (lateral). Figures 202, 203. *Aphidoletes aphidimyza*: 202, wing; 203, male third flagellomere.

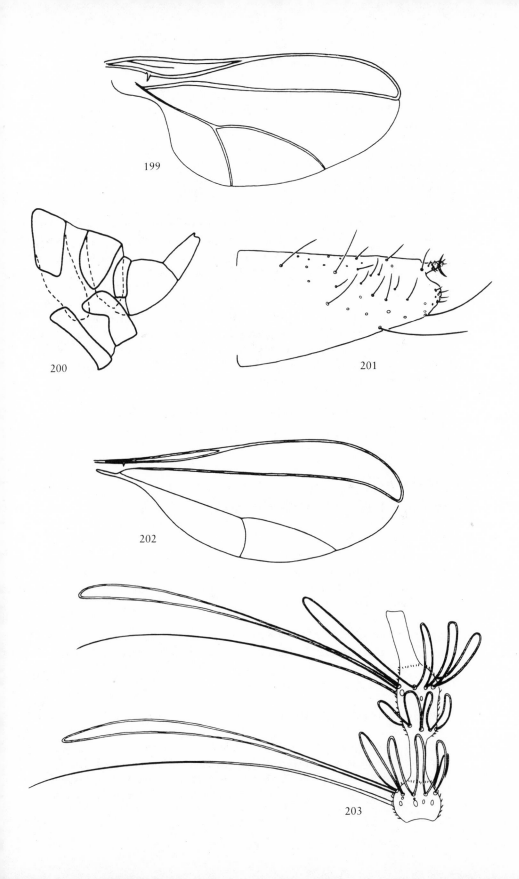

199

200

201

202

203

bahiensis Tavares 1920b: 94, pl. III, figs. 1–3 (*Andirodiplosis*); ♂, ♀, pupa. Syntypes: Bahia and Ilha de Itaparica, Bahia, Brazil; presumed lost. Host: *Andira* sp. (Fabaceae). Distr.: Brazil.

Astrodiplosis

Astrodiplosis Felt 1913b: 217. Type species, *speciosa* Felt (orig. des.).

This genus is known from one conspicuously marked species reared from an irregular stem gall on *Cissus*. The larva originally associated with this species appears to belong to *Neolasioptera* and may be responsible for the gall. The wing (fig. 206) is about 4.5 mm long and strongly marked with brown in cell R_5 and continuing around the wing margin and along Cu. Legs are banded brown and yellow; claws are simple, and the empodia are rudimentary (fig. 205). The ovipositor (fig. 204) is short but tapered, the cerci elongate ovoid, with two sensory setae among the remaining sparse setae. The gonostyli (fig. 207) are swollen and covered with long setae basally, then abruptly constricted before widening slightly toward the apex; they are setulose basally and striate beyond; the hypoproct (fig. 208) has a concave ventral cavity filled with short, pliant setae; and the aedeagus is very narrow, as long as the hypoproct.

speciosa Felt 1913b: 218 (*Astrodiplosis*); ♂, ♀, pupa, larva. Syntypes: Puerto Barrios, Guatemala; Felt Coll. Host: *Cissus* sp. (Vitaceae) (Felt 1921c). Distr.: Guatemala.

Austrodiplosis

Austrodiplosis Brèthes 1914: 154. Type species, *argentina* Brèthes (orig. des.).

This genus is known from one species based on a single female caught in flight. Wings are 1.4 mm long. The tarsal claws (fig. 210) appear to be curved near the base, and the empodia appear to reach the curve of the tarsal claws. The ovipositor (fig. 209) may possibly be recognized again if the sinuous setae on the cerci are not an artifact of preparation.

argentina Brèthes 1914: 154 (*Austrodiplosis*); ♀. Holotype: General Urquiza, Buenos Aires Prov., Argentina; MACN. Host: unknown. Distr.: Argentina.

Bremia

Bremia Rondani 1860: 289. Type species, *Cecidomyia decorata* Loew (orig. des.).

This is a cosmopolitan genus of 17 described species. Many species occur in the Neotropics although none has previously been described. More than 20 species from this region are represented in the USNM collection, of which one is described below to give an example of the bizarre male genitalia found in this group. Another

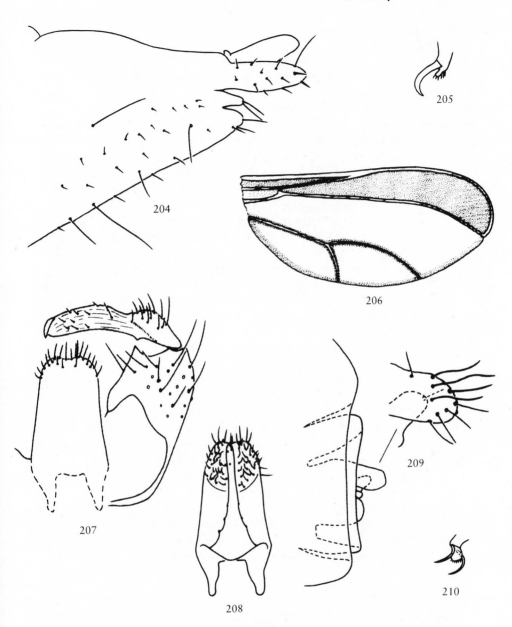

Figures 204–208. *Astrodiplosis speciosa*: **204,** female posterior abdominal segments (lateral); **205,** tarsal claw and empodium; **206,** wing; **207,** male genitalia, the cerci removed (dorsal); **208,** hypoproct and aedeagus (ventral). **Figures 209, 210.** *Austrodiplosis argentina*: **209,** female postabdomen with detail of cerci and hypoproct (lateral); **210,** tarsal claws and empodium.

undescribed species has been recorded from Mexican amber of Oligocene-Miocene age (Gagné 1973a).

Adults can often be found roosting on spider webs. The wing (fig. 211) is 2.0–3.0 mm long, R_1 is remarkably short, M is curved, and R_5 is often darker than the other veins. Male flagellomeres (fig. 216) have the second circumfilum bandlike and closely appressed, the loops of the other two circumfila greatly uneven in length, the longest on the dorsum. Antennal setae are also very long on the dorsum of the flagellomeres. Female flagellomere necks are often setulose. The tarsal claws (fig. 212) are toothed on the forelegs only, strongly bent at basal third, and sometimes wavy beyond the bend. Female cerci (fig. 215) have many short closely adjacent sensory setae postero-ventrally. The male gonocoxite, hypoproct, and aedeagus may be variously lobed as in the following example species.

The larva of an African species is known (Harris 1981a). It is predaceous on dragonfly eggs that occur in gelatinous masses on shrubs over streams in Gabon. Larval antennae are not elongate as in some other predaceous genera, but the head capsule is longer than wide. The terminal segment shows a resemblance to that of *Aphidoletes*, with its two lobes each bearing four setose papillae. The *Bremia* larva lacks ventral pseudopods and has the hind spiracles situated at the end of two dorsocaudal projections.

Bremia mirifica Gagné, new species

Male: Wing, third flagellomere, fore- and midtarsal claws and empodia as in figures 211 and 212. Male genitalia are as in figures 213 and 214: gonocoxites with elongate mesobasal lobe, the lobe setulose except glabrous near apex; hypoproct with baso-lateral arms, tapering abruptly near apical third, the central portion thickly covered ventrally with pliant setae; aedeagus trifurcate from base, the two lateral arms tipped with setae and slightly wider and longer than the central arm.

Holotype: ♂, on spider web, Valle, Colombia, VIII-1977, W. G. Eberhard, deposited in USNM.

Etymology: *mirifica* is a Latin adjective meaning "wonderful."

Remarks: This species is distinct from all other described Western Hemisphere *Bremia* species in the shape of the male genitalia. No other *Bremia* species has such a narrow, long lobe on the gonocoxites and a tripartite aedeagus.

Charidiplosis

Charidiplosis Tavares 1918b: 80. Type species, *concinna* Tavares (orig. des.) = *Charidiplosis triangularis* (Felt).

Figures 211–216. *Bremia mirifica*: 211, wing; 212, fore- and midtarsal claws and empodia; 213, male genitalia (dorsal); 214, aedeagus; 215, female posterior abdominal segments (ventro-lateral); 216, male third flagellomere (setae drawn only to midlength).

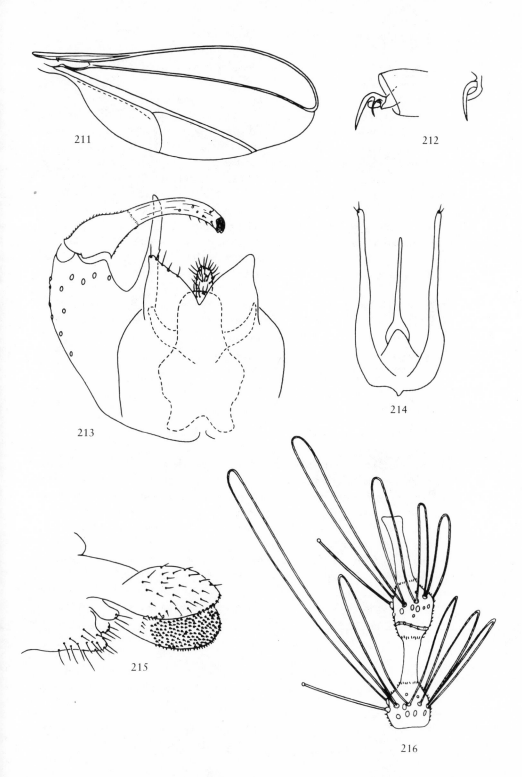

211

212

213

214

215

216

Aphodiplosis Gagné 1973b: 864. Type species, *Clinodiplosis triangularis* Felt (orig. des.). New synonymy.

This genus is known from one species that is widely distributed in the Americas. Larvae have been reared from fungi and cow dung; adults have been repeatedly collected on cacao flowers in the Neotropics (Young 1985). I know of a similar but undescribed species from cacao flowers in Ghana. The male genitalia resemble those of *Monobremia*, an aphid predator (see Harris 1966), but in that genus the mesobasal gonocoxite lobe is longer, and, according to Kieffer (1913a), the ovipositor is short.

The wing (fig. 217) is 2.0–3.0 mm long. The circumfila of the male antennae (fig. 222) are regular to irregular in length. The ovipositor (fig. 219) is elongate, and the eighth tergite is sclerotized and completely setose. The male is unique for its elongate and narrow hypoproct (fig. 220). The larval terminal papillae (fig. 221) are situated in two groups of four, three with setiform papillae, the fourth corniform.

triangularis Felt 1908b: 411 (*Clinodiplosis*); ♂. New combination. Syntypes: Nassau, New York, USA; Felt Coll. Host: mushrooms and decaying organic matter. Distr.: Costa Rica, Brazil; USA (widespread).
concinna Tavares 1918b: 81, pl. IV, figs. 2–12 (*Charidiplosis*); ♂, ♀. New synonym. Syntypes: Bahia, Brazil; presumed lost.

Compsodiplosis

Compsodiplosis Tavares 1909: 14. Type species, *luteoalbida* Tavares (mon.).

Tavares placed five species in *Compsodiplosis*, but indicated (Tavares 1915b) that he used the genus as a catchall category. I am limiting his genus to the type species, a gall maker possibly from *Smilax* in Brazil. Three species (*friburgensis, humilis,* and *tristis*), all from *Styrax* (Styracaceae), are placed in the new genus *Contodiplosis*, and *itaparicana*, reared from an unknown host, is with the unplaced Cecidomyiidi. Tavares's (1909) description of *Compsodiplosis* was accompanied by fine drawings made by Rübsaamen. The male genitalia resemble somewhat those of *Contarinia* with their short, robust gonocoxites, small gonostyli, wide, weakly divided hypoproct, and short, pointed aedeagus. The antenna shows nothing unusual, but the palpus is two-segmented and the empodia are only about half the length of the tarsal claws. The ovipositor resembles in gross aspect that figured for *Contodiplosis* (fig. 224).

luteoalbida (as *luteo-albida*) Tavares 1909: 14, Pl. II, figs. 2–6 (*Compsodiplosis*); ♂, ♀, pupa. Syntypes: Rio Grande do Sul, Brazil; presumed lost. Host: possibly

---

Figures 217–222. *Charidiplosis triangularis*: **217**, wing; **218**, tarsal claw and empodium; **219**, female postabdomen with detail of posterior segments (lateral); **220**, male genitalia (dorsal); **221**, larval posterior segments (dorsal); **222**, male third flagellomere.

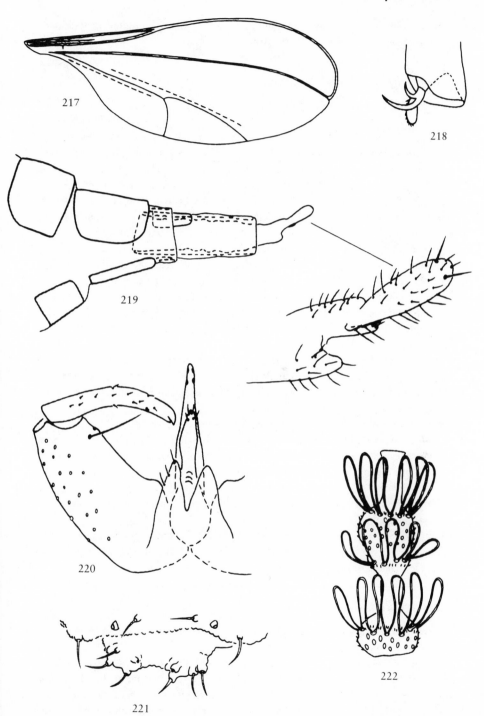

217

218

219

220

221

222

Smilax (Smilacaceae); Tavares thought *luteoalbida* to be an inquiline in galls made by *Asphondylia sulphurea*. Distr.: Brazil.

Contodiplosis Gagné, new genus

Type species, *Compsodiplosis friburgensis* Tavares.

Three species are moved here from Tavares's catchall genus *Compsodiplosis*. All three form complex leaf galls on *Styrax* in Brazil. Tavares (1915b) sketched the male genitalia of all three, and I have been able to study the type series of *C. friburgensis* in detail. The three species differ from *Compsodiplosis* in their robust gonocoxites (fig. 223) with a weak caudomesal lobe and the long, narrow hypoproct. Female antennal nodes and necks are longer and male circumfilar loops of the male are much shorter than Tavares (1909) showed for *Compsodiplosis*. The tarsal claws (fig. 225) are curved beyond midlength as in Compsodiplosis but the empodia are rudimentary rather than half the length of the tarsal claws. Additional evidence that the two genera are distinct is that *Contodiplosis* species form complex leaf galls on *Styrax*, while the one species of *Compsodiplosis* is possibly an inquiline in galls of an *Asphondylia* on *Smilax*. Wings of *C. friburgensis* are 2.0–3.5 mm. long. The ovipositor (fig. 224) is short but much tapered. Pupae have well-developed antennal horns. Larvae are unknown.

Etymology: *Contodiplosis* is a feminine name: *conto* is Greek for "short," in reference to the short loops of the male circumfila; *diplosis* is a suffix commonly used in Cecidomyiidi, meaning "double," with reference to the binodal male flagellomeres.

friburgensis Tavares 1915b: 150, figs. 3–5 (*Compsodiplosis*); ♂, ♀. New combination. Syntypes: Nova Friburgo, Rio de Janeiro, Brazil; presumed lost: specimens of both sexes present in Felt Collection fit Tavares's original description and illustrations, but are labeled "Bahia." Host: *Styrax* sp. (Styracaceae). Distr.: Brazil.

humilis Tavares 1915b: 154, figs. 6, 7 (*Compsodiplosis*); ♂. New combination. Syntypes: Nova Friburgo, Rio de Janeiro, Brazil; presumed lost. Host: *Styrax* sp. (Styracaceae). Distr.: Brazil.

tristis Tavares 1915b: 156, fig. 8 (*Compsodiplosis*); ♂. New combination. Syntypes: Nova Friburgo, Rio de Janeiro, Brazil; presumed lost. Host: *Styrax* sp. (Styracaceae). Distr.: Brazil.

Dactylodiplosis

Dactylodiplosis Rübsaamen 1916a: 452. Type species, *heisteriae* Rübsaamen (mon.).

This genus is known from one described species that forms fuzzy, spherical galls on leaves of *Heisteria*. An undescribed species from unspecified leaf galls of *Protium* (Burseraceae) from Brazil may belong here: its male genitalia (fig. 226) show some resemblance to Rübsaamen's drawings of the type species. The gonocoxites are not

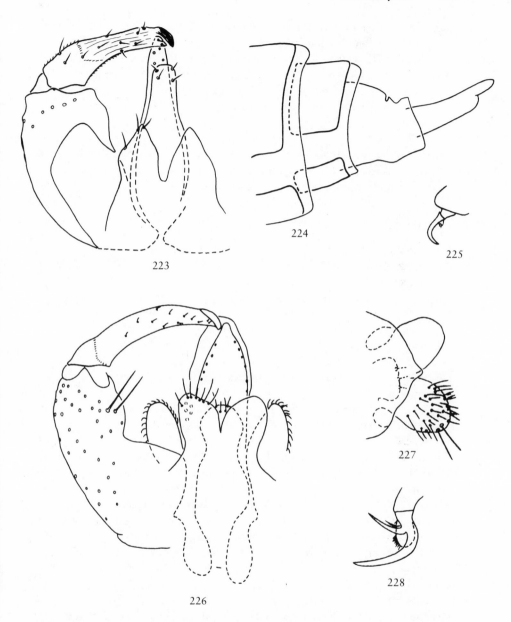

Figures 223–225. *Contodiplosis friburgensis*: **223**, male genitalia (dorsal); **224**, female post-abdomen (lateral); **225**, tarsal claw and empodium. **Figures 226–228.** *Dactylodiplosis* sp. ex *Protium*: **226**, male genitalia (dorsal); **227**, female 10th segment and cerci (dorsal; setae not shown on one cercus); **228**, tarsal claw and empodium.

splayed but have a prominent mesobasal lobe, the aedeagus is large, and the hypoproct is merely rounded and not bilobed. This genus shows some resemblance to *Youngomyia*. The wing is 3 mm long. Claws are strongly curved close to the basal third and have two teeth (fig. 228). The ovipositor is not protrusible; the cerci (fig. 227) are discoid with two long, posterior sensory setae and scattered shorter setae elsewhere. The larval spatula of *D. heisteriae* is reduced and terminates in a single point anteriorly.

heisteriae Rübsaamen 1916a: 454, figs. 24–28 (*Dactylodiplosis*); ♂, ♀, pupa, larva. Syntypes: Auristella, Madre de Dios, Peru, and San Francisco, Amazonas, Brazil; ZMHU. Host: *Heisteria cyanocarpa* (Olacaceae).

Diadiplosis

Diadiplosis Felt 1911h: 54. Type species, *cocci* Felt (orig. des.).

Kalodiplosis Felt 1915b: 229. Type species, *Dicrodiplosis multifila* Felt (orig. des.). New synonymy.

Cleodiplosis Felt 1922a: 1. Type species, *aleyrodici* Felt (orig. des.). New synonymy.

Olesicoccus Borgmeier 1931: 186. Type species, *costalimai* Borgmeier (orig. des.) = *Diadiplosis coccidivora* (Felt). New synonymy.

Phagodiplosis Blanchard 1938: 345. Type species, *haywardi* Blanchard (orig. des.) = *Diadiplosis coccidivora* (Felt). New synonymy.

Ghesquierinia Barnes 1939: 327. Type species, *megalamellae* Barnes (orig. des.). New synonymy.

Vincentodiplosis Harris 1968: 473. Type species, *Lobodiplosis coccidarum* Felt (orig. des.) = *Diadiplosis coccidarum* (Cockerell). New synonymy.

This is a worldwide, chiefly tropical genus with 9 described Neotropical species and 15 other Old World species. I know of one other, undescribed species from the Neotropics. Larvae of this genus are predators of scale insects or, in one case, whiteflies. *Kalodiplosis* and its subjective synonyms (*Cleodiplosis Ghesquierinia, Olesicoccus*, and *Vincentodiplosis*) of Gagné (1973b) are newly combined here with *Diadiplosis* because differences among them are gradual. Of special note, *D. pseudococci* Felt was purposely introduced to Hawaii and *D. multifila* (Felt), a widespread Neotropical species, is known also from Fiji. One species extends as far north as Bermuda.

Adults are small, with a wing length of 1.0–1.5 mm (fig. 230). Eyes (fig. 229) are separated laterally or the facets are very sparse there; facets are usually hexagonal

---

Figures 229–231. *Diadiplosis coccidarum*: 229, head (antennae removed); 230, wing; 231, fore- and midtarsal claws and empodia. Figures 232, 233. *Diadiplosis coccidivora*: 232, male postabdomen (lateral except genitalia dorsal); 233, female postabdomen with detail of posterior segments (dorsolateral). Figures 234–236. *Diadiplosis coccidarum*: 234, larval head (dorsal); 235, spatula with associated papillae; 236, larval terminal segments (dorsal).

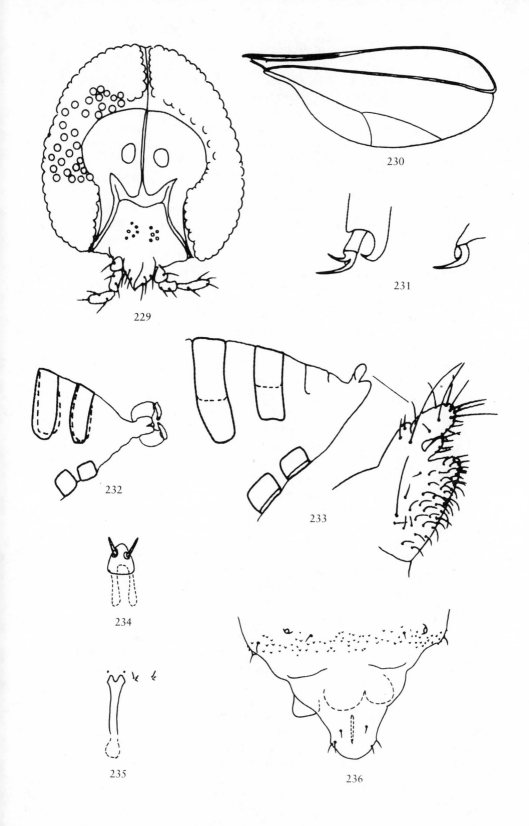

229

230

231

232

233

234

235

236

above and below, circular and farther apart laterally. Claws are bent at the basal third and are toothed or simple (fig. 231). Abdominal sclerites are short and wide (figs. 232, 233), especially in males. The female 10th tergum (fig. 233) has a pair of strong setae and may have lesser ones; female cerci are ovoid and without differentiated apical sensory setae, but in two species, *D. aleyrodici* (Felt) and *D. pulvinariae* (Felt), some dorsal setae are strong and recurved, as are also some setae on the 10th sternum in those and other species. Gonocoxites may have ventroapical lobes; gonostyli are variously shaped, completely setulose, the tooth pectinate; hypoproct entire to deeply lobed; aedeagus variable. Larvae have elongate antennae (fig. 234), a variable spatula (fig. 235), usually pseudopods, a ventral anus, and 4–6 terminal papillae (fig. 236). Ref.: Harris 1968.

aleyrodici Felt 1922a: 1 (*Cleodiplosis*); ♂, ♀ New combination. Syntypes: Panama City, Panama; USNM. Host: *Aleurycus chagentios* (Homoptera: Aleyrodidae). Distr.: Ecuador, Panama.

cocci Felt 1911h: 55 (*Diadiplosis*); ♂, ♀, pupa, larva. Syntypes: St. Vincent; Felt Coll. Host: *Saissetia nigra* (Homoptera: Coccidae). Distr.: Cuba, St. Vincent. Refs.: Harris 1968, Blahutiak and Alayo 1981.

coccidarum Cockerell 1892: 181 (*Diplosis*); ♂, ♀. New combination. Syntypes: Kingston, Jamaica; USNM. Hosts: *Dysmicoccus* sp., *Phenacoccus* sp., *Planococcus citri*, *Pseudococcus adonidum*, and *Saccharicoccus sacchari* (Homoptera: Pseudococcidae); and associated with one or more of *Aleurodes* sp., *Aspidiotus* sp., *Dactylopius* sp., *Orthezia insignis*, and *Parlatoria* sp. Distr.: Colombia, Cuba, Dominica, Dominican Republic, Guyana, Peru, Puerto Rico, St. Vincent, Trinidad, USA (Florida). Ref.: Harris 1968 (as incertae sedis). Note: Harris (1968) did not have the benefit of seeing the type series of this species. The five specimens were recently found glued to a card and labeled only "from TDA Cockerell, Kingston, Jamaica, 3002a." They are in extremely poor condition (the genitalia of all are lost), but they do belong to this genus because the characteristic antennae of both sexes are identical to the synonym *D. coccidarum* Felt, also reared from *Planococcus citri*.

   <u>*coccidarum*</u> Felt 1911f: 195 (*Lobodiplosis*); ♂, ♀. New synonym, new combination. Syntypes: St. Vincent; USNM. Host: *Planococcus citri*. Ref.: Harris 1968 (in *Vincentodiplosis*).

   <u>*cocci*</u> Felt 1913a: 304 (*Karschomyia*); ♂, ♀. Syntypes: Patillas, Puerto Rico; originally in Felt Collection, subsequently lost in return shipment of loan.

coccidivora Felt 1911g: 549 (*Mycodiplosis*); ♂, ♀. New combination. Syntypes: Kingston, Jamaica; USNM (part of series lost in return shipment of loan). Hosts: *Alichtensia* sp., *Coccus* sp., *Eriococcus* sp., *Pulvinaria ficus*, *P. urbicola*, *Saissetia coffeae*, *S. hemisphaerica*, and *S. oleae* (Homoptera: Coccidae, Pseudococcidae). Distr.: widespread from Bermuda, USA (Florida), and Panama, to Argentina. Refs.: Parnell 1966, Harris 1968 (in *Olesicoccus*).

   <u>*costalimai*</u> (as *costa-limai*) Borgmeier 1931: 186 (*Olesicoccus*); ♂, ♀, pupa, larva. Syntypes: São Paulo, Brazil; MZSP and BMNH.

haywardi Blanchard 1938: 345 (*Phagodiplosis*); ♂, ♀. New synonymy, new combination. Syntypes: Concordia, Entre Rios, Argentina; possibly in Blanchard Collection, INTAC.

floridana Felt 1915b: 230 (*Kalodiplosis*); ♂, ♀. New combination. Syntypes: Florida, USA; USNM. Host: *Hypogeococcus festeriana* (Homoptera: Pseudococcidae). Distr.: Cuba, Paraguay, USA (Florida).

multifila Felt 1907c: 19 (*Dichrodiplosis*); ♂. New combination. Holotype: Puerto Rico; USNM. Hosts: *Icerya montserratensis* and *Planococcus citri* (Homoptera: Margarodidae, Pseudococcidae). Distr.: Brazil, Dominica, Dominican Republic, Guyana, Montserrat, Puerto Rico, and Trinidad; also Fiji. Ref.: Harris 1968 (as *Ghesquierinia buscki*).

coccidarum Felt 1911g: 548 (*Dicrodiplosis*); ♀. New synonym, new combination. Holotype: Mayaguez, Puerto Rico; USNM. Host: scale insect. Ref.: Harris 1968 (as incertae sedis). Note: The flagellomeres with their closely spaced, elliptical circumfilar bases, the three-segmented palpus, and the setation of the 10th tergum, two large setae and scattered smaller ones, follow the pattern found in other females of *D. multifila*.

buscki Felt 1914a (*Diadiplosis*); ♂, ♀. Syntypes: Puerto Rico; Felt Coll. Host: scale insect.

pseudococci Felt 1921b: 225 (*Diadiplosis*); ♀, pupa, larva. Syntypes: Kartabo, Guyana; Felt Coll. Hosts: *Dysmicoccus brevipes* and *D. neobrevipes* (Homoptera: Pseudococcidae). Distr.: Brazil, Guatemala, Guyana, Honduras, Jamaica, Mexico; Hawaii (introduced). Ref.: Harris 1968 (in *Vincentodiplosis* and *Diadiplosis*). Note: Nothing in the female of the two synonyms indicates that they are different species; furthermore, both type series were preying on the same host. If the two names are not synonymized, a probably unnecessary name will have to be coined for the later, better known species.

pseudococci Felt 1933: 87 (*Lobodiplosis*); ♂, ♀. New synonymy, new combination. Syntypes: Kunia, Oahu, Hawaii; USNM.

pulvinariae Felt 1912a: 175 (*Mycodiplosis*); ♂, ♀, pupa, larva. New combination. Syntypes: St. Vincent; Felt Coll. Hosts: *Philephedra tuberculosa* and *Protopulvinaria pyriformis* (Homoptera: Coccidae). Distr.: Dominica, Dominican Republic, Guyana, Jamaica, St. Vincent, Trinidad, Venezuela, USA (Florida). Ref.: Harris 1968 (in *Olesicoccus*).

vaupedis Harris 1968: 450 (*Ghesquierinia*); ♂, ♀. New combination. Holotype: ♂; Vaupes, Colombia; BMNH. Host: coccid. Distr.: Colombia.

Enallodiplosis Gagné, new genus

Type species, *Enallodiplosis discordis* Gagné.

This genus is described for a species reared from fruit and curled leaves of *Prosopis tamarugo* (Fabaceae) in Chile. The moderately long antennae, long, pointed mandibles, and long setae of the larvae indicate that this species is predaceous. The com-

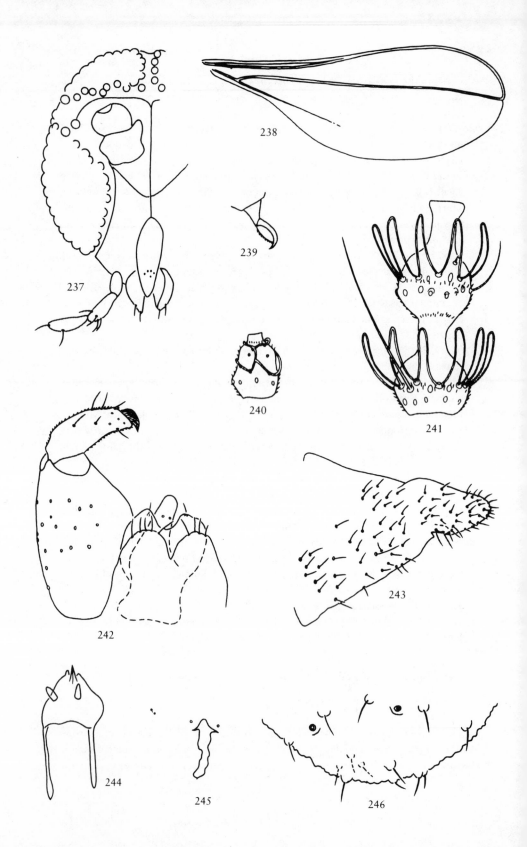

237

238

239

240

241

242

243

244

245

246

pletely setulose gonocoxites are reminiscent of *Diadiplosis*, but the tarsal claws of *Enallodiplosis* are simple and curved beyond midlength, and the empodia attain the curve of the claws (fig. 239). The female is aberrant in Cecidomyiidi for the reduction in the number of flagellomeres from 12 to 7. Adult abdominal terga lack any but the posterior setae; even the anterior pair of trichoid setae normally found on cecidomyiid terga are absent. The wing (fig. 238) is 1 mm long; R_5 is straight, attaining the wing apex; Cu is evanescent before the fork. The eyes (fig. 237) are narrowly discontinuous vertically and conspicuously so laterally. Male flagellomeres (fig. 241) are bifilar, with setulae present only on basal halves of nodes. There are only seven female flagellomeres (fig. 240), each short, narrowed on the apical half, and short-necked. Abdominal terga are weakly sclerotized, without trichoid anterior setae or any other setation other than the posterior row of setae. The female eighth tergite has sparse setae posteriorly; the ovipositor (fig. 243) is short, the 10th tergum is bare, the cerci separate, ovoid, and covered evenly with setae and setulae. The gonocoxites (fig. 242) are short and robust, the gonostyli completely setulose, the hypoproct bilobed and only slightly longer than the cerci; the aedeagus is short, slightly longer than the hypoproct. The pupal abdomen lacks spines dorsally. Larval antennae (fig. 244) are about twice as long as wide, the cephalic apodemes are longer than the head capsule, the spatula (fig. 245) is reduced in size and has only one pair of lateral papillae on each side, only one of the pair setose. All dorsal, pleural, and the eight terminal papillae (fig. 246) of the larva have a long seta and are raised on mamelons. The larval anus is ventral.

Etymology: *Enallopdiplosis* is a feminine name from the Greek *enallo*, meaning "contrary" or "against the grain," and *diplosis*, meaning "double," a suffix commonly used for Cecidomyiidi.

Enallodiplosis discordis Gagné, new species

Description: Head, male and female third flagellomeres, wing, tarsal claw and empodium, male genitalia, female postabdomen, and larval parts as in figures 237–246.

Holotype: ♂, from fruit of *Prosopis tamarugo*, Canchones, Chile, 19-11-1971, Vender, deposited in USNM. Other specimens examined (all in the USNM): 1 ♂, same data as holotype; 9 ♂♂, 7 ♀♀, from *P. tamarugo*, Iquique, Chile; 6 ♀♀, 2 pupal exuviae, 4 larvae, from leaves of *P. tamarugo*, Tarapacá, Pampa del Tamarugal, Chile, II-III-1966, C. Klein K. Two males and one female from Buenos Aires in the Brèthes collection belong to this genus and may belong to this species but are too poorly displayed on the slide mounts to compare them exactly with *E. discordis*.

Figures 237–246. *Enallodiplosis discordis*: **237**, head (left side, antennal flagellomeres removed); **238**, wing; **239**, tarsal claw and empodium; **240**, female third flagellomere; **241**, male third flagellomere; **242**, male genitalia (dorsal); **243**, female posterior abdominal segments (lateral); **244**, larval head; **245**, spatula with associated papillae; **246**, larval terminal segments (dorsal).

Etymology: *discordis* is a Latin adjective and refers to the fact that this species is the only one in the supertribe with fewer than 12 flagellomeres in the female.

Epihormomyia

Epihormomyia Felt 1915a: 155. Type species, *auripes* Felt (mon.).

This genus is known from a single female caught in flight. It was conspicuously colored in life (Felt 1915a), but the colors are now lost or faded on the slide mount. The wing is 3.5 mm long. The flagellomeres (fig. 247) are binodal, the first tarsomeres have a spur (fig. 249), and the tarsal claws (fig. 250) are curved near the basal third and have two teeth. Abdominal tergites are thickly covered with long, pliant scales. The ovipositor (fig. 248) is elongate but not greatly attenuate.

auripes Felt 1915a: 156 (*Epihormomyia*); ♀. Holotype: Malali, Demarara River, Guyana; Felt Coll. Host: unknown. Distr.: Guyana.

Frauenfeldiella

Frauenfeldiella Rübsaamen 1905: 122. Type species, *coussapoae* Rübsaamen (orig. des., as n. g., n. sp.).

This genus is known from one species reared from aerial roots of *Coussapoa*. The wing is about 3 mm long. The antennal circumfila are sinuous. The ovipositor is reportedly short, the cerci large.

coussapoae Rübsaamen 1905: 122 (*Frauenfeldiella*); ♀, larva. Syntypes: Gavea, Rio de Janeiro, and Juruá Miry, Amazonas, Brazil; ZMHU. Host: *Coussapoa* sp. (Moraceae). Distr.: Brazil. Note: The larva associated with the gall was only doubtfully referred to the female.

Gnesiodiplosis

Gnesiodiplosis Tavares 1917b: 162. Type species, *itaparicae* Tavares (orig. des.).

This genus is known from one species taken from elongate rosette bud galls on an undetermined Rubiaceae. The wing is about 3 mm long. The palpus consists of one large segment almost as long as the labella. The male hypoproct (fig. 251) is elongate and narrow but widens toward its apex; the aedeagus is elongate and narrow. The female cerci (fig. 252) are elongate.

itaparicae Tavares 1917b: 162, pl. XI, figs. 2–5 (*Gnesiodiplosis*); ♂, ♀. Syntypes: Bahia and Itaparica, Bahia, Brazil; part presumed lost, but one female sent to Felt by Tavares is deposited in the Felt Collection. Host: an undetermined Rubiaceae. Distr.: Brazil.

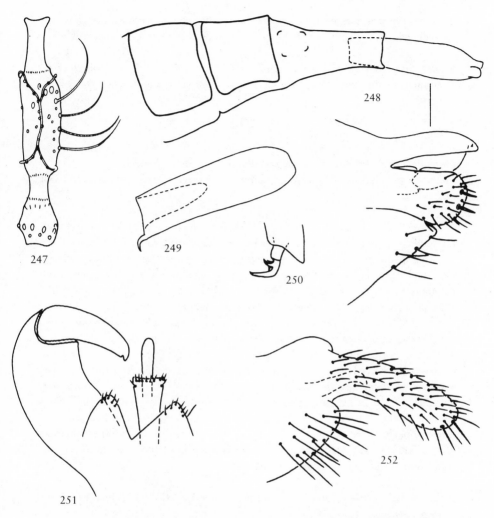

Figures 247–250. *Epihormomyia auripes*: **247,** female third flagellomere; **248,** female post-abdomen with detail of posterior segments (dorsolateral); **249,** first tarsomere; **250,** tarsal claw and empodium. **Figures 251, 252.** *Gnesiodiplosis itaparicae*: **251,** male genitalia (dorsal; redrawn from Tavares 1917b); **252,** female posterior abdominal segments (lateral).

Huradiplosis

Huradiplosis Nijveldt 1968: 61. Type species, *surinamensis* Nijveldt (orig. des.).

This genus is known from adults of one species reared from unspecified leaf galls on *Hura*. The wing is 1.5 mm long. The male gonocoxites are greatly swollen. The short hypoproct and aedeagus and the bifilar male flagellomeres resemble those of some

Contarinia species, but this genus differs from *Contarinia* in that the tarsal claws are strongly bent at midlength and the female postabdomen is not at all protrusible.

surinamensis Nijveldt 1968: 62, figs. 75–81 (*Huradiplosis*); ♂, ♀. Holotype: ♂; Paramaribo, Surinam; ITZA. Host: *Hura crepitans* (Euphorbiaceae). Distr.: Surinam.

Karshomyia

Karshomyia Felt 1908b: 398. Type species, *Mycodiplosis viburni* Felt (orig. des.).
Lobodiplosis Felt 1908b: 397. New synonym. Type species, *Mycodiplosis acerina* Felt (orig. des.).

Although many species have been described from the Holarctic Region under both *Karshomyia* and *Lobodiplosis*, only one has been described from the Neotropics, not counting *K. townsendi*, which is now referred to *Coquillettomyia*. I know of many more undescribed Neotropical species caught in flight and of another species from Dominican amber of Oligocene-Miocene age. Holarctic species have been reared from mushrooms and logs, where they are presumably fungivorous.

Lobodiplosis and *Karshomyia* were previously separated on the basis of differences in the male genitalia, the chief distinctions being whether the gonostyli were modified and whether the gonocoxites had a setose posteroventral lobe. These distinctions, however, are blurred in two undescribed species that I have collected from the United States, so I synonymize the two genera here. The wing (fig. 253) is 2–3 mm long. Tarsal claws may be simple or toothed. The most prominent unifying feature of this genus is that the anterior abdominal tergites (fig. 254) are partially unsclerotized at midlength. The gonopod and aedeagus (figs. 255, 256) are greatly modified and often bizarre. Larvae have a clove-shaped spatula and short, corniform terminal papillae situated on knobs (Mamaev and Krivosheina 1965).

spinulosa Gagné. New name for *Lobodiplosis spinosa* Felt 1909. Host: unknown. Distr.: *Antigua*.
   spinosa Felt 1909: 301 (*Lobodiplosis*); ♂, ♀. New combination; new secondary homonym of *Metadiplosis* (now *Karshomyia*) *spinosa* Felt 1908b. Lectotype, ♂, here designated: St. Johns, Antigua; USNM; paralectotypes, 2 ♀♀ with same data as ♂, do not belong to *Karshomyia* but to an as yet undetermined genus.

Figures 253, 254. *Karshomyia* sp.: 253, wing; 254, female postabdomen (lateral). Figures 255, 256. *Karshomyia spinulosa*: 255, male genitalia (dorsal, ventral lobe of gonocoxite and aedeagus not shown); 256, gonocoxite and aedeagus (ventral). Figures 257, 258. *Neobaeomyza maculata*: 257, female third and fourth flagellomeres; 258, female cercus and hypoproct (lateral).

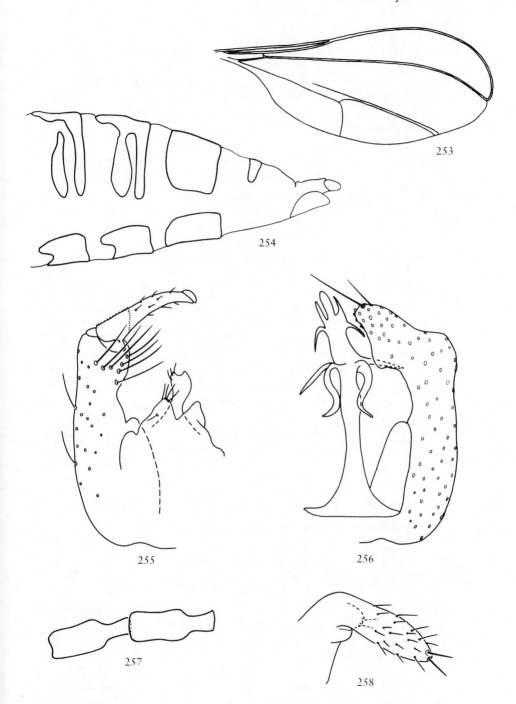

253

254

255

256

257

258

Megaulus

Megaulus Rübsaamen 1916a: 461. Type species, *sterculiae* Rübsaamen (mon.).

This genus is known from one species reared from leaf galls of *Sterculia*. The male is unknown. Female flagellomeres are peculiar for their long necks and long circumfilar loops. The ovipositor is distinctive, fairly long, and with short cerci reminiscent of those of *Contarinia*.

sterculiae Rübsaamen 1916a: 461, figs. 35–37 (*Megaulus*); ♀, pupa, larva. Syntypes: on Rio Acre near San Francisco, and near Cobija, Amazonas, Brazil; ZMHU. Host: *Sterculia* spp. (Sterculiaceae). Distr.: Brazil.

Moehniella

Moehniella Wünsch 1979: 66. Type species, *fernandi* Wünsch (orig. des.).

This genus is known from the female and larva of a species living in fruit of *Mimosa*, probably as an inquiline. The larva was only tentatively associated with the female and appears to belong to *Contarinia*. The wing is about 2 mm long. The antennal flagellomeres are peculiar for their very sinuous circumfila. The wing as originally illustrated appears to have Cu forked from the base or, alternatively, to have a strong M_3 fold and a simple Cu. Because I have not seen either situation elsewhere in the Cecidomyiidi, I am inclined to believe that the drawing is inaccurate. The cerci are elongate, and the hypoproct is deeply bilobed.

fernandi Wünsch 1979: 67, pls. 20–22 (*Moehniella*); ♀, larva. Holotype: ♀; Cerro San Fernando, Colombia; SMNS. Host: *Mimosa leiocarpa* (Fabaceae). Distr.: Colombia.

Neobaeomyza

Neobaeomyza Brèthes 1914: 151. Type species, *maculata* Brèthes (orig. des.).

This genus is known only from the female of one species caught in flight. The type of the single specimen is poorly mounted on a slide, so many of the characters are obscured, but the species should be recognizable when found again. The wing is 1.3 mm long, the flagellomeres (fig. 257) have long necks without setulae, and the ovipositor is nonprotrusible. The cerci (fig. 258) are elongate and have two prominent, long, sensory setae apically and scattered long and short setae elsewhere, none longer than the sensory setae.

maculata Brèthes 1914: 151 (*Neobaeomyza*); ♀. Holotype: General Urquiza, Buenos Aires, Argentina; MACN. Host: unknown. Distr.: Argentina.

Ouradiplosis

Ouradiplosis Felt 1915a: 154. Type species, *aurata* Felt (orig. des.).

This genus is known from one species based on a single female caught in flight. The female is distinct from other known genera for its elongate cerci (fig. 261), which should enable the species to be found again. The type lacks its tarsal claws. The wing (fig. 259) is 4.5 mm long and originally thickly clothed with golden-brown hairs, giving a pronounced color; the slide-mounted specimen no longer has hairs on the wing, but the membrane is dark except for the leading edge. The flagellomeres are as in figure 260. The ovipositor is not protrusible, the cerci (fig. 261) are elongate, and the hypoproct wide with many short setae.

aurata Felt 1915a: 154 (*Ouradiplosis*); ♀. Holotype: Igarape-Açú, Pará, Brazil; CUI. Host: unknown. Distr.: Brazil.

Plesiodiplosis

Plesiodiplosis Kieffer 1912a: 1. Type species, *Clinodiplosis chilensis* Kieffer and Herbst (orig. des.).

This genus is known from a single species caught in flight, but the male genitalia (fig. 262) (redrawn here from Kieffer 1913a) are distinctive. They are reportedly very large, one third the length of the abdomen. The aedeagus is elongate and wider apically than at midlength, and the hypoproct is deeply bilobed. The tarsal claws were described as curved and as long as the empodium. but their exact shape is unknown. The female cerci are reportedly nearly triangular and barely longer than wide as seen from above.

chilensis Kieffer and Herbst 1906: 331 (*Clinodiplosis*); ♂, ♀. Syntypes: vic. Concepción, Chile; presumed lost. Host: unknown. Distr.: Chile. Ref.: Kieffer 1913a.

Resseliella

Resseliella Seitner 1906: 174. Type species, *piceae* Seitner (orig. des.).

This genus is chiefly Holarctic in distribution with 39 known species, 19 from North America. I know of only two species from the Neotropics, both undescribed. One is from stem swellings on *Piper* (Piperaceae) in Costa Rica, the other from a larva intercepted in California in a shipment of Ecuadoran bananas. The genus is unlike most other plant-feeding Cecidomyiidi, particularly the Cecidomyiini, because the tarsal claws are angled at the basal third and are usually toothed. The terminal larval segment (fig. 263) is diagnostic for its two recurved lobes, each ending in a large, corniform papilla.

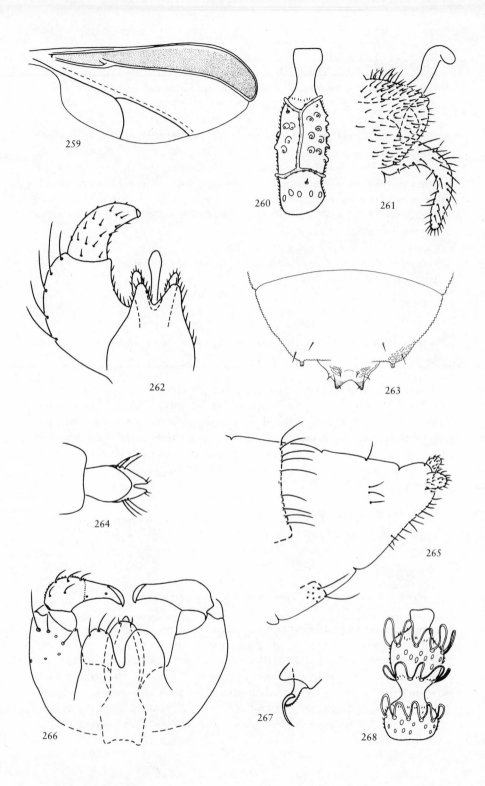

259

260

261

262

263

264

265

266

267

268

Schizodiplosis

Schizodiplosis Kieffer 1912c: 233. Type species, *Clinodiplosis floralis* Kieffer (orig. des.).

This genus is known only from the female of a species caught in flight, but the bulbous, moderately protractile ovipositor (fig. 264) (redrawn here from Kieffer 1913a) may allow this genus to be found again. The flagellomere necks are half as long as the nodes and the circumfila have short loops. The tarsal claws are simple and attain the length of the empodia, but the shape was not specified.

floralis Kieffer and Herbst 1906: 233, pl. I, figs. 5, 6 (*Clinodiplosis*); ♀. Holotype: vic. Concepción, Chile; presumed lost. Host: observed flying in numbers around flowers of a Myrtaceae. Distr.: Chile. Ref.: Kieffer 1913a.

Silvestrina

Silvestrina Kieffer 1912b: 172. Type species, *Arthrocnodax silvestrii* Kieffer (orig. des.) = *Silvestrina cincta* (Felt).

This genus is known from one widely distributed species known from the Americas as well as Africa, Asia, and Europe. Larvae are evidently generalist predators, the adults having been reared in association with mites, scale insects, caterpillars, bees, and beetles, and from stored food products, fresh fruit, and fungi presumably infested with insects or mites. Kieffer's (1913a) illustration of the male genitalia of *S. silvestrii* fit the male of *S. cincta* (Felt). Because other characters of *S. silvestrii*, epecially the characteristic shape of the tarsal claws and empodia, also fit *S. cincta*, the two species are treated as synonyms here. Specimens I have seen from widely separated localities on five continents and from Puerto Rico are identical.

Wings are 1.0–1.5 mm long. Male flagellomeres (fig. 268) have short internodes, necks, and circumfila. Eye facets are farther apart laterally than elsewhere on the eye. Tarsal claws (fig. 267) are bowed near the basal third, and the empodia attain the curve in the tarsal claws. Kieffer (1910a and subseq.) wrote that the claws were twice the length of the empodia, and they are, but his statement should not be taken to mean that the empodia do not reach the bend of the claw. The abdomen is noteworthy for its weak sclerotization and lack of pigmentation. Each of the abdominal tergites

---

Figures 259–261. *Ouradiplosis aurata*: 259, wing; 260, female third flagellomere; 261, female posterior abdominal segments (ventral; setae not shown on one cercus). Figure 262. *Plesiodiplosis chilensis*: male genitalia (dorsal; redrawn from Kieffer 1913a). Figure 263. *Resseliella conicola*, larval terminal segments (dorsal). Figure 264. *Schizodiplosis floralis*, female posterior abdominal segments (dorsal; redrawn from Kieffer 1913a). Figures 265–268. *Silvestrina cincta*: 265, female postabdomen (dorsolateral); 266, male genitalia (dorsal); 267, tarsal claw and empodium; 268, male third flagellomere.

has a single, sparse row of posterior setae (fig. 265) and no other chaetotaxy except for the anterior pair of trichoid sensilla; the sternites have similar setation except for two to three unbroken rows of setae posteriorly. The male genitalia (fig. 266) are compact, the gonocoxites with a short, convex mesobasal lobe, the gonostyli short, bulbous at base, tapering beyond, setulose at least to midlength.

cincta Felt 1907a: 47 (*Cecidomyia*); ♂. Holotype: Albany, New York, USA; Felt Coll. Hosts: insects and mites. Distr.: Brazil, Puerto Rico; Canada, China, Italy, Japan, Republic of South Africa, USA. Refs.: Wolcott 1936, Gagné 1973b.
    macrofila Felt 1907c: 21 (*Cecidomyia*); ♂. Holotype: Las Vegas, New Mexico, USA; Felt Coll.
    silvestrii Kieffer 1910a: 133, figs. I, III (*Arthrocnodax*); ♀. New synonym. Syntypes: Cape Town, Republic of South Africa; presumed lost.

Styraxdiplosis

Styraxdiplosis Tavares 1915b: 147. Type species, *caetitensis* Tavares (orig. des.).
Styracodiplosis Tavares 1918b: 77, emendation.

The type species forms leaf galls on *Styrax*. *Styraxdiplosis*, *Compsodiplosis*, and *Contodiplosis* have all been reared from leaf galls on *Styrax*. Two species are listed here, following Tavares. Both make leaf galls but on different plant families. Tavares (1925) pointed out differences between the somewhat different pupae of the two species, but he knew only the male of *S. caetitensis* and only the female of *S. cearensis*. The two species are not necessarily related. Male flagellomeres have short internodes and short circumfilar loops; the female flagellomeres have very short necks. The tarsal claws are simple but the shape was not specified, so this genus is not in the generic key. The male genitalia have a wide and bilobed hypoproct and unlobed gonocoxites (fig. 269) (redrawn here from Tavares 1918b). They were incorrectly drawn in Tavares 1915b. The ovipositor was described only as moderately protractile.

caetitensis Tavares 1915b: 147, figs. 1, 2 (*Styraxdiplosis*); ♂. Syntypes: Caeteté, near Bahia, Bahia, Brazil; presumed lost. Host: *Styrax* sp. (Styracaceae). Distr.: Brazil. Ref.: Tavares 1918b.
    caetetensis Tavares 1918b: 75, emendation.
cearensis Tavares 1925: 20 (*Styracodiplosis*); ♀, pupa. Syntypes: Ceará, Brazil; presumed lost. Host: *Croton hemiargyreus* (Euphorbiaceae). Distr.: Brazil.

Figure 269. *Styraxdiplosis caetitensis*: male genitalia (dorsal; redrawn from Tavares 1918b). **Figures 270, 271.** *Xenasphondylia albipes*: 270, female postabdomen and detail of tip of ovipositor (dorsolateral); 271, tarsal claw and empodium. **Figures 272–274.** *Youngomyia knabi*: 272, wing; 273, male genitalia (dorsal); 274, male third flagellomere. **Figures 275, 276.** *Youngomyia podophyllae*: 275, male genitalia (dorsal); 276, tarsal claw and empodium.

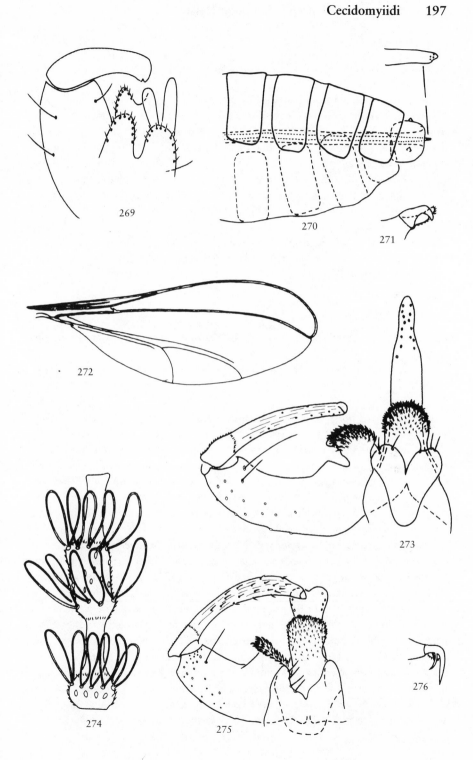

269

270

271

272

273

274

275

276

Xenasphondylia

Xenasphondylia Felt 1915c: 202. Type species, *albipes* (orig. des.).

This genus is known from females of a species reared from an unknown plant. It has until now been placed with the Asphondyliini because of the similarity of the ovipositor shaft to that of *Asphondylia*. The seventh sternite, however, is no larger than the preceding sternites, unlike that of all Asphondyliini, so I regard the ovipositor as a convergent adaptation. The wing is 4 mm long. The tarsal claws (fig. 271) are simple and especially robust at their bases. The eighth tergite is bare, but has two raised, dorsal papillae. The ovipositor is very long, aciculate, pigmented, and rigid, and the cerci are greatly reduced and fused (fig. 270).

albipes Felt 1915c: 202, fig. 5 (*Xenasphondylia*); ♀. Syntypes: San Francisco Mountains, Santo Domingo [presumably Dominican Republic]; USNM. Host: unknown. Distr.: Dominican Republic. Ref.: Möhn 1963a.

Youngomyia

Youngomyia Felt 1908b: 398. Type species, *Dicrodiplosis podophyllae* Felt (orig. des.).
Delphodiplosis Felt 1915a: 155. New synonymy. Type species, *cinctipes* Felt (orig. des.).

This genus is at present known from four species, three from the Western Hemisphere, one from India. The type species of *Youngomyia*, previously known from the United States and newly recorded here from Middle America, is an inquiline in galls formed by *Schizomyia* species on several host plants; the other two species, both newly placed in this genus, are Neotropical and were caught in flight. *Delphodiplosis* is known from a single female bearing a resemblance to the female of the other newly combined species, *Y. knabi*. These two have ovipositors unlike that of *Y. podophyllae*, differences that by themselves would preclude placing these species in the same genus. Yet the male genitalia of *Y. knabi* (fig. 273) and *Y. podophyllae* (fig. 275) have so many points in common as to appear congeneric, so I am confident that the males and females of *Y. knabi* are properly associated. Wings (fig. 272) are 3–4 mm long. The position of the conspicuous but incomplete Rs is variable, situated slightly before to slightly beyond midlength of R_1; for this reason *Youngomyia* occurs in two places in the key to genera. Male flagellomeres are elongate and tricircumfilar (fig. 274). Female circumfilar connectives are numerous, and the flagellomere necks are setulose. The tarsal claws (fig. 276) are strongly bent near the basal third. In *Y. knabi* only the foreclaws are toothed, with an additional, tiny tooth usually visible below the larger; in the other species all claws have two teeth. The female cerci of *Y. podophyllae* (fig. 277) are large, globular, and, on the mesal surface, have more than 100 closely adjacent, short, sensory setae. The cerci of *Y. knabi* (fig. 278) and *Y. cinctipes* (fig.

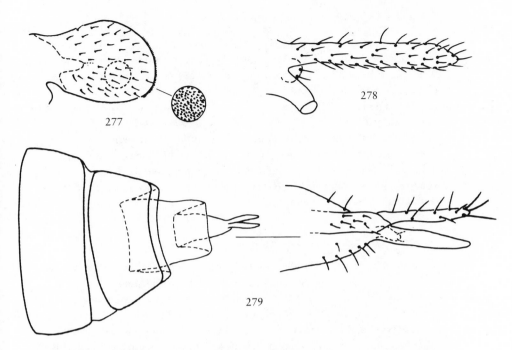

Figure 277. *Youngomyia podophyllae*, female cercus (lateral, with section of mesal surface). **Figure 278.** *Youngomyia knabi*, one female cercus and hypoproct (dorsal). **Figure 279.** *Youngomyia cinctipes*, female postabdomen with detail of cerci and hypoproct (dorsal).

279) are elongate-cylindrical, with no such sensory setae. Gonocoxites (figs. 273, 275) are splayed, each with a large, setose mesobasal lobe; the gonostyli are long, attenuate; the hypoproct is covered on its posterior half with closely appressed spinules; the aedeagus is variably shaped.

cinctipes Felt 1915a: 155 (*Delphodiplosis*); ♀. Holotype: Bartica, Guyana; Felt Coll. Host: unknown. Distr.: Guyana.

knabi Felt 1912d: 154 (*Coquillettomyia*); ♂, ♀. New combination. Lectotype here designated: ♂; Puerto Limón, Costa Rica; USNM (#29246, Felt #a3186); paralectotypes, on slides, one ♂ and two ♀♀, with same data as lectotype, except that one ♂ and one ♀ are in the Felt Collection. In addition to the types series, there are seven pinned specimens, two ♂♂ and five ♀♀, labeled "Port Limon CR" and "*C. knabi*," five of them in the USNM, two in the Felt Collection. Host: unknown, found roosting on spider web. Distr.: Costa Rica.

podophyllae Felt 1907a: 30 (*Dicrodiplosis*); ♂. Holotype: Albany, New York, USA: Felt Coll. Host: inquiline in galls of *Schizomyia* spp. Distr.: Guatemala, Panama; USA (widespread).

podophylli, emendation in Felt 1907b: 126.

Unplaced Species of Cecidomyiidi

itaparicana Tavares 1922: 41, figs. 9–10 (*Compsodiplosis*); ♂, larva. Syntypes: Ilha de Itaparica, Bahia, Brazil; presumed lost. Host: circular leaf spot gall on unknown plant. Distr.: Brazil. Note: Tavares's original figure of the genitalia may eventually allow this species to be placed to genus.

producta Felt 1912a: 177 (*Hyperdiplosis*); ♂, ♀. Syntypes: St. Vincent; Felt Coll. Host: mite galls on inflorescence of *Stachytarpheta jamaicensis* (Verbenaceae). Note: This is probably a predator, but it does not fit into any known genus. The specimens, a male and three females, all mounted on the same slide, are uncleared and too delicate to remount. The male genitalia are mounted laterally, and the separate parts are obscured. The specimens have peglike setae along the antero-ventral surface of the fore femora.

Unplaced Species of Cecidomyiinae

biconica (as *bi-conica*) Cook 1906: 251, fig. 8 (*Cecidomyia*); gall. Syntypes: Santiago de Cuba, Cuba; depos. unknown. Host: unknown. Distr.: Cuba.

bifida Brèthes 1914: 154 (*Trisopsis*); ♂. Holotype: Buenos Aires, Argentina; possibly lost, not found in MACN. Host: unknown. Distr.: Argentina. Note: There is a slide marked "type" in the Brèthes Collection in the MACN, but the male on it cannot be *bifida* because its genitalia do not match those of Brèthes' figure published with his original description. Further, it does not have three eyes (i.e., the eyes are not divided laterally), which one would expect a *Trisopsis* to have. The genitalia are not mounted flat, but generally resemble those of a *Lestodiplosis*. The antennal flagellomeres each have two circumfila. Three other specimens are labeled as this species in Brèthes' collection, but they are labeled BAs [presumably for Buenos Aires], 11-1917, so were evidently collected after publication of *bifida*. These specimens do have "three eyes" but belong not to Trisopsis but to the new genus *Enallodiplosis* described elsewhere in this book.

cayamacensis Cook 1906: 251, pl. 49, fig. 7 (*Cecidomyia*); gall. Syntypes: Cayamas, Santa Clara, Cuba; depos. unknown. Host: unknown [probably Fabaceae]. Distr.: Cuba.

cecropiae Cook 1909: 145 (*Cecidomyia*); gall. Syntypes: Cuba; depos. unknown. Host: *Cecropia obtusa* and *Cecropia* sp. (Moraceae). Distr.: Brazil, Cuba.

fici Cook 1909: 145, fig. 3 (*Cecidomyia*); gall. Syntypes: Cuba; depos. unknown. Host: *Ficus* sp. (Moraceae). Distr.: Cuba.

mazaiana Cook 1906: 252, fig. 9 (*Cecidomyia*); gall. Syntypes: Santiago de Cuba, Cuba; depos. unknown. Host: unknown plant; Cook's figure of gall same as for *torreana* below. Distr.: Cuba.

torreana Cook 1906: 252, fig. 9 (*Cecidomyia*); gall. Syntypes: Santiago de Cuba, Cuba; depos. unknown. Host: unknown plant; Cook's figure of gall same as for *mazaiana* above. Distr.: Cuba.

Unplaced Species of Cecidomyiidae

flavida Blanchard 1852: 350 (*Cecidomyia*); adult. Syntypes: La Serena, Chile; MNHNP? Host: unknown. Distr.: Chile.

furcata Philippi 1865: 631 (*Lasioptera*); adult. Syntypes: Santiago, Chile; MNHNS? Host: unknown. Distr.: Chile.

fuscescens Philippi 1865: 628 (*Cecidomyia*); adult. Syntypes: Chile; MNHNS? Host: unknown. Distr.: Chile.

gollmeri Karsch 1880: 297 (*Cecidomyia*); ♂. Holotype: Caracas, Venezuela; ZMHU? Host: unknown. Distr.: Venezuela.

pallipes Philippi 1865: 630 (*Lasioptera*); adult. Syntypes: San Cristóval, near Santiago, Chile; MNHNS? Host: unknown. Distr.: Chile.

pictipes Williston 1896: 253 (*Diplosis*); ♂ Holotype: St. Vincent; BMNH? Host: unknown. Distr.: St. Vincent.

umbra Walker 1856: 421 (*Cecidomyia*); adult. Type loc.: South America; BMNH? Host: unknown. Distr.: South America.

List of New Taxa and Nomenclatural Changes

Several new taxa are described and many replacement names, synonyms, and new combinations were made in this chapter. New names and changes proposed here are as follows.

New tribes:
    Centrodiplosini
    Lopesiini
New genera:
    Contodiplosis
    Enallodiplosis
New species:
    Bremia mirifica
    Enallodiplosis discordis
    Stomatosema dominicensis
Replacement names:
    Anasphodiplosis for *Anasphondylia* Blanchard
    Bruggmannia pisonioides for *Bruggmanniella pisoniae* Felt
    Calmonia fici for *Asteromyia urostigmatis* Tavares
    Karchomyia spinulosa for *Lobodiplosis spinosa* Felt
    Neolasioptera exeupatorii for *Cecidomyia eupatorii* Cook
    Stephomyia epeugenia for *Stephomyia eugeniae* Tavares
Restored generic names:
    Hemiasphondylia Möhn from *Asphondylia*

Stenodiplosis Reuter from *Contarinia*
Restored specific name:
 Hemiasphondylia mimosa Möhn from *Asphondylia*
New generic synonyms:
 Allocontarinia Solinas under *Stenodiplosis*
 Aphodiplosis Gagné under *Charidiplosis*
 Cleodiplosis Felt under *Diadiplosis*
 Delphodiplosis Felt under *Youngomyia*
 Ghesquierinia Barnes under *Diadiplosis*
 Hemibruggmanniella Möhn under *Bruggmanniella*
 Kalodiplosis Felt under *Diadiplosis*
 Lobodiplosis Felt under *Karshomyia*
 Neurolasioptera Brèthes under *Neolasioptera*
 Olesicoccus Borgmeier under *Diadiplosis*
 Phagodiplosis Blanchard under *Diadiplosis*
 Physalidicola Brèthes under *Neolasioptera*
 Ruebsaamenodiplosis Rübsaamen under *Machaeriobia*
 Toxomyia Felt under *Mycodiplosis*
 Vincentodiplosis Harris under *Diadiplosis*
New specific synonyms:
 Arthrocnodax constricta Felt under *Feltiella insularis*
 Arthrocnodax silvestrii Kieffer under *Silvestrina cincta*
 Asphondylia mimosicola Gagné under *Hemiasphondylia mimosa*
 Charidiplosis concinna Tavares under *Charidiplosis triangularis*
 Dicrodiplosis coccidarum Felt under *Diadiplosis multifila*
 Lobodiplosis coccidarum Felt under *Diadiplosis coccidarum* Cockerell
 Lobodiplosis pseudococci Felt under *Diadiplosis pseudococci* Felt
 Machaeriobia brasiliensis Rübsaamen under *Machaeriobia machaerii*
 Neolasioptera odorati Wünsch under *Neolasioptera cruttwellae*
 Neolasioptera portulacae (Cook 1909) under *Neolasioptera portulacae* (Cook
    1906)
 Phagodiplosis haywardi Blanchard under *Diadiplosis coccidivora*
New combinations:
 Asphondylia crassipalpis Kieffer and Jörgensen to *Rhoasphondylia*
 Autodiplosis iheringi Tavares to *Clinodiplosis*
 Camptoneuromyia petioli Mamaev to *Ledomyia*
 Cecidomyia cocolobae Cook to *Ctenodactylomyia*
 Cecidomyia pisoniae Cook to *Bruggmannia*
 Cleodiplosis aleyrodici Felt to *Diadiplosis*
 Clinodiplosis triangularis Felt to *Charidiplosis*
 Compsodiplosis friburgensis Tavares to *Contodiplosis*
 Compsodiplosis humilis Tavares to *Contodiplosis*
 Compsodiplosis tristis Tavares to *Contodiplosis*
 Contarinia sorghicola (Coquillett) to *Stenodiplosis*

Coquillettomyia knabi Felt to *Youngomyia*
Dichrodiplosis multifila Felt to *Diadiplosis*
Diplosis coccidarum Cockerell to *Diadiplosis*
Eudiplosis bahiensis Tavares to *Clinodiplosis*
Eudiplosis cearensis Tavares to *Clinodiplosis*
Eudiplosis marcetiae Tavares to *Clinodiplosis*
Eudiplosis pulchra Tavares to *Clinodiplosis*
Eudiplosis rubiae Tavares to *Clinodiplosis*
Ghesquierinia vaupedis Harris to *Diadiplosis*
Kalodiplosis floridana Felt to *Diadiplosis*
Karschomyia townsendi Felt to *Coquillettomyia*
Lasioptera cordobensis Kieffer and Jörgensen to *Baccharomyia*
Lasioptera graciliforceps Kieffer and Jörgensen to *Meunieriella*
Lasioptera interrupta Kieffer and Jörgensen to *Baccharomyia*
Lasioptera ornaticornis Kieffer and Jörgensen to *Baccharomyia*
Lestodiplosis picturata Felt to *Clinodiplosis*
Mycodiplosis coccidivora Felt to *Diadiplosis*
Mycodiplosis insularis Felt to *Feltiella*
Mycodiplosis pulvinariae Felt to *Diadiplosis*
Neobaeomyza australis Brèthes to *Mycodiplosis*
Neurolasioptera baezi Brèthes to *Neolasioptera*
Oxasphondylia clavata Tavares to *Stephomyia*
Physalidicola argentata Brèthes to *Neolasioptera*
Toxomyia fungicola Felt to *Mycodiplosis*
Uleella depressa Kieffer to *Bruggmannia*
Uleella globulifex Kieffer to *Bruggmannia*
Uleella lignicola Kieffer to *Bruggmannia*
Uleella longicauda Kieffer to *Bruggmannia*
Uleella longiseta Kieffer to *Bruggmannia*
Uleella micrura Kieffer to *Bruggmannia*
Uleella neeana Kieffer to *Bruggmannia*
Uleella ruebsaameni Kieffer to *Bruggmannia*

6 Plant Hosts and Galls of Neotropical Cecidomyiidae

Hosts

Cecidomyiids are widely represented on higher plants throughout the world and probably occur on every family of Spermatophyta, the seed-bearing plants. Fewer than 10 cecidomyiids are known from the ferns and allies throughout the world, and, to date, none is known from these plants in the Neotropics.

Neotropical gall midges are now known from 86 plant families. As in other parts of the world, the larger the host family the more species of gall midges it appears to have. Families with more than 10 recorded species of gall midges (both described and undescribed) are, in decreasing order, Fabaceae with 105, Asteraceae with 99, Euphorbiaceae and Nyctaginaceae with 24 each, Myrtaceae, Rubiaceae, and Solanaceae with 22 each, and Melastomataceae with 14. These are also some of the largest families of Neotropical plants. The dominant genera of a family change from one region to another, but the larger genera still carry more gall midges in either place. For example, *Baccharis* in the Neotropics is analogous to *Solidago* in the eastern Nearctic in both species richness and the large number of gall midges it carries.

The widespread distribution of Cecidomyiidae among seed-bearing plants is very different from other groups of gall-forming organisms, except perhaps for eriophyid mites (Acarina), which are plentiful in the Neotropics as well as in North America. The Cynipinae (Hymenoptera) that are so numerous in the Holarctic Region on Fagaceae and some Rosaceae are relatively rare in the Neotropics, where for the most part those hosts do not occur. Here, unlike in most parts of the world, Lepidoptera make complex galls, especially on Melastomataceae. Psyllidae (Homoptera), while still a minor gall-making group, appear to be more numerous in the Neotropics than elsewhere.

Galls and Gall Makers

Some plant-feeding gall midges live freely in buds or flowers or even on plant surfaces, but most are found in galls. Cecidomyiid galls are predictable and consistent deformations that occur in response to larval feeding. They may be simple or complex. Simple galls include spots (fig. 281), swellings (fig. 282), and folds (fig. 314) that are not the result of any reorganized plant tissue. Complex galls are the result of fundamentally reorganized plant tissue that results in a structure different from any normally found on a plant (fig. 287). Information on galls in general can be found in Meyer 1987 and Shorthouse and Rohfritsch 1992.

In this book, a few phrases will be used often to characterize some of the simple galls. A "leaf spot" or "leaf blister" is an ovoid to circular area, with one or more larvae in a concavity on the underside of the leaf or inside the leaf tissue. If one or both sides are convex, the gall is called a blister instead of a spot. If the whole leaf is rolled to form a tube, the gall is called a "leaf roll"; if only the margin of the leaf is rolled, it is a "marginal leaf roll." A "vein fold" or "leaf fold" is an invagination of the leaf blade, along a vein or between veins, respectively, that does not affect the whole leaf.

Figures 280–341 show a variety of simple and complex galls. Spherical, fuzzy, complex leaf galls (fig. 312) are one of the commonest kinds of galls on Neotropical plants. Although such galls appear on at least 20 plant families in this region, they are made by a great variety of cecidomyiids.

Cecidomyiidae are responsible for most plant galls, but one cannot always be sure what a gall contains by its appearance alone, especially in the Neotropics. Many complex Lepidoptera galls on Melastomataceae and Psyllidae (Homoptera) galls on, for example, *Schinus* spp. (Anacardiaceae) resemble some formed by cecidomyiids on other groups of plants. The only way to be certain of what forms the galls is to open them when the gall maker is still inside. One should also open more than one gall when trying to determine the gall maker because some may contain hyperparasites or other secondary organisms. For excellent series of photographs of the kinds of galls formed by various insects and mites, see Meyer 1987 and Dreger-Jauffret and Shorthouse 1992.

Cecidomyiid galls will reveal one to several, legless, minute-headed larvae. Full-grown larvae will often be recognizable by a characteristic spatula (fig. 1), although that is lost in some genera. The interior of their galls is usually smooth and clean because gall midge larvae produce no noticeable waste products. The inside surface of galls caused by Asphon-dyliina and by most Alycaulini are lined with a fungal mycelium that remains white while the larva is actively feeding.

Tephritidae (Diptera) and Hymenoptera larvae also have no legs or apparent waste products. Tephritid larvae have only two pairs of spiracles, a prothoracic and a posterior pair, and no head capsule, the only apparent mouthparts being a pair of hooks. Hymenoptera larvae found in galls differ most from gall midges by their fairly conspicuous, opposable mandibles.

Eriophyid mites (Eriophyidae: Acarina) form diverse galls on a great variety of plants. Their galls tend to be made up of greatly convoluted tissue or simple, globular or conical leaf invaginations. The inner surface of their galls is highly convoluted, often covered with fine hairs, and lacks an ovoid chamber like that found in cecidomyiid galls. Mites are always gregarious, seldom as long as 1 mm, and have legs.

Most galls not caused by Diptera, legless Hymenoptera, or mites contain frass or waxy exudate and include several kinds of Homoptera, for example, Aphididae, Coccoidea, and Psyllidae. Most makers of those galls have legs; if not, for example some weevils (Coleoptera), they have prominent head capsules and mandibles.

For Neotropical galls not made by cecidomyiids, Houard 1933 is the best identification guide. It was comprehensive when published and still covers most of the galls known today. The book includes illustrations of many galls whose gall makers are unknown.

How to Use the Keys

The keys that follow include all known published records of damage to Neotropical plants that are attributed to gall midges. They also include unpublished records based on specimens in the collections in the U.S. National Museum of Natural History. Records of damage to unknown host genera are included only for described gall midges.

Entries are arranged under host family and genus in alphabetical order. Listed for each entry is the kind of damage, name of the gall maker, known host(s), distribution, and the original and any other important published reference. Additional notes on discriminating features of the damage accompany an entry when available and judged helpful. The presence of cecidomyiid inquilines is also listed in the notes.

Descriptions of the damage are condensed. Often nothing more is known beyond the gross structure of the gall or the site of the damage. Size of most galls is not important, as in tapered stem swellings that vary in size according to the number of larvae inside. When I have judged size important, I have added measurements if known.

Illustrations are presented of various types of damage but not of every kind. Figure numbers are in parentheses following the character description. No material was available for some, and some of those for which material was available are simple enough to visualize without an illustration. Most illustrations were done from fresh or preserved examples of damage, but some are redrawn or restructured from other works to show galls that are not easily described in words. Most plant parts are shown in actual size or as given in the originals of the redrawn figures. Some details are done in 2×, as noted in the figure legends.

Plant family names in this book all end in "aceae." Forms of family names not based on a genus (e.g., Compositae) are indexed and also appear in parentheses after the correct family name in the major headings. Most scientific plant names are as originally given except corrected for spelling or moved to current genus. Doubtful records are given in quotation marks. Authors of plant names are not provided.

The range of a species is drawn with a broad brush. Neotropical distribution is by country in alphabetical order; non-Neotropical distribution follows after a semicolon. Provinces, departments, states, or territories of the larger countries are given as additional detail. For each previously published cecidomyiid record I provide the best reference or references. The citations will vary in usefulness, but include the original reference and one or more others if they include figures of the gall or are otherwise significant. A reference that includes an illustration of the gall is followed by "(fig.)." An absence of a citation following a record indicates an unpublished record.

Acanthaceae

Aphelandra

Stem swelling. . . . . *Neolasioptera aphelandrae*

    Host: *A. deppeana*. Distr.: El Salvador. Ref.: Möhn 1964a.

Blechum

Unspecified fruit gall . . . . Asphondylia blechi

    Host: *B. pyramidatum*. Distr.: Colombia (Magdalena). Ref.: Wünsch 1979.

Dicliptera

Stem swelling . . . . *Neolasioptera diclipterae*

> Host: *D. assurgens.* Distr.: Colombia (Magdalena). Ref.: Wünsch 1979.

Justicia

Stem swelling . . . . cecidomyiid

> The swelling is a large, irregular aggregation of tiny monothalamous galls, each an ovoid cell surrounded by spongy tissue. Host: *Justicia* sp. Distr.: Brazil (Rio Grande do Sul). Ref.: Tavares 1909 (fig.).

Odontonema

Stem or leaf midvein swelling. . . . . *Neolasioptera odontonemae*

> Host: *O. stricta.* Distr.: El Salvador. Ref.: Möhn 1964a.

Ruellia

Swollen, aborted fruit. . . . . *Asphondylia ruelliae*

> Host: *R. albicaulis.* Distr.: El Salvador. Ref.: Möhn 1959.

Amaranthaceae

Alternanthera

Larva among newly opening, terminal leaves. . . . . *Clinodiplosis* sp.

> Host: *A. philoxeroides* (alligator weed). Distr.: Argentina, Brazil, Uruguay.

Amaranthus

1a   Stem swelling . . . . *Neolasioptera amaranthi*

> Host: *A. spinosus.* Distr.: El Salvador. Ref.: Möhn 1964a.

1b   Unspecified fruit gall . . . . *Asphondylia lopezae*

> Hosts: *A. dubius, A. spinosus,* and *Iresine angustifolia.* Distr.: Colombia (Magdalena). Ref.: Wünsch 1979.

Chamissoa

Unspecified fruit gall .... *Aspondylia evae*

Host: *Chamissoa* sp. Distr.: Colombia (Magdalena). Ref.: Wünsch 1979.

Iresine

1a   Unspecified fruit gall .... *Asphondylia lopezae*

Hosts: *I. angustifolia* and *Amaranthus* spp. Distr.: Colombia (Magdalena). Ref.: Wünsch 1979.

1b   Stem swellings .... *Epilasioptera iresinis, Neolasioptera donamae,* and *Neolasioptera iresinis*

Epilasioptera iresinis. Host: *I. calea.* Distr.: El Salvador. Ref.: Möhn 1964a.

Neolasioptera donamae. Host: *I. angustifolia.* Distr.: Colombia (Magdalena). Ref.: Wünsch 1979.

Neolasioptera iresinis. Host: *I. celosia.* Distr.: El Salvador. Ref.: Möhn 1964b.

Amaryllidaceae

Agave

Larva in leaf, under slight, sometimes discolored concavity ....
.... *Neolasioptera martelli*

Host: *Agave* sp. Distr.: Mexico (Distrito Federal). Ref.: Nijveldt 1967 (fig.).

Anacardiaceae

Anacardium

Leaf blister .... cecidomyiid

Blisters are 2–4 mm in diameter and turn brown to black with age. Host: *A. occidentale.* Distr.: Brazil (Bahia). Refs.: Tavares 1918a, Houard 1933 (fig.).

Mangifera (mango)

1a   Irregular, succulent, often coalesced blisters on flower parts or leaves . . . . . .
. . . . *Erosomyia mangiferae*

Host: *M. indica*. Distr.: Brazil (Bahia), St. Vincent. Ref.: Tavares 1918a (fig.).

1b   Larva under bark of small twigs . . . . *Asynapta mangiferae*

Larvae may be feeding on decaying plant tissue. Host: *M. indica*. Distr.: West Indies. Ref.: Felt 1909.

Schinus

One peculiar gall made by a lepidopteran on *Schinus dependens* in Argentina and Chile is noteworthy here because it has been attributed to Cecidomyiidae. The gall first appears as a tapered stem swelling, which splits when mature to reveal a number of clavate galls (Houard 1933).

1a   Stem swelling . . . . *Bruggmanniella oblita*

This is an irregularly tapered swelling, sometimes to one side of stem, with numerous ovoid cavities. Host: *Schinus* sp. (cannela). Distr.: Brazil (Rio de Janeiro). Refs.: Tavares 1920a, Houard 1933 (fig.).

1b   Leaf galls 2
2a   Leaf or marginal leaf rolls 3
2b   Blister or spheroid galls 4
3a   Leaf entirely rolled, greatly swollen, resembling a pod . . . . cecidomyiid

Host: *Schinus* sp. Distr.: Brazil (Bahia). Refs.: Tavares 1918a, Houard 1933 (fig.).

3b   Marginal leaf roll without conspicuous hypertrophy . . . . cecidomyiid

Host: *Schinus* sp. (cannela). Distr.: Brazil (Rio de Janeiro). Refs.: Tavares 1920a, Houard 1933 (fig.).

4a   Convex blisters 5
4b   Spheroid galls, narrowly attached to leaf. 6
5a   Blister showing on upper side of leaf only . . . . cecidomyiid

The galls are hemispherical, 2.5 mm high, 3 mm large at base. Although attributed to a cecidomyiid, figures in Houard (1933) suggest it may be made by a homopteran. Host: *S. dependens*. Distr.: Argentina (Buenos Aires). Refs.: Tavares 1915a, Houard 1933 (fig.).

5b   Blister showing on both sides of leaf . . . . cecidomyiid

Galls are 5 mm in diameter and smooth. Host: *Schinus* sp. (caroba). Distr.: Brazil (Bahia). Ref.: Tavares 1918a.

6a   Spherical, hairy gall . . . . cecidomyiid

Galls are 3–4 mm in diameter, usually found along the midrib on the lower leaf surface and attached to the leaf by a narrow peduncle. Host: *Schinus* sp. (cannela). Distr.: Brazil (Rio de Janeiro). Refs.: Tavares 1920a, Houard 1933 (fig.).

6b   Ovate or obovate, smooth gall . . . . cecidomyiid

Galls are 5 mm long, smooth, and found on either side of the leaf. Houard (1933, figs. 497, 498) considered a short form of the same gall in the Tavares Collection as different from that originally noted by Tavares. Host: *Schinus* sp. (caroba). Distr.: Brazil (Bahia). Refs.: Tavares 1918a, Houard 1933 (fig.).

Annonaceae

Annona

1a   Stem swelling. . . . . *Asphondylia annonae* and cecidomyiid

Asphondylia annonae. Host: *A. cherimola*. Distr.: El Salvador. Ref.: Möhn 1959.

Cecidomyiid, possibly *A. annonae*. Host: *Annona* sp. Distr.: Brazil (Rio de Janeiro). Refs.: Tavares 1918a, Houard 1933 (fig.).

1b   Conical leaf or twig gall. . . . . cecidomyiid

Host: *Annona* sp. Distr.: Brazil (Rio de Janeiro). Refs.: Tavares 1918a (fig.), Houard 1933 (fig.).

Apiaceae (Umbelliferae)

Gymnophyton

Elongate stem swelling beneath umbel . . . . *Neolasioptera monticola*

Host: *G. polycephalum*. Distr.: Chile (Santiago). Ref.: Kieffer and Herbst 1909.

Apocynaceae

Aspidosperma

1a   Leaf blister.. cecidomyiid

Larvae appear to belong to the tribe Oligotrophini. Hosts: *A. australe*, *A. quebracho-blanco*. Distr.: Argentina and Brazil (Minas Gerais). Ref.: Fernandes et al. 1988 (fig.).

1b   Spheroid bud or stem gall .... *Anasphodiplosis aspidospermae*

Host: *A. quebracho-blanco*. Distr.: Argentina (Buenos Aires, Córdoba). Refs.: Tavares 1915a, Blanchard 1939, Houard 1933 (fig.).

Echites

Complex leaf gall, convex and hairy on one side of leaf, flask-shaped and smooth on the other.. cecidomyiid

Host: *Echites* sp. Distr.: Brazil (Bahia). Refs.: Tavares 1925, Houard 1933 (fig.).

Fernaldia

Larva from flower heads..................................... *Schizomyia* sp.

Host: *Fernaldia* sp. Distr.: El Salvador.

Tabernaemontana

Swollen, closed, aborted flower .... *Asphondylia tabernaemontanae*

Host: *T. amygdalifolia*. Distr.: El Salvador. Ref.: Möhn 1959.

Thevetia

Swollen, closed, aborted flower .... *Asphondylia thevetiae*

An inquiline, *Domolasioptera thevetiae*, was found in the galls (Möhn 1975). Host: *T. plumeriaefolia*. Distr.: El Salvador. Ref.: Möhn 1959.

Araceae

Dieffenbachia

Root swelling . . . . cecidomyiid

> Host: *Dieffenbachia* sp. Distr.: Mexico.

Heteropsis

Subcylindrical leaf gall. . . . . cecidomyiid

> Galls have a single larva and are 10 mm long, 2 mm in diameter, narrowed apically, and inserted principally on the abaxial surface of the midrib. Host: *H. salicifolia*. Distr.: Brazil (Rio de Janeiro). Ref.: Rübsaamen 1907.

Philodendron

Root swelling . . . . cecidomyiid

> Galls are up to 20 mm long by 5–6 mm diameter. Larvae belong to the tribe Oligotrophini. Host: *Philodendron* sp. Distr.: Brazil (Rio de Janeiro), Costa Rica. Ref.: Rübsaamen 1908.

Araliaceae

Gilibertia

Leaf petiole or vein swellings. . . . . *Brachylasioptera gilibertiae*

> Host: *G. arborea*. Distr.: El Salvador. Ref.: Möhn 1964b.

Arecaceae

Raphia

Larva beneath seed coat . . . . *Contarinia* sp.

> Host: *R. farinifera*. Distr.: French Guiana.

Aristolochiaceae

Aristolochia

1a   Larva in flowers. . . . . *Clinodiplosis* sp.

Host: *A. cymbifera*. Distr.: Brazil.

1b   Petiole and leaf vein swelling............................. cecidomyiid

Galls are glabrous and elongate, up to 18 mm long and 4 mm in diameter. Host: *A. bicolor*. Distr.: Brazil (Amazonas). Ref.: Rübsaamen 1905.

Asteraceae (Compositae)

Ageratina

Spheroid outgrowth at node (280)............................... cecidomyiid

Host: *A. mairetiana*. Distr.: Mexico (Morelos).

Ageratum

Enlarged, closed, aborted flower............................. *Asphondylia* sp.

Host: *A. conyzoides*. Distr.: Brazil (Paraná).

Ambrosia

1a   Larva in flower .... *Contarinia* sp.

Host: *A. tenuifolia*. Distr.: Argentina (Tucumán).

1b   Bud galls... 2
2a   Green, leafy bud gall........................ *Rhopalomyia ambrosiae*

Figures 280–293. Galls of Asteraceae. **280,** Stem gall, *Ageratina mairetiana* (Morelos, Mexico): cecidomyiid. **281–293,** *Baccharis.* **281,** Leaf galls, *B. salicifolia* (Córdoba, Argentina): *Geraldesia* sp. **282,** Stem gall, *Baccharis* sp. (Buenos Aires, Argentina): *Baccharomyia* sp. **283,** Bud galls, *B. eupatorioides* (redrawn from Kieffer and Herbst 1909): *Scheueria agglomerata.* **284,** Bud galls, *B. eupatorioides* (redrawn from Kieffer and Herbst 1909): *Scheueria longicornis.* **285,** Bud galls, *B. dracunculifolia* (Corientes, Argentina): *Asphondylia* sp. **286,** Aggregated bud galls, *B. salicifolia* (redrawn from Jörgensen 1917): *Rhopalomyia globifex.* **287,** Stem galls, *B. salicifolia* (Córdoba, Argentina): *Rhoasphondylia crassipalpis.* **288,** *B. dracunculifolia* (Entre Ríos, Argentina): *Rhoasphondylia friburgensis.* **289,** Bud galls, *B. dracunculifolia* (Minas Gerais, Brazil): cecidomyiid. **290,** Leaf galls, *Baccharis* sp. (redrawn from Houard 1933): cecidomyiid. **291,** Stem gall, *B. trimera* (Entre Ríos, Argentina): *Baccharomyia ramosina.* **292,** Stem gall, *B. spartiodes* (Neuquen, Argentina): *Baccharomyia* sp. **293,** Stem gall, *B. artemisioides* (Neuquen, Argentina): *Neolasioptera heterothalami.*

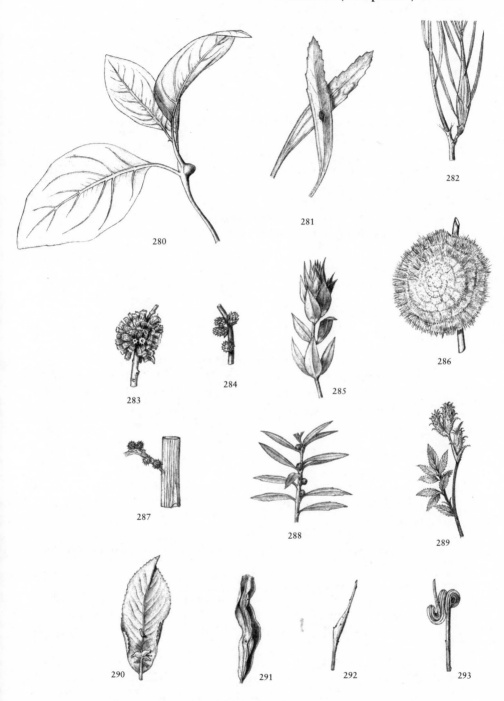

280

281

282

283

284

285

286

287

288

289

290

291

292

293

This species causes a foreshortening of the stem. Host: *A. artemisiifolia.* Distr.: USA (Florida). Refs.: Gagné 1975c, 1989 (fig.).

2b   White-fuzzy, ovoid bud galls 3
3a   Larval cell lined with a white fungus .... *Asphondylia ambrosiae*

Hosts: *A. artemisiifolia* and *Ambrosia* spp. Distr.: USA (Florida; California and Texas). Refs.: Gagné 1975c, 1989 (fig.).

3b   Larval cell green, not lined with a white fungus........................
.... *Contarinia partheniicola*

Hosts: *A. artemisiifolia* and *Ambrosia* spp. Distr.: USA (Florida; California and Texas). Ref.: Cockerell 1900, Gagné 1989 (fig.).

Baccharis

1a   Leaf galls 2
1b   Stem, bud, or flower galls. 6
2a   Leaf roll .... cecidomyiid

The halves of the leaf roll are swollen and the whole leaf resembles a pod. Although attributed to a cecidomyiid by Tavares (1917b: 124), I have seen similar galls on *Baccharis* in Argentina made by Psyllidae. Hosts: *Baccharis* spp. Distr.: Brazil (São Paulo). Ref.: Houard 1933 (fig.).

2b   Leaf blister or complex outgrowth of leaf 3
3a   Simple leaf blisters (281) .... *Geraldesia* spp.

Blisters are 3–5 mm in length and contain one larva each. Pupation occurs in the gall. Several species, all presumably belonging to *Geraldesia*, are involved:

Geraldesia cumbrensis. Host: *B. trinervis.* Distr.: El Salvador. Ref.: Möhn 1975.

Geraldesia spp. Hosts: *B. dracunculifolia*, *B. salicifolia* sp., and *Baccharis* spp. Distr.: Argentina (Buenos Aires, Córdoba, Tucumán), Brazil (Bahia, Minas Gerais). Refs.: Tavares 1915a, 1917b.

3b   Complex outgrowths of leaf 4
4a   Cylindrical gall, tapered apically, at oblique angle to leaf................
.... *Rhoasphondylia* sp.

This species is only tentatively determined as a *Rhoasphondylia* on the basis of a single pupa. Galls are inconspicuous, about 4 mm long and the same color and texture as the leaf. They may be the same as those referred to by Tavares (1917b: 124) from *Baccharis* sp. in Brazil. Host: *B. dracunculifolia*. Distr.: Brazil (Minas Gerais).

4b   Spherical or flask-shaped galls....................................5a
5a   Spherical gall (288) ....*Rhoasphondylia friburgensis*

Galls are usually on the upper or lower midribs but also occur on the stems. Hosts: *B. dracunculifolia*, *B. schultzii*, and *B. trinervis*. Distr.: Brazil (Bahia), El Salvador. Refs.: Tavares 1917b, Möhn 1960a.

5b   Flask-shaped gall (290) ....cecidomyiid

Galls are ca. 6–7 mm long and glabrous. Host: *Baccharis* sp. Distr.: Brazil (Bahia). Ref.: Tavares 1917b.

6a   Spheroid, pedicellate stem galls...................................7
6b   Bud galls, flower galls, or tapered stem swellings.....................8
7a   Gall with scalelike growths, situated on side of stem (287) ....
....*Rhoasphondylia crassipalpis*

Host: *B. salicifolia*. Distr.: Argentina (Córdoba, Mendoza, San Juan). Refs.: Kieffer and Jörgensen 1910, Jörgensen 1917, Tavares, 1915a.

7b   Gall smooth, situated on upper or lower leaf midrib surface or on stem (288)
....*Rhoasphondylia friburgensis*

See couplet 5a.

8a   Larvae in flowers..9
8b   Larvae in other plant parts 11
9a   Larva free among florets in flower head..................*Contarinia* sp.

The flower head may be enlarged and contain gall-making psyllids. Hosts: *B. dracunculoides*, *B. pingraea*, and *B. salicifolia*. Distr.: Argentina (Buenos Aires, San Luis), Brazil (Minas Gerais).

9b   Flower galls...10
10a   Swollen flower receptacle ....*Neolasioptera rostrata*

Hosts: *B. myrsinites*, *B. spartioides*, and other *Baccharis* spp. in USA. Distr.:

Argentina (Mendoza), Dominican Republic; USA (southern states). Ref.: Gagné and Boldt 1989.

10b   Cylindrical floret gall . . . . *Asphondylia baccharis*

This gall is reportedly 2.5–3.0 mm long, black, thin-walled, and covered with hairs. Host: *Baccharis* sp. Distr.: Chile. Ref.: Kieffer and Herbst 1905.

11a   Leafy or hairy bud galls 12
11b   Stem or bud swellings with at most a few scattered leaves 15
12a   Galls simulating flowers, spherical in aggregate (283, 286) . . . .
         . . . . *Rhopalomyia globifex* and *Scheueria agglomerata*

Rhopalomyia globifex. The aggregate gall of this and the next species may be 25–30 mm in diameter, but the hairs or scales covering the surface appear to be longer than on galls made by *S. agglomerata*. Host: *B. salicifolia*. Distr.: Argentina (Mendoza, San Juan). Refs.: Kieffer and Jörgensen 1910, Jörgensen 1917 (fig.).

Scheueria agglomerata. See remarks under *Rhopalomyia globifex*. Host: *B. eupatorioides*. Distr.: Chile (Concepción). Ref.: Kieffer and Herbst 1909 (fig.).

12b   Leafy rosette galls or hairy, tapered bud swelling. 13
13a   Hairy, tapered bud swelling. . . . . *Baccharomyia cordobensis*

Galls are 8–10 mm long and 6–7 mm in diameter with a single larval chamber. Host: *B. coridifolia*. Distr.: Argentina (Córdoba). Refs.: Kieffer and Jörgensen 1910; Jörgensen 1917 (fig.), Houard 1933 (fig.).

13b   Leafy rosette galls 14
14a   Gall with bracts all of same size, appressed (284) . . . . *Scheueria longicornis*

Branch buds form elongate, tight rosettes, 8–10 mm long by 6–7 mm in diameter. Host: *B. eupatorioides*. Distr.: Chile (Concepción). Refs.: Kieffer and Herbst 1909, Houard 1933 (fig.).

14b   Leaves at base of gall much longer and larger than those inside, the rosette loose, more open (285, 289) . . . .
         . . . . . . *Asphondylia* sp., *Dasineura* spp., *Rhopalomyia* sp., and cecidomyiid

Asphondylia sp. The outer leaves of the galls are pointed. Host: *B. dracunculifolia*. Distr.: Argentina and Brazil (São Paulo). Refs.: Tavares 1917b (fig.), Houard 1933 (fig.).

Dasineura chilensis. The outermost leaves are wide and enclose the gall. Host: *B. rosmarinifolia.* Distr.: Chile (Concepción). Refs.: Kieffer and Herbst 1909, Houard 1933 (fig.).

Dasineura subinervis. The gall is reportedly reminiscent of that of *Dasineura chilensis.* Host: *B. rosmarinifolia.* Distr.: Chile (Valparaíso). Ref.: Kieffer and Herbst 1911.

Rhopalomyia sp. The innermost leaves are linear and tightly fit together to enclose a large larva feeding at the base of the gall. Outermost leaves harbor free-ranging larvae of *Clinodiplosis* sp. and *Lestodiplosis* sp. Host: *B. capitalensis.* Distr.: Argentina (Jujuy).

Cecidomyiid. Gall similar to figure 289. Host: *B. elaegnoides.* Distr.: Brazil (Rio de Janeiro). Ref.: Tavares 1917b.

15a   Spindleform, elongated and twisted internodal swelling (293). . . . .
. . . . *Neolasioptera heterothalami*

Host: *B. artemisioides* and *B. spartioides.* Distr.: Argentina (Mendoza, San Juan). Refs.: Kieffer and Jörgensen 1910, Jörgensen 1917 (fig.).

15b   Spindleform, variable swelling, not twisted. 16
16a   Internodal stem swellings, occasionally at apex of plant, usually with several larvae in tunnels (282, 291, 292) . . . .
. . . . *Baccharomyia* spp. and *Neolasioptera lathami*

These galls are variable in size and hardness. Besides the localities and hosts listed below, generally similar galls have been reported from or recently found in Argentina (Buenos Aires, La Pampa), Brazil (Minas Gerais), and Chile on *B. dracunculifolia, B. pingraea, B. serrulata,* and *Baccharis* spp. Refs.: Kieffer and Jörgensen 1910, Kieffer and Herbst 1911, Jörgensen 1917 (fig.), Houard 1933 (fig.).

Baccharomyia interrupta. Host: *B. subulata.* Distr.: Argentina (Mendoza). Refs.: Kieffer and Jörgensen 1910, Jörgensen 1917.

Baccharomyia ornaticornis. Host: *B. salicifolia.* Distr.: Argentina (Córdoba). Refs.: Kieffer and Jörgensen 1910, Tavares 1915a (fig.). Jörgensen 1917 (fig.).

Baccharomyia ramosina. The gall is usually soft and spongy. It may sometimes occur on buds also (Houard 1933, fig.). Host: *B. trimera.* Distr.: Argentina (Entre Ríos) and Brazil (Rio de Janeiro). Ref.: Tavares 1917b.

Neolasioptera lathami. Hosts: *B. halimifolia* and *Baccharis* spp. Distr.: USA (Florida; SE USA). Ref.: Gagné 1971.

16b   Apical, one-celled galls . . . . *Calopedilla herbsti* and cecidomyiids

Galls are elongate ovoid, woody, 6–10 mm long by 3–4 mm in diameter, smooth, and sometimes with one or more leaves emanating from apex, usually at the tips of branches, more seldom in leaf axils. Kieffer and Herbst (1909, 1911) described (without figures) two other generally similar galls from Chile on the same host, and Kieffer (1908) reported a similar gall on *B. bogotensis* from Colombia. Host: *B. rosmarinifolia.* Distr.: Chile (Concepción). Refs.: Kieffer, 1903, Kieffer and Herbst 1906 (fig.), Houard 1933 (fig.).

Bidens

Tapered swelling of stem, petiole, or leaf vein . . . . *Asphondylia anaceliae*

Hosts: *B. squarrosa* and *B. refracta.* Distr.: El Salvador. Refs.: Möhn 1959, 1960a.

Borrichia

Swollen pair of leaves, joined at base . . . . *Asphondylia borrichiae*

Host: *B. frutescens.* Distr.: Mexico, USA (Florida; North Carolina). Refs.: Rossi and Strong 1990 (fig.)., Stiling et al. 1992.

Brickellia

Swollen achene. . . . . *Neolasioptera brickelliae*

Host: *B. pacayensis.* Distr.: El Salvador. Ref.: Möhn 1964b.

Calea

1a   Flower or fruit galls 2
1b   Stem galls 3
2a   Swollen flower . . . . *Asphondylia caleae*

Host: *C. integrifolia.* Distr.: El Salvador. Ref.: Möhn 1959.

2b   Swollen achenes . . . . *Neolasioptera caleae* and *Neolasioptera parvula*

Achenes infested with larvae of these species show only a slight swelling to differentiate them externally from uninfested fruit. Host: *C. integrifolia.* Distr.: El Salvador. Ref.: Möhn 1964b.

3a   Round stem gall, pointed at apex . . . . *Asphondylia zacatechichi*

Host: *C. zacatechichi.* Distr.: El Salvador. Ref.: Möhn 1960a.

3b   Tapered stem swellings . . . . *Neolasioptera* spp.

Neolasioptera lapalmae and *Neolasioptera serrata.* Host: *C. zacatechichi.* Distr.: El Salvador. Refs.: Möhn 1964a,b.

Neolasioptera olivae. Host: *C. caracasana.* Distr.: Colombia (Magdalena). Ref.: Wünsch 1979.

Chaetospira

Smooth, rounded leaf or bud gall . . . . *Rhoasphondylia sanpedri*

Leaf galls are as large as 3 mm diameter and variously colored green, red, brown, and violet. Host: *C. funcki.* Distr.: Colombia (Magdalena). Ref.: Wünsch 1979.

Chromolaena

1a   Larvae in flower heads 2
1b   Larvae in galls on leaf or stem 5
2a   Larvae in achenes 3
2b   Larvae free in petal tubes or among achenes 4
3a   Larva in greatly swollen achene . . . . *Asphondylia corbulae*

Galls are 2–3 mm in diameter, and the larval chamber is lined with a white fungus. Two or three galls may form on each flower head, and few if any seeds on the infested head develop. Hosts: *C. odorata, Eupatorium* sp., and *Fleischmannia microstemon.* Distr.: Trinidad. Ref.: Gagné 1977b, Cruttwell (unpub.).

3b   Larva in normal-appearing achene . . . . *Neolasioptera frugivora*

Hosts: *C. odorata* and *Fleischmannia microstemon.* Distr.: Trinidad. Ref.: Gagné 1977b, Cruttwell (unpub.).

4a   Larva in petal tube .... *Dasineura corollae*

Larvae are found singly in the tube of young flowers. Host: *C. odorata*. Distr.: Trinidad. Ref.: Gagné 1977b, Cruttwell (unpub.).

4b   Larva among achenes.................... *Contarinia* sp. near *perfoliata*

One to several larvae appear on each head and prevent seed set. Host: *C. odorata*. Distr.: Trinidad. Ref.: Gagné 1977b, Cruttwell (unpub.).

5a   Woody, conical leaf gall (295) .... *Clinodiplosis eupatorii*

Galls usually occur on the leaf upper surface. Each contains one larva. Hosts: *C. ivaefolia*, *C. odorata*, and *Eupatorium* sp. Distr.: Brazil (Pará), Costa Rica, St. Vincent, Trinidad. Refs.: Gagné 1977b, Cruttwell (unpub.).

5b   Stem or bud galls.. 6
6a   Tapered stem swelling....................... *Neolasioptera cruttwellae*

Hosts: *C. ivaefolia* and *C. odorata*. Distr.: Colombia (Magdalena), Trinidad. Refs.: Gagné 1977b, Wünsch 1979, Cruttwell (unpub.).

6b   Enlarged buds or pyriform stem galls.............................. 7
7a   Hollow pyriform gall on stem or buds .... *Perasphondylia reticulata*

Trotteria lapalmae is an inquiline in these galls (Möhn 1975). Hosts: *C. odorata* and *Eupatorium* sp. Distr.: Bolivia, Brazil (Pará), El Salvador, Mexico (Vera Cruz), Trinidad. Refs.: Möhn 1960a, Gagné 1977b, Cruttwell (unpub.).

7b   Swollen bud...................................... *Clinodiplosis* spp.

Larvae live gregariously among the leaves forming the gall. Hosts: *C. odorata* and *Fleischmannia microstemon*. Distr.: Trinidad, Costa Rica. Ref.: Gagné 1977b, Cruttwell (unpub.).

Chrysanthemum

Conical gall on leaf, stem, or bud.................. *Rhopalomyia chrysanthemi*

Galls are about 2 mm long and usually project obliquely from the host tissue. Hosts: *Chrysanthemum* spp., including cultivars. Distr.: Argentina and Colombia; worldwide, wherever chrysanthemums are grown. Refs.: Ahlberg 1939, Barnes 1948 (fig.), Mallea et al. 1983 (fig.), Gagné 1989 (fig.).

Chuquiraga

1a   Irregular, tapered swelling on stem, petiole, or main leaf ribs . . . .
. . . . cecidomyiid

Galls are rough, spongy, and have many larval cavities. Host: *C. tomentosa*.
Distr.: Brazil (Rio Grande do Sul). Ref.: Tavares 1909 (fig.).

1b   Subcylindrical, whitish yellow, complex leaf gall, usually on lower surface. .
. . . . cecidomyiid

Host: *C. tomentosa*. Distr.: Brazil (Rio Grande do Sul). Ref.: Tavares 1909
(fig.).

Conyza

1a   Stem swelling . . . . *Neolasioptera erigerontis*

Host: *C. canadensis*. Distr.: El Salvador; USA (widespread). Refs.: Felt 1907a,
Möhn 1964b, Gagné 1989 (fig.).

1b   Ovoid leaf blister (294) . . . . *Asteromyia modesta*

The galls usually show some purple discoloration. Unlike other *Asteromyia*
spp., *A. modesta* does not have a symbiotic fungus in its galls. Host: *C. canadensis*. Distr.: Argentina (Buenos Aires) and Brazil (Minas Gerais, Rio de
Janeiro); USA (widespread). Refs.: Felt 1907b, Gagné 1989 (fig.).

Eupatorium

1a   Swollen, closed flower. . . . . *Asphondylia corbulae*

Galls are 2–3 mm in diameter, and the larval chamber is lined with a white
fungus. Two or three galls may form on each flower head, and few if any seeds
on the infested head develop. Hosts: *Chromolaena odorata*, *Eupatorium* sp.,
and *Fleischmannia microstemon*. Distr.: El Salvador, Trinidad. Refs.: Möhn
1960a, Gagné 1977b.

1b   Swollen stems or leaf parts. 2
2a   Swollen stems . . . . *Isolasioptera* spp. and *Neolasioptera* spp.

Isolasioptera eupatoriensis. See also couplet 4a. Hosts: *E. morifolium* and
Eupatorium sp. Distr.: El Salvador. Ref.: Möhn 1964b.

Isolasioptera palmae Host: *E. collinum*. Distr.: El Salvador. Ref.: Möhn 1975.

Neolasioptera dentata. Host: *Eupatorium* sp. Distr.: El Salvador. Ref.: Möhn 1964b.

Neolasioptera exeupatorii. Host: *E. villosum*. Distr.: Cuba. Ref.: Cook 1909 (fig.).

Possibly *Neolasioptera* sp. Host: *Eupatorium* sp. Distr.: Brazil (Bahia). Ref.: Tavares 1917b.

2b   Leaf swellings. 3
3a   Leaf petiole swelling. . . . . cecidomyiid

The swelling occurs at the base of the petiole, is 10 mm by 7 mm, glabrous, green or violaceous, with an ellipsoidal larval cavity. Host: *E. laevigatum*. Distr.: Brazil (Rio de Janeiro). Ref.: Tavares 1917b.

3b   Leaf vein or laminar swelling. 4
4a   Swollen leaf vein . . . . *Isolasioptera* spp.

Isolasioptera eupatoriensis. See couplet 2a.

Isolasioptera sp. This is an apparently undescribed species. Host: *E. adenophorum*. Distr.: Mexico (Morelos).

4b   Laminar blister . . . . *Geraldesia eupatorii*

The gall is ovoid, 3–4 mm long by 1.5 mm wide; the surface is glabrous, rugose and green. The adult escapes from the upper leaf surface. Host: *Eupatorium* sp. Distr.: Brazil (Rio de Janeiro), El Salvador. Refs.: Tavares 1917b, Möhn 1975.

Fleischmannia

1a   Larva in swollen bud . . . . *Clinodiplosis* spp.

Two species of *Clinodiplosis* were reared from larvae living gregariously among the bud leaves. Hosts: *F. microstemon* and *Chromolaena odorata*. Distr.: Costa Rica, Trinidad. Ref.: Gagné 1977b.

1b   Larvae in flower head. 2
2a   Larva in greatly swollen achene. . . . . *Asphondylia corbulae*

Galls are 2–3 mm diameter, and the larval chamber is lined with a white fungus. Two or three may form on each flower head, and few if any remaining seeds develop. Hosts: *F. microstemon*, *Eupatorium* sp., and *Chromolaena odorata*. Distr.: El Salvador, Trinidad. Refs.: Möhn 1960a, Gagné 1977b.

2b   Larva in normal-appearing achene . . . . *Neolasioptera frugivora*

Hosts: *F. microstemon* and *Chromolaena odorata*. Distr.: Trinidad. Ref.: Gagné 1977b.

Flourensia

Larva in flower head . . . . *Clinodiplosis* sp.

Host: *F. hirta*. Distr.: Argentina (La Rioja).

Heliopsis

Larva in flower head. . . . . *Dasineura* sp.

Host: *H. buphthalmoides*. Distr.: Guatemala.

Heterothalamus

See couplet 15a under *Baccharis*, *B. spartioides*.

Hymenostephium

1a   Closed and swollen flower . . . . *Asphondylia salvadorensis*

Host: *H. cordatum*. Distr.: El Salvador. Ref.: Möhn 1959.

1b   Swollen achene . . . . *Neolasioptera compositarum*

Hosts: *H. cordatum* and *Tridax procumbens*. Distr.: El Salvador, Guatemala. Ref.: Möhn 1964b.

Hymenoxys

Swollen flower receptacle. . . . . *Asphondylia* sp.

Lestodiplosis sp. was reared from infested flower heads. Host: *H. robusta*. Distr.: Argentina (Jujuy).

Melanthera

1a   Closed, swollen flower .... *Asphondylia duplicornis*

Host: *M. aspera.* Distr.: Colombia (Magdalena). Ref.: Wünsch 1979.

1b   Bud or stem galls... 2
2a   Swollen bud .... *Asphondylia melantherae*

Infested buds have many leaves growing from them and contain one to five larval cells. Host: *M. nivea.* Distr.: El Salvador. Ref.: Möhn 1959.

2b   Tapered stem swellings........................... *Neolasioptera* spp.

Neolasioptera melantherae. Host: *M. nivea.* Distr.: El Salvador. Ref.: Möhn 1964a.

Neolasioptera mincae. Host: *M. aspera.* Distr.: Colombia (Magdalena). Ref.: Wünsch 1979.

Mikania

1a   Cylindrical, spinelike gall on stem or leaf .... *Clinodiplosis* sp.

This gall may be similar to that formed by *Clinodiplosis eupatorii* on *Chromolaena odorata*, although those on *Mikania* were reported from stems as well as leaves. Hosts: *M. guaco* and *Mikania* sp. Distr.: Brazil (Rio de Janeiro). Ref.: Rübsaamen 1907.

1b   Stem or leaf galls, none spinelike 2
2a   Spheroid, complex outgrowths of stem or leaf 3
2b   Simple leaf blister or tapered swelling of stem, petiole, or veins.......... 4
3a   Mammiform swelling along side of stem................... cecidomyiid

Galls are 2.5–4 mm in diameter, whitish, and rough. Host: *M. hirsutissima.* Distr.: Brazil (Rio Grande do Sul). Refs.: Tavares 1909 (fig.), Houard 1933 (fig.).

3b   Subspherical leaf gall .... *Asphondylia ulei*

Galls are 5 mm in diameter and densely covered with long hairs. Host: *Mikania* sp. Distr.: Brazil (Amazonas). Ref.: Rübsaamen 1907.

4a   Leaf blister.. cecidomyiid

Blisters are circular to elliptical and apparent on both leaf surfaces. Host: *Mikania* sp. Distr.: Brazil (Rio de Janeiro). Ref.: Rübsaamen 1908.

4b   Tapered swellings of stems, petioles, and leaf veins 5
5a   Swellings on petiole or leaf veins . . . . *Alycaulus* spp.

    Alycaulus mikaniae. Host: *Mikania* sp. Distr.: Brazil (Amazonas). Refs.: Rübsaamen 1916a (fig.), Houard 1933 (fig.).

    Alycaulus trilobatus. An inquiline, *Domolasioptera acuario*, was also reported from these galls in Colombia. Hosts: *M. cordifolia*, *M. micrantha*, and *Mikania* sp. Distr.: Colombia (Magdalena), El Salvador. Refs.: Möhn 1964a, Wünsch 1979.

5b   Stem swellings . . . . *Asphondylia moehni* and cecidomyiids

    Asphondylia moehni. Galls are irregular, spongy, and usually to one side of the stem. Host: *M. guaco*. Distr.: Brazil (Rio Grande do Sul). Refs.: Tavares 1909 (fig.), Houard 1933 (fig.).

    Cecidomyiids. Several other stem swellings made by unidentified gall midges have been reported: two from *Mikania* sp. in Brazil (Rübsaamen 1908, Tavares 1909 [fig.], Houard 1933 [fig.]) and one from *M. lindleyana* in Brazil (Rübsaamen 1907).

Parthenium

Swollen male floret . . . . *Asphondylia* sp.

    Host: *P. hysterophorus*. Distr.: Argentina (Tucumán).

Piptocarpha

Hemispherical leaf gall on median vein. . . . . cecidomyiid

    Galls are 12 mm diameter and are more apparent on the lower surface than on the upper. The surface is covered with dense stellate hairs, and the larval cavity is large. Host: *Piptocarpha* sp. Distr.: Brazil (Rio de Janeiro). Ref.: Rübsaamen 1908.

Porophyllum

Tapered stem swelling . . . . *Zalepidota ituensis*

Galls are 10 mm long, up to 6 mm wide, longitudinally striate, and contain one larval cavity lined with characteristic white fungus. Host: *Porophyllum* sp. Distr.: Brazil (São Paulo). Ref.: Tavares 1917b.

Senecio

Tapered stem swellings . . . . . . *"Janetiella" montivaga* and *Neolasioptera senecionis*

"Janetiella" montivaga. Galls are succulent and reddish. Host: *S. mendocinus.* Distr.: Argentina (Mendoza). Refs.: Kieffer and Jörgensen 1910, Jörgensen 1917 (fig.), Houard 1933 (fig.).

Neolasioptera senecionis. Host: *S. kermesinus.* Distr.: El Salvador. Ref.: Möhn 1964a.

Sinclairia

Larva in flower head . . . . *Dasineura* sp.

Host: *S. vagans.* Distr.: Guatemala.

Tithonia

Budlike, many-chambered gall on stem . . . . *Asphondylia tithoniae*

Host: *T. rotundifolia.* Distr.: El Salvador. Ref.: Möhn 1960a.

Tridax

Swollen achene . . . . *Neolasioptera compositarum*

Hosts: *T. procumbens* and *Hymenostephium cordatum.* Distr.: El Salvador, Guatemala. Ref.: Möhn 1964b.

Trixis

Spheroid, apically pointed gall on bud or leaf . . . . *Asphondylia trixidis*

Host: *T. radialis.* Distr.: El Salvador. Ref.: Möhn 1959.

Verbesina

Tapered swelling on stem, petiole, or leaf veins . . . . *Neolasioptera verbesinae*

Hosts: *V. fraseri*, *V. sublobata*, and *V. turbacensis*. Distr.: El Salvador. Ref.: Möhn 1964a.

Vernonia

1a	Spheroid or tapered stem or petiole swellings	2
1b	Swollen buds or achenes.	3
2a	Spheroid stem or petiole swelling	*Asphondylia* sp.

Galls are succulent and several may coalesce; they are thick-walled and monothalamous. Host: *V. polyanthes*. Distr.: Brazil (Minas Gerais).

2b   Tapered stem swelling . . . . *Neolasioptera vernoniensis*

Host: *Vernonia* sp. Distr.: Brazil (Minas Gerais), El Salvador. Ref.: Möhn 1964a.

3a   Swollen achene. . . . . *Asphondylia herculesi*

Host: *V. canescens*. Distr.: El Salvador. Ref.: Möhn 1959.

3b   Swollen bud . . . . *Asphondylia ajallai*

Hosts: *V. canescens* and *V. patens*. Distr.: El Salvador. Ref.: Möhn 1959, 1960a.

Zexmenia

1a   Swollen achene . . . . *Asphondylia zexmeniae*

Host: *Z. frutescens*. Distr.: El Salvador. Ref.: Möhn 1960a.

1b   Elongate-ovoid, abaxial leaf vein gall. . . . . *Ameliomyia zexmeniae*

Host: *Z. frutescens*. Distr.: El Salvador. Ref.: Möhn 1960a.

Berberidaceae

Berberis

Artichoke-like bud gall . . . . cecidomyiid

The gall consists of a great number of closely appressed, modified leaves; a single larva lives in the center. The gall maker is an undescribed species of

Oligotrophini that fits in no known genus. Hosts: *B. darwinii* and *B. empetrifolia*. Distr.: Argentina (Rio Negro).

Bignoniaceae

In addition to the gall midges listed below under the various genera, *Neolasioptera salvadorensis* was reared from stem and tendril swellings on an undetermined Bignoniaceae in El Salvador (Möhn 1964a).

Adenocalymma

Swelling on larger leaf veins . . . . *Neolasioptera samariae*

> Host: *A. dugandii*. Distr.: Colombia (Magdalena). Ref.: Wünsch 1979.

Arrabidaea

Woody, tapered swelling on stems, petioles, or leaf veins . . . . cecidomyiid

> Tavares (1918a: 57, 1925: 9) distinguished between a very large, 35 mm wide, stem swelling with many larval cells and a smaller, 5–6 mm wide, more ellipsoidal swelling of the branches of the inflorescence, petioles, and leaf veins with a central larval cell. Hosts: *A. coleocalyx* and *Arrabidaea* sp. Distr.: Brazil (Rio de Janeiro). Ref.: Houard 1933 (fig.).

Bignonia

Tapered stem swelling with many larval cavities . . . . cecidomyiid

> Host: *Bignonia* sp. Distr.: Brazil (Rio Grande do Sul). Refs.: Tavares 1909 (fig.), Houard 1933 (fig.).

Crescentia

Larva in flowers . . . . *Prodiplosis* spp.

> Two different species have been reared from larvae found in flowers. Host: *C. alata*. Distr.: Nicaragua. Ref.: Maes 1990.

Godmania

Aborted fruit . . . . *Asphondylia godmaniae*

Host: *G. aesculifolia.* Distr.: El Salvador. Ref.: Möhn 1959.

Lundia

Spheroid gall on leaves or inflorescence .... cecidomyiid

Host: *Lundia* sp. Distr.: Brazil (Bahia). Ref.: Tavares 1925.

Pithecoctenium

Swollen tendril .... *Neolasioptera salvadorensis*

Host: *P. echinatum.* Distr.: El Salvador. Ref.: Möhn 1975.

Pseudocalymma

Swollen stems, tendrils, petioles, or leaf veins .... various Alycaulini

Brachylasioptera rotunda. This species was reared from stem, petiole, and vein swellings. Host: *P. macrocarpa.* Distr.: El Salvador. Ref.: Möhn 1964b.

Lobolasioptera media. This species was reared from vein swellings. Host: *P. macrocarpa.* Distr.: El Salvador. Ref.: Möhn 1964b.

Neolasioptera grandis. This species was reared from stem and petiole swellings. Host: *P. macrocarpa.* Distr.: El Salvador. Ref.: Möhn 1964a.

Neolasioptera pseudocalymmae. This species was reared from stem, tendril, and petiole swellings. Host: *P. macrocarpa.* Distr.: El Salvador. Ref.: Möhn 1964a.

Tabebuia

Spheroid gall at base of leaf .... cecidomyiid

Host: *T. ochracea.* Distr.: Brazil (Minas Gerais). Ref.: Fernandes et al. 1988.

Boraginaceae

Cordia

1a   Tapered stem swelling .... *Neolasioptera cordiae* and *Neolasioptera* sp.

Neolasioptera cordiae. Host: *C. cana.* Distr.: El Salvador. Ref.: Möhn 1964a.

Neolasioptera sp. This unidentified cecidomyiid is tentatively referred to this genus. Host: *Cordia* sp. Distr.: Brazil (Rio de Janeiro). Ref.: Tavares 1925.

1b   Bud, flower, or leaf galls. 2

2a   Fuzzy bud gall . . . . cecidomyiid

Host: *C. sellowiana.* Distr.: Brazil (Minas Gerais). Ref.: Fernandes et al. 1988 (fig.).

2b   Flower and leaf galls. 3

3a   Closed and swollen flower . . . . *Asphondylia cordiae*

Host: *C. alba* and *C. dentata.* Distr.: El Salvador. Ref.: Möhn 1959.

3b   Round leaf gall . . . . cecidomyiid

Two inquilines, *Meunieriella cordiae* (Möhn 1975) and *Meunieriella gairae* (Wünsch 1979), were reared from the gall. Hosts: *C. alliodora* and *C. subtruncata.* Distr.: Colombia (Magdalena), El Salvador. Refs.: Möhn 1975, Wünsch 1979.

Heliotropium

Tapered stem gall . . . . *Neolasioptera tridentifera*

Host: *H. curassavicum.* Distr.: Argentina (Mendoza, San Juan). Refs.: Kieffer and Jörgensen 1910, Jörgensen 1917 (fig.).

Tournefortia

1a   Tapered stem swellings. . . . . . . *Neolasioptera tournefortiae* and cecidomyiids

Neolasioptera tournefortiae. Host: *T. volubilis.* Distr.: El Salvador. Ref.: Möhn 1964a.

Cecidomyiid. Host: *T. angustiflora.* Distr.: Brazil (Amazonas). Ref.: Rübsaamen 1916a.

Cecidomyiid. Host: *T. pohlii.* Distr.: Brazil (Rio de Janeiro). Ref.: Tavares 1922.

1b   Distorted flowers or fruit 2

2a   Distorted and swollen fruit. . . . . *Asphondylia tournefortiae*

Hosts: *T. angustiflora* and *T. volubilis.* Distr.: Brazil (Amazonas), El Salvador. Refs.: Rübsaamen 1916, Möhn 1960a.

2b   Swollen and closed flower .... cecidomyiid

Host: *T. pohlii.* Distr.: Brazil (Rio de Janeiro). Ref.: Tavares 1922.

Bromeliaceae

Besides the following, there are in the USNM hard blister galls made by an undetermined cecidomyiid on an unidentified bromel.

Tillandsia

Swollen aerial root (as in 299) .... *Neolasioptera* sp.

Host: *Tillandsia* sp. Distr.: Bolivia, Costa Rica, Mexico.

Burseraceae

Bursera

Closed, swollen flower .... *Burseramyia burserae*

An inquiline, *Camptoneuromyia burserae,* has been reared from the gall (Möhn 1975). Host: *B. simaruba.* Distr.: El Salvador. Refs.: Möhn 1960a, 1975.

Protium

1a   Swollen, rolled leaf margin .... cecidomyiid

Host: *P. heptaphyllum.* Distr.: Brazil (Bahia). Ref.: Tavares 1922.

1b   Variously shaped leaf blade galls...................................... 2
2a   Globular, closed leaf gall, showing equally on both leaf surfaces (296) .....
.... *Dasineura braziliensis*

These galls may distort the leaf. An inquiline, *Meunieriella insignis,* was also reared from these galls. Host: *P. heptaphyllum.* Distr.: Brazil (Bahia). Ref.: Tavares 1922.

2b   Variously shaped galls on one side of the leaf, the other side showing an opening (297)..cecidomyiids

Six such galls were described by Tavares (1922) and 40 others were described and illustrated by Houard (1933). Three are shown in figure 297. Many are apparently caused by Psyllidae, but some are attributable to cecidomyiids. Hosts: *Protium* spp. Distr.: Cuba to Paraguay.

Cactaceae

Cereus

Variable stem swelling....................................*Neolasioptera cerei*

The swelling may extend for a length of 25 mm and is many-celled, the cells oriented perpendicularly to the axis of the stem. Host: *C. setaceus*. Distr.: Brazil (Rio de Janeiro). Ref.: Rübsaamen 1905.

Eriocereus

Larva in young shoot......................................*Contarinia* sp.

Host: *E. (Harrisia) martinii*. Distr.: Argentina (Formosa).

Opuntia

Aborted fruit...*Asphondylia betheli*

Hosts: *Opuntia* spp. Distr.: Mexico (widespread, S to Oaxaca); USA (southwestern). Refs.: Cockerell 1907, Mann 1969.

Caryocaraceae

Caryocar

1a   Slightly swollen bud............................*Prodiplosis floricola*

Hosts: *C. brasiliense*; also on Rosaceae in North America. Distr.: Brazil (São Paulo); USA. Ref.: Gagné 1986.

1b   Spheroid, hairy leaf gall (298)............................cecidomyiid

Galls show on both sides of the leaf. Host: *C. brasiliense*. Distr.: Brazil (Minas Gerais). Ref.: Tavares 1922.

Caryophyllaceae

Arenaria

Bud gall. . . . . *Asphondylia yukawai*

Hosts: *Arenaria* sp. and *Melochia* spp. (Sterculiaceae). Distr.: Colombia (Magdalena). Ref.: Wünsch 1979.

Chenopodiaceae

Atriplex

Stem swelling (300) . . . . *Neolasioptera* sp.

Host: *Atriplex* sp. Distr.: Argentina (Mendoza).

Chenopodium

Larva in flower bud . . . . *Prodiplosis longifila*

This species feeds on many plant families. Larvae are gregarious and kill the buds. Hosts: *C. ambrosioides* (wormseed); also alfalfa, castorbean, citrus, green beans, potato, tomato, and wild cotton. Distr.: Colombia, Ecuador, Peru (Lambayeque), USA (Florida). Refs.: Diaz B. 1981, Gagné 1986, Peña et al. 1989.

Suaeda

1a   Ovoid gall at tip of branch (301). . . . . *Asphondylia swaedicola*

The gall is large, 10 mm long, 6–8 mm diameter, with a few leaves near middle, and a single larval chamber. Host: *S. divaricata*. Distr.: Argentina (Mendoza, San Juan). Refs.: Kieffer and Jörgensen 1910, Jörgensen 1917 (fig.).

1b   Spherical stem swelling . . . . cecidomyiid

The gall is 3–4 mm long, 3–4 mm diameter, with thin walls and a single larval chamber. Host: *S. divaricata*. Distr.: Argentina (Mendoza). Refs.: Kieffer and Jörgensen 1910, Jörgensen 1917 (fig.).

Clusiaceae (Guttiferae)

Clusia

1a   Fusiform, pointed bud gall . . . . *Uleia clusiae*

The gall is about 15 mm high, 6 mm diameter and contains one yellowish larva. Host: *Clusia* sp. Distr.: Brazil (Amazonas). Ref.: Rübsaamen 1905.

1b   Leaf blister . . . . cecidomyiid

Host: *C. lanceolata*. Distr.: Brazil (Rio de Janeiro).

Kielmeyera

Swollen, aborted flower bud. . . . . *Arcivena kielmeyerae*

Hosts: *K. rosea* and *Kielmeyera* spp. Distr.: Brazil (Minas Gerais, São Paulo). Ref.: Gagné 1984.

Vismia

1a   Cylindrical leaf invagination on under side of leaf. . . . . cecidomyiid

Galls are 3–4 mm long and open as a slit on the upper side of the leaf. Host: *V. guianensis*. Distr.: Brazil (Bahia). Refs.: Tavares 1922, Houard 1933 (fig.).

2a   Marginal leaf roll . . . . cecidomyiid

Host: *V. guianensis*. Distr.: Brazil (Bahia). Refs.: Tavares 1922, Houard 1933 (fig.).

Combretaceae

Combretum

1a   Enlarged, irregularly swollen bud . . . . *Houardodiplosis rochae*

Host: *C. leprosum*. Distr.: Brazil (Ceará). Ref.: Tavares 1925 (fig.).

1b   Tapered swelling of tendril . . . . *Neolasioptera combreti*

Host: *C. farinosum*. Distr.: El Salvador. Ref.: Möhn 1964a.

Connaraceae

Connarus

Stem swelling . . . . cecidomyiid

Host: *C. uleanus*. Distr.: Brazil (Rio de Janeiro). Ref.: Rübsaamen 1905.

Convolvulaceae

Ipomoea

1a   Deformed, rolled leaf . . . . cecidomyiid

·Host: *Ipomoea* sp. Distr.: Brazil (Rio de Janeiro). Ref.: Rübsaamen 1907.

1b   On other plant parts or a simple petiole swelling 2
2a   Swollen stem . . . . cecidomyiid

Swellings can be up to four times normal stem diameter. Host: *Ipomoea* sp. Distr.: Brazil (Rio de Janeiro). Ref.: Rübsaamen 1907.

2b   Larvae in bud or petiole 3
3a   Larva in flower bud . . . . *Schizomyia ipomoeae*

Camptoneuromyia meridionalis is an inquiline in this gall (Felt 1910b). Host: *Ipomoea* sp. Distr.: St. Vincent. Refs.: Felt 1910a,b.

3b   Larva in elongate-ovoid bud or swollen petiole . . . . *Asphondylia convolvuli*

Host: *I. arborescens*. Distr.: El Salvador. Ref.: Möhn 1960a.

Merremia

1a   Swollen tendril . . . . *Neolasioptera merremiae*

Host: *M. umbellata*. Distr.: El Salvador. Ref.: Möhn 1964a.

1b   Withered, unopened flower.................. *Salvalasioptera merremiae*

Host: *M. umbellata*. Distr.: El Salvador. Ref.: Möhn 1975.

Cucurbitaceae

Citrullus and *Cucumis*

Swollen, aborted flower .... *Asphondylia* sp.

Camptoneuromyia sp. is an inquiline in the flowers. Hosts: *Citrullus* sp. (watermelon) and *Cucumis* sp. (melon). Distr.: Peru (Lambayeque). Ref.: Korytkowski and Llontop 1967.

Cucumis (see also under *Citrullus*)

With fungus in or under skin on flower end of fruit .... *Neolasioptera hibisci*

The fungus is usually associated with a wound in the fruit. Hosts: *Cucumis melo* (melon); also *Hibiscus moscheutos* (Malvaceae) in USA. Distr.: Honduras; USA. Refs.: Felt 1907b, Gagné 1989.

Rytidostylis

Swollen, closed, flower bud .... *Asphondylia* sp.

Host: *R. brevisetosa*. Distr.: Venezuela (Aragua).

Dilleniaceae

Curatella

1a   Swollen, closed flower with fungus-lined solitary larval cell..............
.... *Asphondylia curatellae*

Host: *C. americana*. Distr.: El Salvador. Ref.: Möhn 1959.

1b   Closed flower with gregarious larvae...................... cecidomyiid

The gall maker is an undescribed species of Cecidomyiidi. An inquiline, *Domolasioptera curatellae*, has been reared from these galls. Host: *C. americana*. Distr.: El Salvador. Ref.: Möhn 1975.

Davilla

1a   Leaf blister . . . . cecidomyiid

Galls are 3–4 mm in diameter, convex above, flat below. Host: *D. rugosa.*
Distr.: Brazil (Rio de Janeiro). Ref.: Tavares 1922.

1b   Swollen, aborted flower . . . . cecidomyiid

Host: *D. rugosa.* Distr.: Brazil (Rio de Janeiro). Ref.: Tavares 1922.

Ericaceae

Pernettya

1a   Irregular stem swelling, 5–50 mm long . . . . *Xyloperrisia azarae*

The gall has many larval cells. Host: *P. furens* (Kieffer and Herbst 1905).
Distr.: Chile (Concepción). Ref.: Kieffer 1903.

1b   Enlarged, spherical, apical or axilliary bud . . . .
      . . . . *Pernettyella longicornis* and cecidomyiid

Pernettyella longicornis. Galls are 12 mm long, 7 mm diameter. Host: *P. fu-rens.* Distr.: Chile (Concepción). Ref.: Kieffer and Herbst 1909 (fig.).

Cecidomyiid. Galls are 8 mm long, 7 mm wide. Host: *P. phillyreaefolia.*
Distr.: Argentina (Córdoba). Ref.: Tavares 1915a (fig.).

Erythroxylaceae

Erythroxylum

1a   Aborted, deformed fruit . . . . *Asphondylia erythroxylis*

Host: *E. mexicanum.* Distr.: El Salvador. Ref.: Möhn 1959.

1b   Leaf or stem galls 2
2a   Enlarged, swollen bud . . . . cecidomyiid

Host: *E. frangulifolium.* Distr.: Brazil (Minas Gerais). Ref.: Fernandes et al.
1988 (fig.).

2b   Swollen stem internodes or leaf galls 3

3a   Tapered stem swelling. . . . . *Neolasioptera erythroxyli*

      Host: *E. mexicanum.* Distr.: El Salvador. Ref.: Möhn 1964a.

3b   Spheroid, fuzzy leaf gall. . . . . cecidomyiid

      Host: *E. coelophlebium.* Distr.: Brazil (Minas Gerais). Ref.: Fernandes et al. 1988 (fig.).

Euphorbiaceae

Acalypha

Besides the two galls keyed, an unspecified leaf gall on *Acalypha unibracteata* made by an undetermined Cecidomyiidi, has been recorded from El Salvador with reference to *Meunieriella acalyphae*, an inquiline (Möhn 1975).

1a   Simple, tapered swelling of branch or leaf midrib . . . . cecidomyiid

      Host: *Acalypha* sp. Distr.: Peru (Loreto). Ref.: Rübsaamen 1905.

1b   Complex, cylindrical, hairy gall on leaf or stem. . . . . cecidomyiid

      Host: *Acalypha* sp. Distr.: Brazil (Rio de Janeiro). Refs.: Tavares 1918a, Houard 1933 (fig.).

Alchornea

Fusiform, multicelled stem swelling. . . . . cecidomyiid

      Host: *Alchornea* sp. Distr.: Brazil (Amazonas). Ref.: Rübsaamen 1905.

Caperonia

Leaf fold . . . . cecidomyiid

      Host: *Caperonia* sp. Distr.: Brazil (Amazonas). Ref.: Rübsaamen 1905.

Colliguaja

1a   Fusiform stem swelling with many individual cells with solitary larvae . . . . .
. . . . *Promikiola rubra*

This gall appears at the basal region of the inflorescence and is twice the normal stem diameter. Host: *C. odorifera*. Distr.: Chile (Valparaíso). Ref.: Kieffer and Herbst 1911.

1b   Fusiform stem swelling with single chamber and gregarious larvae . . . .
. . . . *Riveraella* spp.

Riveraella colliguayae. The gall is 30 mm long by 15 mm diameter, thin-walled, with small scattered leaves emerging from surface. Host: *C. odorifera*. Distr.: Chile (Valparaíso). Refs.: Philippi 1873 (fig.), Kieffer and Herbst 1911.

Riveraella sp. The gall is 10–15 mm long by 5–8 mm diameter. Host: *C. odorifera*. Distr.: Chile. Ref.: Kieffer and Herbst 1911.

Croton

1a   Irregular, fusiform stem swellings . . . . cecidomyiids

Host: *C. buxifolius*. Distr.: Brazil (Rio de Janeiro). Ref.: Rübsaamen 1905.

Host: *C. migrans*. Distr.: Brazil (Minas Gerais). Ref.: Tavares 1922.

1b   Leaf galls 2
2a   Complex, spheroid galls with stellate hairs. . . . .
. . . . *Styraxdiplosis cearensis* and cecidomyiids

Styraxdiplosis cearensis. Host: *C. hemiargyreus*. Distr.: Brazil (Ceará). Refs.: Tavares 1925 (fig.), Houard 1933 (fig.).

Cecidomyiid. Galls are on lower leaf surface. Host: *Croton* sp. Distr.: Brazil (Rio de Janeiro). Ref.: Rübsaamen 1905.

Cecidomyiid. Galls are on upper leaf surface. Host: *Croton* sp. Distr.: Brazil (Rio de Janeiro). Ref.: Rübsaamen 1905.

2b   Simple leaf galls. 3
3a   Upwardly rolled leaf edge . . . . cecidomyiid

Host: *Croton* sp. Distr.: Brazil (Rio de Janeiro). Ref.: Rübsaamen 1905.

3b   Blister gall . . . . cecidomyiid

Galls are elliptical and visible on upper surface. Host: *C. argentinus*. Distr.: Argentina (Córdoba). Refs.: Tavares 1915a, Houard 1933 (fig.).

Dalechampia

Without seeing the galls reported under this host, it is not clear to me whether they are tissue swellings or complex outgrowths.

Cecidomyiid. Ovoid gall on branch, petiole or leaf lamina, 7 mm long by 5 mm wide, covered in part with dense, white-gray pile. Host: *Dalechampia* sp. Distr.: Brazil (Rio de Janeiro). Ref.: Rübsaamen 1905.

Meunieriella dalechampiae. Stem gall, up to 10 mm long by 3 mm diameter, covered with dense, yellowish stiff hairs. Host: *D. filicifolia*. Distr.: Brazil (Rio de Janeiro). Ref.: Rübsaamen 1905.

Hevea

1a   Deformed, thickened leaflets . . . . *Contarinia* sp.

Larvae are gregarious, but it is not clear from the original description whether they are inside or outside the leaves. Host: *H. brasiliensis*. Distr.: Brazil (Amazonas). Ref.: Rübsaamen 1907.

1b   Fusiform stem swelling, 15 mm long, 9 mm in diameter. . . . . cecidomyiid

Host: *H. brasiliensis*. Distr.: Brazil (Amazonas). Ref.: Rübsaamen 1907.

Hura

Leaf blister . . . . cecidomyiid

Blisters are circular, ca. 2 mm diameter, visible on both sides of leaf, variably discolored. Host: *H. crepitans*. Distr.: Brazil (Amazonas). Ref.: Rübsaamen 1907. Note: Nijveldt (1968) described *Huradiplosis surinamensis* from an unspecified leaf gall on *H. crepitans*.

Jatropha

Swollen, closed flower. . . . . *Asphondylia jatrophae*

Host: *J. curcas*. Distr.: El Salvador. Ref.: Möhn 1959.

Mabea

Enlarged, lignified bud. . . . . cecidomyiid

> Host: *Mabea* sp. Distr.: Brazil (Amazonas). Ref.: Rübsaamen 1907.

Manihot

1a   Elongate-cylindrical, smooth leaf gall (302) . . . . *Iatrophobia brasiliensis*

> In addition to the gall maker, two inquilines have been reared from the galls: *Ledomyia manihot* from *M. utilissima* in St. Vincent (Felt 1912e) and *Meunieriella lantanae* from *M. utilissima* in Surinam (Nijveldt 1968). Hosts: *M. dichotoma*, *M. palmata*, *M. tripartita*, and *M. utilissima*. Distr.: Brazil, Colombia, Costa Rica, Guyana, Peru, St. Vincent, Tobago, Trinidad. Refs.: Rübsaamen 1907, 1916a (fig.); Tavares 1918a (fig.); Houard 1933 (fig.); Korytkowski and Sarmiento 1967 (as *Hyperdiplosis*).

1b   Globular, rugose leaf gall. . . . . *Schizomyia manihoti*

> Host: *M. utilissima*. Distr.: Brazil (Ceará), Colombia (Magdalena). Refs.: Tavares 1925 (fig.), Wünsch 1979.

Phyllanthus

Larva in fruit . . . . *Asphondylia siccae*

> Host: *P. distichus*. Distr.: Jamaica. Ref.: Felt 1916.

Ricinus

Larva in flower bud . . . . *Prodiplosis longifila*

> This species feeds on many plant families. Larvae are gregarious and kill the buds. Hosts: *R. communis* (castorbean); also alfalfa, citrus, green beans, potato, tomato, wild cotton, and wormseed. Distr.: Colombia, Ecuador, Peru (Lambayeque), USA (Florida). Refs.: Diaz B. 1981, Gagné 1986, Peña et al. 1989.

Sapium

1a   Enlarged, deformed fruit or flower. . . . . cecidomyiid

> Host: *S. hippomane*. Distr.: Brazil (Amazonas). Ref.: Rübsaamen 1908.

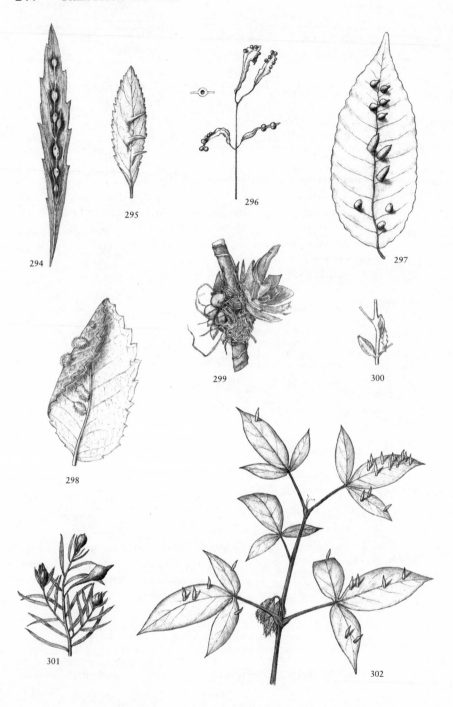

294

295

296

297

298

299

300

301

302

1b   Smooth, spheroid leaf gall with depressed center. . . . . cecidomyiid

> Hosts: *S. biglandulosum* and *S. hippomane*. Distr.: Brazil (Amazonas, Minas Gerais). Refs.: Rübsaamen 1908, Fernandes et al. 1988 (fig.).

Tragia

Enlarged, swollen terminal bud . . . . cecidomyiid

> Host: *T. dodecandra*. Distr.: Argentina (Córdoba). Refs.: Tavares 1915a, Houard 1933 (fig.).

Fabaceae (Leguminosae)

Fabaceae are divided into three subfamilies, the Caesalpinioideae, the Faboideae, and the Mimosoideae. Gall midges may segregate themselves among these three subfamilies, but the evidence so far is very thin. Many kinds of galls made by gall midges are known on the subfamily Caesalpinioideae, on *Bauhinia, Brownea, Cassia, Cercidium, Copaifera, Cynometra, Hymenaea,* and *Swartzia,* but none of the gall formers has yet been named. In contrast, many cecidomyiids are known from the subfamily Faboideae, on *Aeschynomene, Andira, Canavalia, Crotalaria, Cyamopsis, Dalbergia, Desmodium, Erythrina, Geoffroea, Gliricidia, Indigofera, Lonchocarpus, Machaerium, Medicago,* and *Phaseolus.* Many *Asphondylia* species have been reared from these, chiefly from swollen flowers or pods, and several species of *Neolasioptera* are known from stem swellings, but it is not known whether the species of these two genera form distinctive groups here. Two species, *Asphondylia cabezasae* and *A. websteri,* have more than one host genus in this subfamily, and *A. websteri* is known from a Nearctic *Parkinsonia,* a Ceasalpinioideae. *Machaerium* is exclusive host to several species of the tribe Anadiplosini. Otherwise, the gall midges on Faboideae belong to monotypic genera and nothing more can be made of them here. As for the subfamily Mimosoideae, the genera *Acacia, Anadenanthera, Calliandra, Enterolobium, Inga, Mimosa, Piptadenia, Pithecellobium, Prosopis,* and *Schrankia* are hosts

---

Figures 294–302. Galls of Asteraceae, Bromeliaceae, Burseraceae, Caryocaraceae, Chenopodiaceae, and Euphorbiaceae. **294,** Leaf galls, *Conyza canadensis* (Minas Gerais, Brazil); *Asteromyia modesta.* **295,** Leaf galls, *Chromolaena odorata* (Trinidad): *Clinodiplosis eupatorii.* **296,** Leaf galls with one in cross section, *Protium heptaphyllum* (redrawn from Houard 1933): *Dasineura braziliensis.* **297,** Three kinds of leaf galls, *Protium* sp. (redrawn from Houard 1933): cecidomyiids. **298,** Leaf galls, *Caryocar brasiliense* (Minas Gerais, Brazil): cecidomyiid. **299,** Aerial root galls, Bromeliaceae (Oaxaca, Mexico): *Neolasioptera* sp. **300,** Stem swellings, *Atriplex* sp. (Mendoza, Argentina): *Neolasioptera* sp. **301,** Apical bud swellings, *Suaeda divaricata* (San Juan, Argentina): *Asphondylia swaedicola.* **302,** Leaf galls, *Manihot tripartita* (Minas Gerais, Brazil): *Iatrophobia brasiliensis.*

to gall midges. Only the genus *Hemiasphondylia*, forming flower and pod galls on three genera of this subfamily, shows an apparent specificity to the subfamily.

In addition to the gall midges listed under the various genera below, three species have been described from undetermined Fabaceae. These are *Autodiplosis parva* from leaf rolls in Brazil (Bahia) (Tavares 1916, Houard 1933 [fig.]) and *Anadiplosis caetetensis* and *A. procera* from spherical, fuzzy, leaf or twig galls in Brazil (Bahia) (Tavares 1920c, Houard 1933 [fig.]).

Acacia

1a   Irregularly tapered stem or petiole galls . . . . cecidomyiids

Cecidomyiids. Four stem galls attributed to gall midges are listed and some illustrated in Houard 1933. They range from smooth galls with a tunnel in the pith or several longitudinal tunnels through the gall tissue, to irregular warty galls with ovoid cells along the horizontal axis. A stem gall attributed to an undetermined Cecidomyiidi was mentioned by Möhn (1975) with reference to *Meunieriella acaciae*, an inquiline. Hosts: *A. cavenia, A. eburnia, A. riparioides*, and *Acacia* spp. Distr.: Argentina (Buenos Aires, Corrientes, and Santa Fé), El Salvador. Refs.: Tavares 1915a, Möhn 1975.

1b   Leaf galls 2
2a   Swollen, joined pair of leaflets (303) . . . . cecidomyiid

Hosts: *A. jurema* and *Acacia* sp. Distr.: Argentina (Corrientes), Brazil (Bahia). Ref.: Tavares 1925.

2b   Conical gall on petiole . . . . cecidomyiid

Galls are 5–9 mm long and about 5 mm diameter at their base. They sometimes occur in masses and contort the leaves. Host: *Acacia* spp. Distr.: Brazil (Minas Gerais, Rio de Janeiro). Ref.: Rübsaamen 1905.

Aeschynomene

Stem swelling . . . . *Neolasioptera aeschynomensis*

Host: *A. montevidensis*. Distr.: Argentina (Buenos Aires). Ref.: Brèthes 1918 (fig.).

Anadenanthera

Leaflet modified into lenticular gall (304) . . . . cecidomyiid

Host: *A. peregrina*. Distr.: Brazil (Minas Gerais).

Andira

1a   Closed flower, swollen at base . . . . *Asphondylia andirae*

Host: *A. inermis*. Distr.: El Salvador. Ref.: Möhn 1959.

1b   Stem and leaf galls. 2
2a   Spheroid, long-haired gall on stem or petiole . . . . cecidomyiid

Host: *A. parvifolia*. Distr.: Brazil (Minas Gerais). Ref.: Fernandes et al. 1988 (fig.).

2b   Leaf galls without long hair 3
3a   Swollen petiole . . . . cecidomyiid

Galls are spongy and may be many-chambered. Host: *Andira* sp. (angelim). Distr.: Brazil (Rio de Janeiro). Refs.: Tavares 1920b, Houard 1933 (fig.).

3b   Galls on veins or blade of leaf 4
4a   Circular leaf blister . . . . cecidomyiid

Galls are 3 mm diameter and encircled by a lighter area. Host: *Andira* sp. Distr.: Brazil. Ref.: Rübsaamen 1907.

4b   Spheroid or cylindrical leaf galls 5
5a   Galls with circular or slitlike opening on one leaf surface 6
5b   Galls closed on both surfaces. 7
6a   Gall ovoid-petiolate on lower leaf surface, cylindrical and lipped on upper (306) . . . . cecidomyiid

Host: *A. parvifolia*. Distr.: Brazil (Minas Gerais).

6b   Simple, elongate fold of leaf surface . . . . *Andirodiplosis bahiensis*

Host: *Andira* sp. (angelim). Distr.: Brazil (Bahia). Refs.: Tavares 1920b (fig.), 1922 (fig.); Houard 1933 (fig.).

7a   Gall hemispherical, showing equally on both sides of leaf . . . . . cecidomyiid

Galls are usually situated on the veins. Hosts: *A. frondosa* and *Andira* sp. Distr.: Brazil (Rio de Janeiro). Refs.: Rübsaamen 1905, 1907.

7b   Gall showing principally on one side of leaf................ cecidomyiid

Host: *Andira* sp. Distr.: Brazil (Bahia). Ref.: Tavares 1922 (fig.), Houard 1933 (fig.).

Bauhinia

1a   Tapered stem swelling .... cecidomyiid

Stem internodes are greatly foreshortened, three to four times normal diameter, and covered with dense spines but no leaves. Host: *Bauhinia* sp. Distr.: Brazil (Amazonas). Ref.: Rübsaamen 1907.

1b   Spheroid stem or leaf galls....................................... 2
2a   Pulpy stem or leaf gall, without long hairs .... cecidomyiid

Galls are 4 mm diameter. Host: *Bauhinia* sp. Distr.: Brazil (Rio de Janeiro). Ref.: Tavares 1920b.

2b   Leaf gall with dense hairs (305) .... cecidomyiid

Host: *B. forficata*. Distr.: Brazil (Minas Gerais). Ref.: Fernandes et al. 1988 (fig.).

Brownea

Cylindrical, nearly petiolate leaf gall............................ cecidomyiid

Galls are ca. 7 mm long and may be found in large aggregates. Host: *Brownea* sp. Distr.: Brazil (Amazonas). Refs.: Rübsaamen 1905, Houard 1933 (fig.).

Calliandra

Swollen flower .... cecidomyiid

The gall maker is an undescribed Cecidomyiidi. An inquiline, *Camptoneuromyia calliandrae*, was reared from the galls. Host: *C. houstoniana*. Distr.: El Salvador. Ref.: Möhn 1975.

Canavalia

Swollen flower bud .... *Asphondylia canavaliae*

Host: *C. maritima*. Distr.: Colombia (Magdalena). Ref.: Wünsch 1979.

Cassia

Swollen bud, often causing stem to bend (307) . . . . cecidomyiid

>   Host: *C. aphylla*. Distr.: Argentina (Salta).

Leaf fold . . . . cecidomyiid

>   Host: *Cassia* sp. Distr.: Brazil (Rio de Janeiro). Refs.: Tavares 1920b, Houard 1933 (fig.).

Cercidium

Leaf roll. . . . . cecidomyiid

>   Host: *Cercidium* sp. Distr.: Argentina (Córdoba).

Copaifera

1a   Spheroid stem or petiole gall. . . . . cecidomyiid

>   Host: *C. langsdorfii*. Distr.: Brazil (Minas Gerais). Ref.: Fernandes et al. 1988 (fig.).

1b   Ovoid, pointed leaf gall . . . . cecidomyiid

>   Host: *C. langsdorfii*. Distr.: Brazil (Minas Gerais). Ref.: Fernandes et al. 1988 (fig.).

Crotalaria

1a   Swollen leaf midvein. . . . . *Asphondylia palaciosi*

>   Host: *C. longirostrata*. Distr.: El Salvador. Ref.: Möhn 1959.

1b   Flower or fruit galls 2
2a   Fruit greatly swollen at base. . . . . *Asphondylia cabezasae*

>   Hosts: *C. mucronata*, *Desmodium nicaraguense*, and *Gliricidia sepium*. Distr.: El Salvador. Ref.: Möhn 1959.

2b   Unspecified flower gall. . . . . cecidomyiid

>   The undescribed gall maker belongs to the Cecidomyiidi. An inquiline, *Camp-*

toneuromyia crotalariae, was reared from the galls. Host: *C. longirostrata*. Distr.: El Salvador. Ref.: Möhn 1975.

Cyamopsis

Swollen, deformed pod............................... *Asphondylia websteri*

Host: *C. tetragonoloba* (guar) and *Phaseolus* sp.; also *Medicago* and *Parkinsonia* in Mexico and USA. Distr.: Dominican Republic, Honduras; Mexico (northern), USA (southwestern). Refs.: Gagné and Wuensche 1986, Gagné and Woods 1988.

Cynometra

1a   Rolled leaf edge... cecidomyiid

Affected area may be thickened and ridged; sometimes whole leaf may be deformed. Host: *Cynometra* sp. Distr.: Brazil (Amazonas). Ref.: Rübsaamen 1905.

1b   Flask-shaped leaf gall on upper part of leaf .... cecidomyiid

Galls are 6 mm high. Host: *Cynometra* sp. Distr.: Brazil (Amazonas). Ref.: Rübsaamen 1905.

Dalbergia

1a   Stem, bud, or pod galls... 2
1b   Leaf galls 4
2a   Swollen pod, the side walls thickened .... cecidomyiid

Larvae occur in separate cells. Host: *D. monetaria*. Distr.: Brazil (Amazonas). Ref.: Rübsaamen 1907.

2b   Stem or bud galls... 3
3a   Cylindrical bud gall, up to 30 mm long..................... cecidomyiid

Host: *Dalbergia* sp. Distr.: Brazil (Santa Catarina). Ref.: Rübsaamen 1905.

3b   Tapered stem swelling .... cecidomyiid

Host: *Dalbergia* sp. Distr.: Brazil (Rio de Janeiro). Ref.: Rübsaamen 1907.

4a   Tapered petiole or vein swelling .... cecidomyiid

Host: *Dalbergia* sp. Distr.: Brazil (Rio de Janeiro). Ref.: Rübsaamen 1907.

4b   Complex leaf galls. 5
5a   Spheroid, closed gall (310) . . . . *Uleella dalbergiae*

Galls are red or brown, as long as 5 mm; the gall illustrated in figure 310 is only tentatively assigned to *Uleella dalbergiae*. Host: *Dalbergia* sp. Distr.: Brazil (Rio de Janeiro). Ref.: Rübsaamen 1907.

5b   Pouchlike leaf gall open to one surface. . . . . cecidomyiids

Several generally similar galls have been described from the leaves, but only one kind has been illustrated. Hosts: *Dalbergia* spp. Distr.: Brazil (Bahia). Refs.: Rübsaamen 1905, 1907; Tavares 1922 (fig.); Houard 1933 (fig.).

Desmodium

Closed, swollen flower . . . . *Asphondylia cabezasae*

Hosts: *D. nicaraguense*, *Crotalaria mucronata*, and *Gliricidia sepium*. Distr.: El Salvador. Ref.: Möhn 1960a.

Enterolobium

Swollen, aborted flower . . . . *Hemiasphondylia enterolobii*

Host: *E. cyclocarpum*. Distr.: Costa Rica. Ref.: Gagné 1978a.

Erythrina

Rolled leaflet. . . . . cecidomyiid

Host: *E. mulungu*. Distr.: Brazil (Bahia). Ref.: Tavares 1920b.

Geoffraea

1a   Spherical stem bud gall covered with modified leaves (308). . . . .
. . . . *Allodiplosis crassa*

Galls are thin-walled, 12 mm diameter, and have a single spherical larval chamber. The gall illustrated in figure 308 is only tentatively attributed to *Allodiplosis crassa*. Hosts: *G. decorticans* and *Geoffraea* sp. Distr.: Argentina (Mendoza, Salta, Tucumán). Refs.: Kieffer and Jörgensen 1910, Tavares 1915a (fig.), Jörgensen 1917 (fig.), Houard 1933 (fig.).

1b   Irregular bud or stem swellings not covered with modified leaves 2

2a   Tapered, spongy stem swelling . . . . cecidomyiid

> Host: *G. decorticans*. Distr.: Argentina (Córdoba). Refs.: Tavares 1915a (fig.), Houard 1933 (fig.).

2b   Irregular, hard, terminal or lateral bud swelling. . . . . cecidomyiid

> Host: *G. decorticans*. Distr.: Argentina (Córdoba). Refs.: Tavares 1915a, Houard 1933 (fig.).

Gliricidia

Closed, swollen flower . . . . *Asphondylia cabezasae*

> Host: *G. sepium*, *Crotalaria mucronata*, and *Desmodium nicaraguense*. Distr.: El Salvador. Ref.: Möhn 1960a.

Hymenaea

In addition to the gall listed below, an unspecified leaf gall, formed by an undescribed Cecidomyiidi on *H. courbaril* in El Salvador, has been mentioned with regard to an inquiline, *Copinolasioptera salvadorensis*, that was reared from the gall (Möhn 1975).

Leaf blister . . . . cecidomyiid

> The gall is 3–4 mm in diameter. Hosts: *H. courbaril* and *Hymenaea* sp. Distr.: Brazil (Minas Gerais), French Guiana. Refs.: Houard 1924 (fig.), 1933 (fig.).

Indigofera

1a   Swollen fruit. . . . . *Asphondylia indigoferae*

> Three inquilines have been reared from these galls: *Trotteria lindneri* (Möhn 1963b), *Camptoneuromyia crotalariae* and *Domolasioptera baca* (Möhn 1975). Host: *I. suffruticosa*. Distr.: El Salvador. Ref.: Möhn 1960a.

1b   Stem swelling. . . . . *Neolasioptera indigoferae*

> Host: *I. suffruticosa*. Distr.: El Salvador. Ref.: Möhn 1964a.

Inga

1a   Larva in pod . . . . cecidomyiid

Host: *I. ingoides*. Distr.: Guadeloupe.

1b   Galls on bud, stem, or leaf... 2
2a   Rosette bud gall .... cecidomyiid

Host: *I. strigillosa*. Distr.: Brazil (Amazonas). Ref.: Rübsaamen 1907.

2b   Stem or leaf galls.. 3
3a   Tapered, swollen stems or leaf veins.............................. 4
3b   Complex leaf galls.. 5
4a   Swelling on mid or lateral veins .... *Neolasioptera ingae*

Hosts: *I. leptoloba* and *I. spuria*. Distr.: El Salvador. Ref.: Möhn 1964a. Similar galls noted by Tavares (1920b) from Brazil on *Inga* spp. may be this or another species of *Neolasioptera*.

4b   Stem swelling.. cecidomyiid

An inquiline, *Meunieriella ingae*, was reared from the galls. Host: *I. leptoloba*. Distr.: El Salvador. Ref.: Möhn 1975.

5a   Fuzzy spherical gall (312) .... cecidomyiid

Hosts: *I. ingoides* and *Inga* spp. Distr.: Brazil (Minas Gerais, Santa Catarina). Refs.: Rübsaamen 1907, Tavares 1920b, Fernandes et al. 1988 (fig.).

5b   Flask-shaped gall not covered with hairs................... cecidomyiid

Host: *Inga* sp. Distr.: Brazil (Amazonas). Ref.: Rübsaamen 1907.

Lonchocarpus

1a   Closed, slightly swollen flower................ *Asphondylia lonchocarpi*

Host: *L. minimiflorus*. Distr.: El Salvador. Ref.: Möhn 1960a.

1b   Leaf galls 2
2a   Swollen leaf vein...................................... cecidomyiid

The gall maker is undescribed, but two inquilines, *Meunieriella armeniae* and *Meunieriella lonchocarpi*, have been reared from the galls. Host: *Lonchocarpus* sp. Distr.: El Salvador. Ref.: Möhn 1975.

2b   Podlike leaf gall...................................... cecidomyiid

Host: *L. guilleminianus*. Distr.: Brazil (Minas Gerais). Ref.: Fernandes et al. 1988 (fig.).

Machaerium

In addition to the galls keyed below, an unspecified leaf gall was reported from *Machaerium biovulatum* in El Salvador with reference to an inquiline, *Meunieriella machaerii*, reared from the galls (Möhn 1975).

1a   Spherical, fuzzy galls (as in 310), usually on leaves, but found also on branches and buds . . . . Anadiplosini

Anadiplosis pulchra. Host: originally given as *Mimosa* sp. with the common name bico de pato, which means "duck's beak." The common name is a reference to the particular shape of a *Machaerium* fruit, so this host was presumably a species of *Machaerium*. Distr.: Brazil (Rio de Janeiro). Refs.: Tavares 1916, Houard 1933 (fig.).

Anadiplosis venusta. Host: *Machaerium* sp. Distr.: Brazil (Rio de Janeiro). Refs.: Tavares 1916 (fig.), Houard 1933 (fig.).

Anadiplosis sp. Host: *Machaerium angustifolium*. Distr.: Brazil (Minas Gerais). Ref.: Fernandes et al. 1987 (fig.).

Machaeriobia machaerii. Host: *Machaerium* sp. Distr.: Brazil (Santa Catarina). Ref.: Rübsaamen 1907.

1b   Stem and leaf swellings or rolled leaves 2
2a   Tightly rolled young leaf . . . . cecidomyiid

Host: *M. caratinganum*. Distr.: Brazil (Minas Gerais).

2b   Tapered stem or leaf vein swellings. 3
3a   Rough-textured stem swelling. . . . . cecidomyiid

Host: *Machaerium* sp. Distr.: Brazil (Rio de Janerio). Refs.: Tavares 1920b, Houard 1933 (fig.).

3b   Smooth leaf vein swelling . . . . cecidomyiid

Host: *Machaerium* sp. Distr.: Brazil (Rio de Janeiro). Refs.: Tavares 1920b, Houard 1933 (fig.).

Medicago

Larva in flower bud . . . . *Prodiplosis longifila*

This species feeds on many plant families. Larvae are gregarious and kill the buds. Hosts: *M. sativa* (alfalfa); also castorbean, citrus, green beans, potato, tomato, wild cotton, and wormseed. Distr.: Colombia, Ecuador, Peru (Lambayeque), USA (Florida). Refs.: Diaz B. 1981, Gagné 1986, Peña et al. 1989.

Mimosa

1a   Leaf blister . . . . cecidomyiid

A presumed inquiline, *Meunieriella lonchocarpi*, was reared from the galls. Host: *Mimosa* sp. Distr.: El Salvador. Ref.: Möhn 1975.

1b   Flower and pod galls 2
2a   Swollen, deformed flower or young pod . . . . *Hemiasphondylia mimosae*

Two inquilines, *Moehniella fernandi* and *Camptoneuromyia* sp., have been reared from these galls in Colombia (Magdalena). Hosts: *M. albida*, *M. leiocarpa*, and *Prosopis juliflora*. Distr.: Colombia (Magdalena). Refs.: Möhn 1960a, Wünsch 1979.

2b   Deformed pod . . . . *Tavaresomyia mimosae*

Host: *M. caesalpinifolia*. Distr.: Brazil (Ceará). Ref.: Tavares 1925.

Phaseolus

1a   Elongate, tapered stem swelling . . . . *Neolasioptera phaseoli*

Host: *Phaseolus* sp. Distr.: El Salvador. Ref.: Möhn 1975.

1b   Larvae in buds or pods 2
2a   Larva in flower bud . . . . *Prodiplosis longifila*

This species feeds on many plant families. Larvae are gregarious and kill the buds. Hosts: *Phaseolus* sp. (green beans); also alfalfa, castorbean, citrus, potato, tomato, wild cotton, and wormseed. Distr.: Colombia, Ecuador, Peru (Lambayeque), USA (Florida). Refs.: Diaz B. 1981, Gagné 1986, Peña et al. 1989.

2b   Larva in pod . . . . *Asphondylia websteri*

Hosts: *Phaseolus* sp. and *Cyamopsis tetragonoloba*; also *Medicago* and *Parkinsonia* in Mexico and USA. Distr.: Dominican Republic, Honduras; Mexico (northern), USA (southwestern). Ref.: Gagné and Wuensche 1986, Gagné and Woods 1988.

Piptadenia

1a   Tapered petiole swelling. . . . . cecidomyiid

Hosts: *P. communis* and *P. gonoacantha* (now considered synonymous). Distr.: Brazil (Minas Gerais, Rio de Janeiro). Ref.: Rübsaamen 1908.

1b   Spheroid galls on leaflets . . . . cecidomyiids

Rübsaamen (1908) described five kinds of galls from leaflets of *P. communis*. Although they may be distinct, they were not illustrated, so it is not possible to distinguish among them here. Fernandes et al. (1988) described and illustrated a fuzzy one. Hosts: *P. communis* and *P. gonoacantha*. Distr.: Brazil (Minas Gerais, Rio de Janeiro, Santa Catarina). Refs.: Rübsaamen 1908, Fernandes et al. 1988 (fig.).

Pithecellobium

1a   Aborted, deformed pod. . . . . *Asphondylia* sp.

Hosts: *P. pallens* and *P. keyensis*. Distr.: USA (Florida); Mexico (Nuevo Leon).

1b   Conical or cylindrical leaf gall . . . . cecidomyiid

Host: *P. glomeratum*. Distr.: Brazil (Amazonas). Ref.: Rübsaamen 1908. Note: In addition to this leaf gall, several more were found and illustrated by Houard (1924, 1933) from *P. pedicellare* collected in French Guiana. He attributed them to "insects," but they are probably formed by cecidomyiids.

Figures 303–312. Galls of Fabaceae. 303, Leaf galls, *Acacia* sp. (Corrientes, Argentina): cecidomyiid. 304, Leaf galls, *Anadenanthera peregrina* (Minas Gerais, Brazil): cecidomyiid. 305, Leaf galls, *Bauhinea forficata* (Minas Gerais, Brazil): cecidomyiid. 306, Leaf galls, *Andira* sp. (Minas Gerais, Brazil): cecidomyiid. 307, Stem galls, *Cassia* sp. (Salta, Argentina): cecidomyiid. 308, Bud galls, *Geoffraea* sp. (Salta, Argentina): *Allodiplosis crassa*. 309, Bud galls, *Prosopis flexuosa* (La Rioja, Argentina): *Rhopalomyia prosopidis*. 310, Leaf galls, *Dalbergia* sp. (Minas Gerais, Brazil): *Uleela dalbergiae*. 311, Stem galls, *Prosopis* sp. (Salta, Argentina): *Tetradiplosis* sp. 312, Leaf galls, *Inga* sp. (Minas Gerais, Brazil): cecidomyiid.

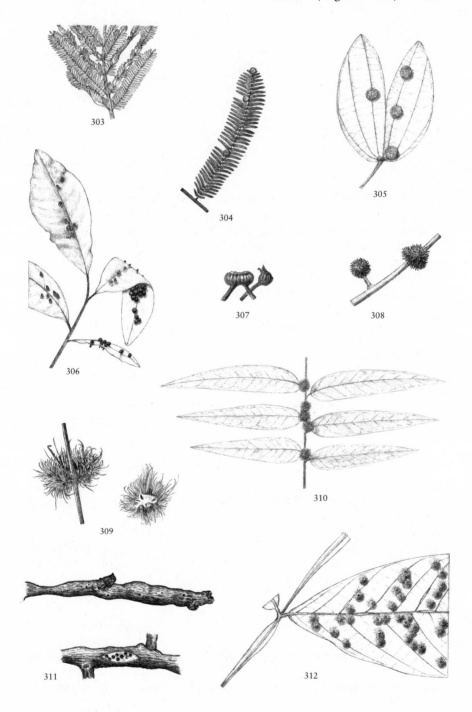

303

304

305

306

307

308

309

310

311

312

Prosopis

1a   Fruit gall .... *Hemiasphondylia mimosae*

Hosts: *P. juliflora* and *Mimosa* spp. Distr.: Colombia (Magdalena), El Salvador. Ref.: Wünsch 1979.

1b   Bud or stem galls... 2

2a   Spherical, spongy, bud gall covered with narrow, stiff leaves (309)........
.... *Rhopalomyia prosopidis*

Galls are polythalamous, 10–25 mm in diameter. Hosts: *P. alba*, *P. alpataco*, *P. campestris*, and *P. flexuosa*; it is possible that various hosts yield different species of *Rhopalomyia*. Distr.: Argentina (Mendoza, San Luis). Refs.: Kieffer and Jörgensen 1910, Jörgensen 1917 (fig.), Houard 1933 (fig.).

2b   Stem swellings (311)..
.... *Liebeliola prosopidis*, *Tetradiplosis sexdentatus*, and cecidomyiid

Many kinds of stem swellings have been described from various *Prosopis* species, and three cecidomyiids have been reported from them. Galls are variable in size, color, and texture.

Liebeliola prosopidis. The gall is spheroid and much greater in diameter than its stem. A probable inquiline, *Meunieriella graciliforceps*, was also reared from these galls. Host: *P. strombulifera*. Distr.: Argentina (Mendoza, San Juan). Refs.: Kieffer and Jörgensen 1910: 428 (fig.), Houard 1933 (fig.).

Tetradiplosis sexdentatus. Galls (311) are rough-textured with ovoid chambers. Hosts: *P. alpataco* and *P. campestris*. Distr.: Argentina (Mendoza). Refs.: Kieffer and Jörgensen 1910 (fig.), Jörgensen 1917 (fig.), Houard 1933 (fig.).

Cecidomyiid. This swelling occurs at the base of lateral branches. The larval chamber lies along the pith. Host: *Prosopis* sp. Distr.: Argentina (Santa Fé, Córdoba). Refs.: Tavares 1915a (fig.), Houard 1933 (fig.).

Swartzia

1a   Complex, biconical leaf gall on lower leaf surface............ cecidomyiid

The gall occurs on the lower leaf surface. The larval cell is located at the base of the gall. Host: *S. stipulifera*. Distr.: Brazil (Amazonas). Refs.: Rübsaamen 1908 (fig.), Houard 1933 (fig.).

1b   Simple, concave leaf spot gall on lower leaf surface. . . . . cecidomyiid

>   Host: *Swartzia* sp. Distr.: Brazil (Amazonas). Ref.: Rübsaamen 1908 (fig.).

Fagaceae

Nothofagus

Rosette bud gall. . . . . *Rhopalomyia nothofagi*

>   Host: *N. obliqua.* Distr.: Chile (Los Lagos). Refs.: Gagné 1974, Madrid 1974 (fig.).

Quercus (oak)

Hemispherical leaf galls. . . . . *Polystepha* spp.

>   *Polystepha lapalmae.* Host: *Q. hondurensis.* Distr.: El Salvador. Ref.: Möhn 1960b.

>   *Polystepha salvadorensis.* Host: *Q. eugeniaefolia.* Distr.: El Salvador. Ref.: Möhn 1960b.

Flacourtiaceae

Carpotroche

Spherical, hairy leaf gall . . . . cecidomyiid

>   Galls are up to 10 mm diameter, one-celled, and occur along veins on the underside of the leaf. Host: *C. longifolia.* Distr.: Brazil (Amazonas). Ref.: Rübsaamen 1905.

Casearia

1a   Spherical, peduncled, terminal bud gall. cecidomyiid

>   Individual galls are 2 mm diameter but can form a large globular mass of 15 mm diameter mixed with modified leaves. Host: *Casearia* sp. Distr.: Brazil (Rio de Janeiro). Ref.: Rübsaamen 1905.

1b   Larva in fruit gall, the fruit smaller than normal. . . . . *Asphondylia* sp.

>   Host: *C. bauquitana.* Distr.: El Salvador. Ref.: Möhn 1960a.

Tetrathylacium

Spherical, brownish, ridged leaf gall .... cecidomyiid

> The gall is 2–3 mm diameter and shows on both sides of leaf. Host: *Tetrathylacium* sp. Distr.: Peru (Loreto). Ref.: Rübsaamen 1908.

Xylosma

Larva in fruit .... *Asphondylia xylosmatis*

> Infested fruit are smaller than normal. Two inquilines, *Camptoneuromyia xylosmatis* and *Domolasioptera adversaria*, have been reared from the galls (Möhn 1975). Host: *X. flexuosa*. Distr.: El Salvador. Ref.: Möhn 1959.

Heliconiaceae

Heliconia

Larva from unspecified chamber in stem .... cecidomyiid

> Larvae in cocoons were intercepted at Miami, Florida, USA. Host: *Heliconia* sp. Distr.: Costa Rica.

Lamiaceae (Labiatae)

Gardoquia

Spheroid stem enlargement (313) .... Dasineura gardoquiae

> Galls are hollow, smooth, and woody and have a large central chamber with a single, yellowish larva. Host: *G. gilliesii*. Distr.: Chile (Concepción). Refs.: Kieffer and Herbst 1905, 1906 (fig.).

Hyptis

1a   Slightly thickened fruit .... *Asphondylia hyptis*

> Host: *H. verticillata*. Distr.: El Salvador. Ref.: Möhn 1960a.

1b   Stem swelling .... *Neolasioptera hyptis*

> Host: *H. suaveolens*. Distr.: El Salvador. Ref.: Möhn 1964a.

Mentha

Swollen leaf or marginal leaf roll (314)...................... *Clinodiplosis* sp.

Host: *Mentha* sp. Distr.: Jamaica.

Ocimum

Swollen flower .... *Asphondylia ocimi*

Host: *O. micranthum*. Distr.: El Salvador. Ref.: Möhn 1959.

Salvia

1a   Stem swelling................................. *Neolasioptera salviae*

Host: *S. occidentalis*. Distr.: El Salvador. Ref.: Möhn 1964a.

1b   Bud or leaf galls.. 2
2a   Swollen bud................................. *Sciasphondylia salviae*

Host: *Salvia* sp. Distr.: El Salvador. Ref.: Möhn 1960a.

2b   Hemispherical, fleshy leaf gall (315) .... *Asphondylia* sp.

Host: *S. leucantha*. Distr.: Mexico (Morelos).

Teucrium

Fuzzy stem swelling .... *Neolasioptera baezi*

Host: *T. inflatum*. Distr.: Argentina (Entre Ríos). Ref.: Brèthes 1922.

Lauraceae

In addition to the gall midges listed below under the various genera, *Macroporpa ulei* was reared from complex, conical leaf vein galls on an undetermined Lauraceae from Peru (Madre de Dios) (Rübsaamen 1916a [fig.], Houard 1933 [fig.]).

Goeppertia

Hemispherical stem swelling.................................... cecidomyiid

Swellings are about 3 mm diameter, rugose, finely haired, woody, each with a single, ovoid larval cavity, and usually found in aggregate. Host: *G. hirsuta*. Distr.: Brazil (Rio Grande do Sul). Refs.: Tavares 1909 (fig.), Houard 1933 (fig.).

Nectandra

1a   Stem swellings .... cecidomyiids

Several differently shaped galls have been recorded, but in insufficient detail to separate here. Hosts: *Nectandra* spp. Distr.: Brazil (Rio Grande do Sul). Refs.: Tavares 1909, 1921; Houard 1933 (fig.).

1b   Blister or spheroid leaf galls (316) ....
.... "*Oligotrophus*" *nectandrae* and cecidomyiids

"*Oligotrophus*" *nectandrae*. The original description is unclear as to whether the gall is complex as are those below or is merely a simple blister. Host: *N. megapotamica*. Distr.: Paraguay. Ref.: Kieffer 1910b.

Cecidomyiids. Several different galls have been described, from flattened, parasol-like to spherical, and hairy or smooth. Hosts: *Nectandra* spp., incl. cannela. Distr.: Brazil (Rio Grande do Sul, Santa Catarina). Refs.: Rübsaamen 1908, Tavares 1921, Houard 1933 (fig.).

Ocotea

1a   Leaf blister.. cecidomyiid

Host: *Ocotea* sp. Distr.: Brazil. Ref.: Rübsaamen 1908.

1b   Stem swelling.. cecidomyiid

A sciarid reared from this gall and described as *Microcerata ocoteae* Köhler is unlikely to be the gall maker (Pritchard 1951). Host: *O. acutifolia*. Distr.: Argentina (Buenos Aires). Ref.: Köhler 1932 (fig.).

Liliaceae

Yucca

1a   Larva in unspecified leaf gall .... *Clinodiplosis* sp.

Host: *Yucca* sp. Distr.: Peru.

1b   Larva in fruit . . . . *Contarinia* sp.

Host: *Yucca* sp. Distr.: Mexico.

Loganiaceae
Buddleja

Swollen, aborted flower . . . . *Asphondylia buddleia*

Hosts: *B. americana* and *B. racemosa*. Distr.: El Salvador; USA (Texas). Refs.: Felt 1935, Painter 1935 (fig.), Möhn 1960a.

Loranthaceae
Phthirusa

Larva from unspecified damage, probably in flower . . . . *Asphondylia* sp.

Host: *P. pyrifolia*. Distr.: Trinidad.

Psittacanthus

Larva from unspecified damage, probably in flower. . . . . *Schizomyia* sp.

Host: *P. robustus*. Distr.: Brazil.

Struthanthus

1a   Swollen fruit and/or multichambered stem swelling. . . . . . . *Asphondylia* spp.

Asphondylia parasiticola. Reared from many-chambered stem swellings and enlarged fruit, which were much longer and thicker than normal fruit. Host: *S. marginatus*. Distr.: El Salvador. Ref.: Möhn 1960a.

Asphondylia struthanthi. Reared from swollen fruit. Host: *Struthanthus* sp. Distr.: Brazil (Ceará). Ref.: Rübsaamen 1916a.

1b   Leaf blister . . . . cecidomyiid

Host: *S. vulgaris*. Distr.: Brazil. Ref.: Arduin et al. 1991.

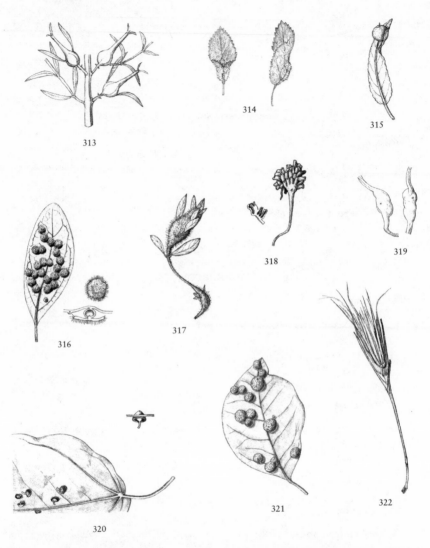

Figures 313–322. Galls of Lamiaceae, Lauraceae, Lythraceae, Malvaceae, Moraceae, Nyctaginaceae, Orchidaceae, and Poaceae. **313,** Stem galls, *Gardoquia gillesii* (redrawn from Kieffer and Herbst 1906): *Dasineura gardoquiae*. **314,** Leaf galls, *Mentha* sp. (Jamaica): *Clinodiplosis* sp. **315,** Leaf gall, *Salvia leucantha* (Morelos, Mexico): *Asphondylia* sp. **316,** Leaf galls with enlargement and cross section, *Nectandra* sp. (redrawn from Houard 1933): cecidomyiid. **317,** Stem galls, *Cuphea* sp. (Minas Gerais, Brazil): *Neolasioptera* sp. **318,** Bud galls with cross section, undetermined Malvaceae (redrawn from Houard 1933): *Metasphondylia squamosa*. **319,** Aerial root galls, *Cattleya* sp. (Central America): *Clinodiplosis cattleyae* and *Neolasioptera* sp. **320,** Leaf galls with one in lateral view, *Ficus* sp. (redrawn from Houard 1933): *Calmonia fici*. **321,** Leaf galls, *Pisonia* sp. (U.S. Virgin Is.): *Bruggmannia* sp. **322,** Stem gall, *Aristida gyrans* (Florida, USA): *Chilophaga gyrantis*.

Lythraceae

Cuphea

1a   Swollen, aborted flower .... *Asphondylia* sp.

Hosts: *C. appendiculata, C. cyanea, C. hookeriana, C. nitidula,* and *C. pinetorum.* Distr.: Mexico, Panama.

1b   Stem swelling covered with sticky hairs (317)........... *Neolasioptera* sp.

Host: *Cuphea* sp. Distr.: Brazil (Minas Gerais).

Malpighiaceae

In addition to the gall midges listed below under the various genera, *Macroporpa peruviana* was reared from fuzzy, spherical leaf galls on an undetermined Malpighiaceae in Peru (Madre de Dios) (Rübsaamen 1916a [fig.], Houard 1933 [fig.]). Larvae of a *Clinodiplosis* species were found in gall tissue surrounding the central larval cavity of these galls.

Banisteria

Unspecified leaf gall, possibly a blister .... cecidomyiid

The gall, from a plant in the Munich Herbarium, was illustrated in Küster 1903 to show cellular structure peripheral to the gall. Host: *Banisteria* sp. Distr.: South America (unspecified).

Byrsonima

1a   Conical leaf gall .... cecidomyiid

Host: *B. verbascifolia.* Distr.: Brazil. Refs.: Tavares 1921, Houard 1933.

1b   Unopened, aborted flower .... cecidomyiids

One larva of an undescribed Cecidomyiidi generally occurs on each flower, but gregarious larvae of an unknown cecidomyiid are occasionally found also. Host: *B. crassa.* Distr.: Brazil (Minas Gerais, Rio de Janeiro).

Heteropteris

Swollen, aborted flower...................................... *Contarinia* sp.

Host: *Heteropteris* sp. Distr.: Brazil (Rio de Janeiro).

Stigmaphyllon

Elongate-ovoid leaf gall. . . . . cecidomyiid

Host: *S. lalandianum.* Distr.: Brazil (Minas Gerais). Ref.: Fernandes et al. 1988.

Tetrapteris

Spherical leaf gall. . . . . cecidomyiid

Host: *Tetrapteris* sp. Distr.: Brazil. Refs.: Rübsaamen 1929 (fig.), Houard 1933 (fig.).

Malvaceae

In addition to the gall midges listed below under the various genera, *Metasphondylia squamosa* was reared from aggregated, cylindrical bud galls (318) on an undetermined Malvaceae in Brazil (Bahia) (Tavares 1918a [fig.]), Houard 1933 [fig.]).

Gossypium

1a   Larva in stems. . . . . *Asynapta gossypii*

Larvae feed in the soft tissue of the bark and developing wood and may be feeding only on rot. Host: *Gossypium* sp. Distr.: Barbados. Refs.: Coquillett 1905, Barnes 1949.

1b   Larva in flowers or flower buds. 2
2a   Larva in flower. . . . . *Contarinia gossypii*

The gregarious larvae feed among the flower parts. When full grown the larvae drop to the soil, and the flowers rot and die. Hosts: *Gossypium* spp. Distr.: Antigua, Barbados, Colombia (Bolívar, Valle), Montserrat, St. Croix, Tortola. Refs.: Felt 1908a, Callan 1940, Barnes 1949.

2b   Larva in flower bud. . . . . *Prodiplosis longifila*

This species feeds on many plant families. Larvae are gregarious and kill the buds. Hosts: *Gossypium* spp. (wild cotton); also alfalfa, castorbean, citrus, green beans, potato, tomato, and wormseed. Distr.: Colombia, Ecuador, Peru (Lambayeque), USA (Florida). Refs.: Diaz B. 1981, Gagné 1986, Peña et al. 1989.

Malvaviscus

1a Enlarged, swollen, closed flower gall . . . . *Asphondylia malvavisci*

  Host: *M. arboreus.* Distr.: El Salvador. Ref.: Möhn 1959.

1b Stem swelling . . . . *Neolasioptera malvavisci*

  Host: *M. arboreus.* Distr.: El Salvador. Ref.: Möhn 1975.

Sida

1a Bud gall . . . . *Asphondylia sidae*

  Hosts: *S. acuta* and *S. rhombifolia.* Distr.: Colombia (Magdalena). Ref.: Wünsch 1979.

1b Tapered stem swelling . . . . *Neolasioptera sidae*

  Host: *S. rhombifolia.* Distr.: El Salvador. Ref.: Möhn 1964a.

Melastomataceae

This family is host to many Lepidoptera that form complex galls similar to those made by some cecidomyiids on other plant families.

Clidemia

Hairy, spherical gall on twigs and leaves . . . . cecidomyiid

  Host: *Clidemia* sp. Distr.: Brazil (Minas Gerais). Refs.: Tavares 1917a (fig.), Houard 1933 (fig.).

Marcetia

Rosette bud gall . . . . *Clinodiplosis marcetiae*

  Host: *Marcetia* sp. Distr.: Brazil (Rio de Janeiro). Refs.: Tavares 1917a, Houard 1933 (fig.).

Miconia

In addition to the galls keyed below, there is a reference to a foliar excrescence in Frauenfeld 1858 on *M. trinervis*.

1a   Galls on fruit or flowers..2
1b   Galls on stems or leaves..3
2a   Deformed fruit ....cecidomyiid

Infested fruit are cylindrical, 10–12 mm long by 1–2 mm wide, velvety, and contain two to four larvae. Hosts: *M. auriculata*, *M. ibaguensis*, and *Miconia* sp. Distr.: Brazil (Amazonas), Peru (San Martín). Ref.: Rübsaamen 1907.

2b   Flowers transformed into a spheroid mass ....cecidomyiid

Galls are 5–9 mm wide by 4–5 mm high. Host: *Miconia* sp. Distr.: Brazil. Ref.: Rübsaamen 1907.

3a   Tapered stem swelling ....cecidomyiid

Galls are multichambered, up to 22 mm long by 10 mm wide, and hard and woody. Host: *Miconia* sp. Distr.: Brazil (Rio de Janeiro). Refs.: Rübsaamen 1907, Tavares 1917a, Houard 1933 (fig.).

3b   Complex leaf galls......................................cecidomyiids

Several different kinds of galls have been described but not illustrated. These are spherical or cylindrical, smooth or hairy, and variously placed on the leaf veins, blade, and petiole and on twigs. Hosts: *Miconia* spp. Distr.: Brazil. Refs.: Rübsaamen 1907, Tavares 1925, Houard 1933 (fig.).

Mouriri

Swollen flower gall.................................possibly *Asphondylia* sp.

Host: *M. ulei*. Distr.: Brazil (Amazonas). Ref.: Rübsaamen 1908.

Ossaea

Spheroid leaf gall.....................................*Lopesia brasiliensis*

Galls are about 5 mm diameter, reddish to brown, with dense hair. They are situated on principal veins and occasionally on the petiole and can deform the leaf when numerous. Host: *Ossaea* sp. Distr.: Brazil (Rio de Janeiro, Santa Catarina). Ref.: Rübsaamen 1908.

Tamonea (Miconia?)

Spherical, hairy leaf gall . . . . cecidomyiid

A similar gall on a *Tibouchina* is caused by a species of *Rochadiplosis*. Host: *Tamonea* sp. Distr.: Brazil (Rio Grande do Sul). Ref.: Tavares 1909 (fig.).

Tibouchina

1a   Elongate, hairy swelling of leaf vein, petiole, or stem . . . . cecidomyiid

Swellings contain a single larva. Host: *T. granulosa*. Distr.: Brazil (Rio de Janeiro). Ref.: Rübsaamen 1908.

1b   Blister or spherical leaf galls 2
2a   Blister gall on upper leaf surface. . . . . cecidomyiid

The gall is about 3 mm long and 2 mm wide and has a slitlike opening on the lower leaf surface. Host: *Tibouchina* sp. Distr.: Brazil (Rio de Janeiro). Refs.: Tavares 1917a, Houard 1933 (fig.).

2b   Spherical hairy leaf gall . . . . *Rochadiplosis tibouchinae*

Host: *Tibouchina* sp. Distr.: Brazil (Rio de Janeiro). Ref.: Tavares 1917a (fig.).

Tococa

Hairy swelling of midvein . . . . *Clinodiplosis* sp.

Host: *Tococa* sp. Distr.: Brazil (Pará). Ref.: Rübsaamen 1908.

Meliaceae

Guarea

1a   Spherical, white-haired, leaf gall . . . . *Guarephila albida*

Galls are only 1.5–2 mm diameter, thin-skinned, have a rounded larval cavity, and are situated on the underside of the leaf. Host: *Guarea* sp., possibly *G. trichilioides*. Distr.: Brazil (Rio Grande do Sul). Refs.: Tavares 1909 (fig.), Houard 1933 (fig.).

1b   Elongate, many-celled twig or stem gall . . . . *Asphondylia guareae*

Host: *Guarea* sp. Distr.: El Salvador. Ref.: Möhn 1959.

Trichilia

Rounded, many-celled twig or stem gall .... *Asphondylia trichiliae*

Host: *Trichilia* sp. Distr.: El Salvador. Ref.: Möhn 1959.

Menispermaceae

Cissampelos

Stem swelling .... *Neolasioptera cissampeli*

Host: *C. pareira*. Distr.: El Salvador. Ref.: Möhn 1964a.

Monimiaceae

Mollinedia

Irregularly tapered stem swelling................................. cecidomyiid

Host: *M. elegans*. Distr.: Brazil (Rio Grande do Sul). Ref.: Tavares 1909 (fig.).

Peumus

Stem swelling... cecidomyiids

Galls are 10–12 mm long by 3 mm diameter and contain one or more elongate cells, each with a single, yellow larva. Larvae of two different species have been taken from these galls. Host: *P.* (as *Boldoa*) *boldus*. Distr.: Chile (Concepción). Refs.: Kieffer and Herbst 1905, 1906.

Siparuna

Larvae in or on flowers .... cecidomyiids

Larvae of *Asynapta*, *Clinodiplosis*, *Dasineura*, and an unidentified Oligo-trophini have been found in male flowers of *Siparuna* spp. Females of *Asynapta* sp. have been seen covered with pollen and apparently ovipositing on male flowers. Hosts: *Siparuna* spp. Distr.: Bolivia. Refs.: Feil and Renner 1991.

Moraceae

Cecropia

1a   Leaf blister . . . . *"Cecidomyia" cecropiae*

Hosts: *C. obtusa* and *Cecropia* sp. Distr.: Brazil (Amazonas), Cuba. Refs.: Cook 1909, Rübsaamen 1905.

1b   Petiole or vein swelling. . . . . cecidomyiid

Host: *Cecropia* sp. Distr.: Brazil (Amazonas). Ref.: Rübsaamen 1905.

Chlorophora

Swollen, closed flower bud with gregarious larvae . . . . . Clinodiplosis chlorophorae

Host: *C. tinctoria*. Distr.: Brazil (Rio de Janeiro). Ref.: Rübsaamen 1905.

Coussapoa

Spheroid gall at extremity of aerial root . . . . *Frauenfeldiella coussapoae*

Galls are 3.5 mm long to 1.5 mm wide and may be found in aggregates of 30 mm diameter. Host: *Coussapoa* sp. Distr.: Brazil (Amazonas, Rio de Janeiro). Ref.: Rübsaamen 1905.

Ficus (fig)

1a   Deformed fruit (syconia) . . . . *Ficiomyia perarticulata*

Galls, each with one larva, are pocket-shaped outgrowths of the receptacle. A fruit may contain many galls. Host: *F. citrifolia*. Distr.: Dominica (possibly a different species of *Ficiomyia*), USA (Florida). Ref.: Roskam and Nadel 1990 (fig.).

1b   Leaf galls 2
2a   Marginal leaf roll . . . . cecidomyiid

Host: *Ficus* sp. Distr.: Brazil (Bahia). Refs.: Tavares 1917b, Houard 1933 (fig.).

2b   Blister or complex leaf galls. 3
3a   Circular leaf blister . . . . cecidomyiids

Several such galls have been noted, from almost flat to hemispherical. They are smooth and contain one larva of the gall maker. An inquiline, *Meunieriella lucida*, has been reared from the galls (Möhn 1975). Hosts: *Ficus* spp. Distr.: Brazil (Amazonas), Colombia (Amazonas), Cuba, El Salvador. Refs.: Cook 1909 ("*Cecidomyia*" *fici*) (fig.), Rübsaamen 1907, Houard 1933 (fig.).

3b   Complex leaf galls (320) . . . . *Calmonia fici* and *Calmonia urostigmatis*

Galls are top-shaped or spheroid and show on both leaf surfaces. Host: *Ficus* sp. Distr.: Brazil (Bahia). Refs.: Tavares 1917b, Houard 1933 (fig.).

Pourouma

Circular leaf blister or spherical, pedunculate gall . . . . cecidomyiids

Host: *P. cuspidata*. Distr.: Brazil (Amazonas). Ref.: Rübsaamen 1908.

Sorocea

Swollen, deformed stem apex . . . . *Bruggmanniella braziliensis*

Galls are large and multichambered. Host: *S. ilicifolia*. Distr.: Brazil (Rio Grande do Sul). Refs.: Tavares 1909 (fig.), Houard 1933 (fig.).

Myrsinaceae

Ardisia

Larva in fruit, the fruit remaining small . . . . *Asphondylia ardisiae*

Host: *A. paschalis*. Distr.: El Salvador. Ref.: Möhn 1959.

Myrsine

1a   Leaf galls 2
1b   Bud or stem galls 3
2a   Subspherical, pedicellate leaf gall . . . . *Bruggmannia braziliensis*

Galls are ca. 2.5 mm diameter and occur on the underside of the leaf. Host: *Myrsine* sp. Distr.: Brazil (Rio Grande do Sul). Refs.: Tavares 1906, 1909 (fig.). Note: Tavares (1909) described and illustrated an additional spherical leaf gall, but it is not clear exactly how it differs from the one made by *Bruggmannia braziliensis*.

2b   Hemispherical leaf gall apparent on both leaf surfaces . . . . cecidomyiid

Host: *Myrsine* (as *Rapanea*) sp. Distr.: Brazil (São Paulo). Ref.: Tavares 1922.

3a   Ovate or elongate blister gall on stem . . . . cecidomyiid

Host: *Myrsine* sp. Distr.: Brazil (Rio Grande do Sul). Ref.: Tavares 1909 (fig.).

3b   Rosette bud gall . . . . cecidomyiid

Host: *Myrsine* sp. Distr.: Brazil (Rio Grande do Sul). Ref.: Tavares 1909 (fig.).

Myrtaceae

In addition to the gall midges under the various genera below, two species have been described from leaf galls on unidentified Myrtaceae in Brazil. These galls are reillustrated here (redrawn from Houard 1933) because they may be found again. One is a clublike gall (fig. 323) formed by *Stephomyia clavata* (Tavares 1920b), the other an amorphous to hemispherical gall (fig. 324) formed by *Anasphondylia myrtacea* (Tavares 1920b).

Eugenia

1a   Deformed fruit . . . . *Dasineura eugeniae*

Host: *E. buxifolia*. Distr.: USA (Florida). Ref.: Felt 1912c.

1b   Bud, stem, and leaf galls 2
2a   Bud gall made up of elongate rosette of leaves . . . . cecidomyiid

Galls vary between 20 and 30 mm in length to about one-third as wide. A single larva is enclosed in an ovoid, central cell. Host: *Eugenia* sp. Ref.: Rübsaamen 1907.

2b   Stem and leaf galls 3
3a   Tapered swelling of stem (325), petiole, or midrib . . . . cecidomyiid

Swellings contain one or more ovoid larval cells. Host: *E. uniflora* and *Eugenia* sp. Distr.: Brazil (Bahia), Peru (San Martín). Refs.: Rübsaamen 1907, Tavares 1921, Houard 1933 (fig.).

3b   Leaf rolls, blisters, or complex leaf or stem galls 4

4a   Simple leaf roll, sometimes associated with hypertrophy. 5
4b   Leaf blisters or complex outgrowths of leaf or stem 6
5a   Simple leaf roll or marginal leaf roll, without swelling . . . . cecidomyiid

Larvae are gregarious inside cavity. Hosts: *E. rotundifolia* and *E. uniflora*. Distr.: Brazil (Bahia). Refs.: Tavares 1921, Houard 1933 (fig.).

5b   Both sides of leaf base swollen and recurved toward upper surface to midvein
. . . . cecidomyiid

The central larval cavity contains an orange larva. Host: *Eugenia* sp. Distr.: Brazil (Bahia). Refs.: Tavares 1921, Houard 1933 (fig.).

6a   Cylindrical outgrowths of leaves and stems (326, 327) . . . . *Stephomyia* spp.

Eugenia spp. show a great variety of such galls, which differ in shape, hairiness, and size, from 3 to 10 mm in length. Each contains a single larval cavity. Several varieties of galls were listed from Brazil by Rübsaamen (1907) and Tavares (1909). Galls made by the two described *Stephomyia* species are as follows:

Galls are ellipsoidal, glabrous, 5–6 mm high. *Stephomyia epeugenia*. Host: *Eugenia* sp. Distr.: Brazil (Rio de Janeiro). Refs.: Tavares 1916 (fig.), 1921, 1922 (fig.).

Galls are ellipsoidal, hairy, and 4–5 mm high. *Stephomyia eugeniae*. Host: *E. buxifolia*. Distr.: USA (Florida). Ref.: Felt 1913c.

6b   Simple blister or complex hemispheroid leaf galls 7
7a   Blister gall. . . . . cecidomyiids

Many slight variations of biconvex, one-celled, blister galls have been recorded. They are visible especially from the top surface of the leaf. The adults escape from beneath the leaf. An inquiline, *Meunieriella eugeniae*, has been reared from the galls (Möhn 1975). Hosts: *E. uniflora* and *Eugenia* sp. Distr.: Brazil (Rio de Janeiro), El Salvador. Refs.: Rübsaamen 1907, Tavares 1921.

7b   Hemispheroid gall set into concave depression, showing as a blister on reverse surface of leaf . . . . cecidomyiid

Host: *Eugenia* sp. Distr.: Brazil (Rio de Janeiro). Ref.: Tavares 1921.

Myrceugenia

Swollen apical or lateral bud (329). . . . . "*Oligotrophus*" *eugeniae*

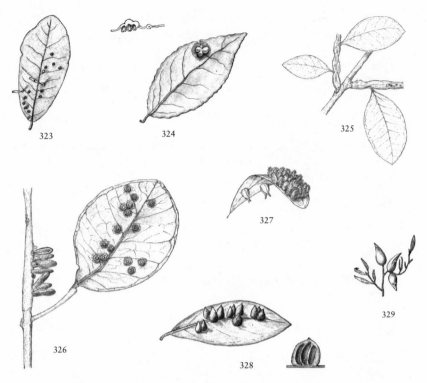

Figures 323–329. Galls of Myrtaceae. **323,** Leaf galls, undetermined Myrtaceae (redrawn from Houard 1933): *Stephomyia clavata*. **324,** Leaf galls with cross section, undetermined Myrtaceae (redrawn from Houard 1933): *Anasphondylia myrtaceae*. **325,** Stem galls, *Eugenia uniflora* (Rio de Janeiro, Brazil): cecidomyiid. **326,** Stem and leaf galls, *Eugenia rotundifolia* (Rio de Janeiro): *Stephomyia* spp. **327,** Two kinds of leaf galls, *Eugenia axilaris* (Florida, USA): *Stephomyia* spp. **328,** Leaf galls with one in cross section, 2×, *Myrcia itambensis* (Minas Gerais, Brazil): cecidomyiid. **329,** Bud galls, *Myrceugenia stenophylla* (redrawn from Houard 1933): *"Oligotrophus" eugeniae*.

Hosts: *M. stenophylla* and possibly *M. ferruginea*; that on the latter may be a different kind of gall (its gall maker was not described), but one cannot be certain from the original description. Distr.: Chile (Concepción, Valparaíso). Refs.: Kieffer and Herbst 1909, 1911.

Myrcia

1a   Rolled leaf margin . . . . cecidomyiid

Host: *M. itambensis*. Distr.: Brazil (Minas Gerais). Ref.: Fernandes et al. 1988 (fig.).

1b   Ovoid, complex leaf gall or simple, spheroid leaf swelling 2
2a   Ovoid, complex gall along midvein (328) . . . . cecidomyiid

> Host: *M. itambensis*. Distr.: Brazil (Minas Gerais). Ref.: Fernandes et al. 1988 (fig.).

2b   Simple spheroid swelling, causing leaf deformation . . . . cecidomyiid

> Galls resemble those of *Anasphondylia myrtacea* from an unknown Myrtaceae (324). Host: *M. itambensis*. Distr.: Brazil (Minas Gerais). Ref.: Fernandes et al. 1988 (fig.).

Myrciaria

1a   Multichambered swelling of apical bud . . . . cecidomyiid

> Host: *Myrciaria* sp. Distr.: Brazil (São Paulo). Ref.: Tavares 1921, Houard 1933 (fig.).

1b   Rolled young leaf . . . . cecidomyiid

> Host: *M. floribunda*. Distr.: Brazil (Rio de Janeiro).

Pimenta

Swollen, misshapen, unopened flower bud . . . . *Schizomyia* sp.

> Host: *P. dioica*. Distr.: Jamaica. Ref.: Parnell and Chapman 1966 (fig.).

Psidium

1a   Rolled leaf margin . . . . cecidomyiid

> Host: *Psidium* sp. (araçá mirim). Distr.: Brazil (Bahia). Refs.: Tavares 1921, Houard 1933 (fig.).

1b   Outgrowths of leaf 2
2a   Spherical gall, up to 12 mm diameter, along edge of leaf . . . . . . cecidomyiid

> Baer (1907) reported that many country people believed horseflies came from these galls. Host: *Psidium* sp. Distr.: Brazil (Goias). Ref.: Houard 1933 (fig.).

2b   Horn-shaped outgrowths of leaf lamina 3
3a   Blister gall with elongate, filiform process . . . . cecidomyiid

The gall occurs on the lower leaf surface. Larvae live in the convexity below the 2–3 mm long process. Host: *Psidium* sp. Distr.: Brazil (Bahia). Refs.: Tavares 1921 (fig.), Houard 1933 (fig.).

3b   Elongate-ovoid gall.................................... cecidomyiid

The gall, 3–4 mm long, surrounds a single larval cell and is open apically. Host: *Psidium* sp. Distr.: Brazil (Amazonas). Refs.: Rübsaamen 1908, Houard 1933 (fig.).

Nyctaginaceae

Boerhavia

Swollen flower or fruit............................. *Asphondylia boerhaaviae*

Several cecidomyiid inquilines have been reared from these galls: *Camptoneuromyia exigua*, *Camptoneuromyia boerhaaviae*, and *Trotteria salvadorensis*, all from El Salvador and the last two from Colombia also. Hosts: *B. coccinea*, *B. diffusa*, *B. erecta*, and *Commicarpus scandens*. Distr.: Colombia (Magdalena), El Salvador, Jamaica. Refs.: Möhn 1959, Wünsch 1979 (fig.), Freeman and Geoghagen 1987, 1989.

Commicarpus

Swollen flower or fruit............................. *Asphondylia boerhaaviae*

Hosts: *C. scandens* and *Boerhavia* spp. Distr.: Jamaica. Ref.: Freeman and Geoghagen 1987, 1989.

Neea

Spheroid, usually hairy, complex galls on stem or leaf.......... *Bruggmannia* spp.

Eight species of *Bruggmannia* (*depressa*, *globulifex*, *lignicola*, *longicauda*, *longiseta*, *micrura*, *neeana*, and *ruebsaameni*) have been described from such galls on *Neea* species in Brazil (Amazonas, Rio de Janeiro, Santa Catarina). No illustrations have ever been made of the galls, so it is difficult to characterize them now without a comprehensive collection of new material. Refs.: Rübsaamen 1908, Houard 1933 (fig.).

Pisonia

1a   Hemispherical (321) or blister leaf galls ....
.... *Bruggmannia* spp. and *Pisphondylia salvadorensis*

Three species of *Bruggmannia* and one species of *Pisphondylia* have been reared from such galls on *Pisonia*. Their galls differ in details of color, hairiness, and size. In the absence of a comprehensive study they are difficult to characterize. The described species are:

Bruggmannia pisoniae. Host: *Pisonia* sp. Distr.: Cuba. Ref.: Cook 1909 (fig.).

Bruggmannia pisonifolia. Host: *P. nigricans.* Distr.: St. Vincent. Ref.: Felt 1912b.

Bruggmannia pustulans. An inquiline, *Meunieriella pisoniae,* has been reared from its galls (Möhn 1975). Host: *P. micranthocarpa.* Distr.: El Salvador. Ref.: Möhn 1960b.

Pisphondylia salvadorensis. Host: *P. micranthocarpa.* Distr.: El Salvador. Ref.: Möhn 1960b.

Another record of a blister leaf gall, possibly made by a *Bruggmannia,* is from *Pisonia* sp. in Brazil (Bahia). Refs.: Tavares 1922 (fig.), Houard 1933 (fig.).

1b   Galls on flowers, buds, or stems 2
2a   Subconical to irregular stem swellings. . . . .
. . . . *Bruggmannia pisonioides* and *Pisoniamyia* spp.

Three cecidomyiids have been described from stem galls that have not been illustrated or well characterized. These are:

Bruggmannia pisonioides. This gall is an irregular elevation of the stem. Host: *P. nigricans.* Distr.: St. Vincent. Ref.: Felt 1912a (as *pisoniae*).

Pisoniamyia mexicana. Galls are subconical. Host: *P aculeata.* Distr.: Mexico (Tamaulipas). Ref.: Felt 1911e.

Pisoniamyia similis. Galls are round and found in aggregate. Host: *P. micranthocarpa.* Distr.: El Salvador. Ref.: Möhn 1960b.

2b   Galls on flowers or buds. 3
3a   Platelike growth on swollen flower base . . . . *Pisoniamyia armeniae*

Host: *P. micranthocarpa.* Distr.: El Salvador. Ref.: Möhn 1960b.

3b   Conical terminal or lateral bud gall . . . . cecidomyiid

Galls as reported are found in aggregate and form a large mass. Host: *Pisonia* sp. Distr.: Brazil (Bahia). Refs.: Tavares 1922, Houard 1933 (fig.).

Tricycla

1a   Spherical, rough, often aggregated bud gall . . . . *Rhopalomyia tricyclae*

       Host: *T. spinosa*. Distr.: Argentina (Mendoza). Refs.: Kieffer and Jörgensen 1910, Jörgensen 1917 (fig.).

1b   Tapered stem swelling . . . . cecidomyiid

       Host: *T. spinosa*. Distr.: Argentina (Mendoza). Ref.: Jörgensen 1916 (fig.).

Olacaceae

Heisteria

All four kinds of galls keyed below were taken on *Heisteria cyanocarpa* in Brazil (Acre) and described and illustrated by Rübsaamen (1916a).

1a   Marginal leaf roll . . . . cecidomyiid
1b   Blister or spheroid leaf galls. 2
2a   Hairy, spherical leaf gall . . . . *Dactylodiplosis heisteriae*
2b   Glabrous spherical or blister leaf galls 3
3a   Spheroid, tapered gall, showing on one side of leaf . . . . cecidomyiid
3b   Blister leaf gall, apparent on both sides of leaf . . . . cecidomyiid

Ximenia

Elongate, usually single-celled stem swelling . . . . *Asphondylia ximeniae*

       Host: *X. americana*. Distr.: El Salvador. Ref.: Möhn 1959.

Oleaceae

Forestiera

Cylindrical, hairy leaf gall. . . . . cecidomyiid

       This undescribed gall midge belongs to the tribe Oligotrophini. Host: *F. segregata*. Distr.: USA (Florida).

Onagraceae

Jussiaea

1a   Swollen or deformed seed pod . . . . *Asphondylia vincenti*

An inquiline, *Camptoneuromyia siliqua*, has been reared from these galls. Hosts: *J. angustifolia*, *J. linifolia*, and *J. suffruticosa*. Distr.: El Salvador, Puerto Rico, St. Vincent. Refs.: Felt 1911b, Needham 1941 (fig.), Möhn 1960a.

1b   Tapered stem swelling. . . . . *Asphondylia rochae*

Galls are more or less regular, as much as 30 mm long and 5 mm wide, and have numerous larval cavities. Host: *Jussiaea* sp. Distr.: Brazil (Ceará). Refs.: Tavares 1918b, Houard 1933 (fig.).

Orchidaceae

Swollen aerial root (319) . . . . *Clinodiplosis cattleyae* and *Neolasioptera* sp.

Clinodiplosis cattleyae. Larvae of this species are only occasionally found in the galls, so they could be secondary feeders. Hosts: *Cattleya* spp., *Epidendrum* spp., *Laelia* spp., and orchids. Distr.: Mexico to Brazil, Jamaica. Ref.: Molliard 1903 (fig.).

Neolasioptera sp. These larvae are most often found in the galls. Hosts: *Cattleya* spp., *Epidendrum* spp., *Laelia* spp., *Oncidium* spp., and orchids. Distr.: Mexico south to Brazil, Peru, Jamaica.

Passifloraceae

Passiflora

Hemispherical leaf gall . . . . cecidomyiid

Host: *P. coccinea*. Distr.: Brazil (Amazonas). Ref.: Rübsaamen 1908.

Phytolaccaceae

Achatocarpus

1a   Leaf blister . . . . *Atolasoptera calida*

Host: *A. nigricans*. Distr.: El Salvador. Ref.: Möhn 1975.

1b   Swollen fruit . . . . *Asphondylia achatocarpi*

Host: *A. nigricans*. Distr.: El Salvador. Ref.: Möhn 1959.

Rivina

Fruit gall . . . . cecidomyiid

An inquiline, *Trotteria rivinae*, has been reared from the gall. Host: *R. humilis*. Distr.: Colombia (Magdalena). Ref.: Wünsch 1979.

Pinaceae

Pinus (pine)

Larva in pitch extruded from the plant . . . . *Cecidomyia reburrata*

Larvae live completely immersed in pitch produced in wounds that may be accompanied by swellings of host tissue. Host: *P.caribaea* and *P.taeda* in USA Distr.: Cuba; USA (Maryland to Florida). Refs.: Gagné 1978c (fig.), Gagné 1989 (fig.).

Piperaceae

Peperomia

Enlarged, fleshy bud. . . . . cecidomyiid

Galls are about 7 mm long, surrounded with several rows of bractlike leaves. A central cavity encloses one larva. Host: *P. controversa*. Distr.: Brazil (Amazonas). Ref.: Rübsaamen 1908.

Piper

1a   Rosette bud gall . . . . cecidomyiid

Host: *Piper* sp. Distr.: Brazil (Rio de Janeiro). Ref.: Rübsaamen 1908.

1b   Stem or leaf galls, not rosettes. 2
2a   Stem galls 3
2b   Leaf galls 6
3a   Horn-shaped, complex gall . . . . cecidomyiid

Galls are 20–35 mm long, 10 mm large at base, and tapering to apex, partly hairy, with basal, ovoid larval cell. Host: *Piper* sp. Distr.: Brazil (Rio de Janeiro). Ref.: Rübsaamen 1908.

3b   Simple stem swellings 4

4a   Knobby, irregular, spongy swelling . . . .
. . . . *Zalepidota piperis* and *Zalepidota tavaresi*

Host: *Piper* sp. Distr.: Brazil (Rio de Janeiro, São Paulo). Refs.: Tavares 1909 (fig.), Tavares 1925 (fig.).

4b   Smooth, tapered swellings 5
5a   Elongate, gradually tapered swelling . . . . . . *Heterasphondylia salvadorensis*

Galls are 2–7 cm long and twice as wide as normal stem diameter, with many, ovoid larval cavities perpendicular to plant axis. A similar gall was reported from Brazil (Rio de Janeiro) by Rübsaamen (1908). Host: *Piper* sp. Distr.: El Salvador. Ref.: Möhn 1960a.

5b   Spherical, abruptly tapered swelling . . . . *Resseliella* sp.

The gall interior is woody; larvae live in elongate tunnels. Host: *P. hispidum.* Distr.: Costa Rica.

6a   Circular blister . . . . cecidomyiid

Host: *Piper* sp. Distr.: Brazil (Rio de Janeiro). Ref.: Rübsaamen 1908.

6b   Spheroid galls inserted on petioles or main veins 7
7a   Smooth, globular gall . . . . cecidomyiid

Host: *Piper* sp. Distr.: Brazil (Rio de Janeiro). Ref.: Rübsaamen 1908.

7b   Velvety, ovoid gall . . . . cecidomyiid

Host: *Piper* sp. Distr.: Brazil (Rio de Janeiro). Ref.: Rübsaamen 1908.

Poaceae (Gramineae)

The imbricated stem galls of grasses (322) are the result of the foreshortening of the internodes so that the leaves are crowded together.

Aristida

Imbricated stem gall (322) . . . . *Chilophaga gyrantis*

Host: *A. gyrans.* Distr.: USA (Florida). Ref.: Gagné and Stegmaier 1971 (fig.).

Bambusa

Imbricated stem gall... cecidomyiid

Galls can be 15 cm long by 8 cm diam. Two separate galls were described but differ only in the color of the leaves. Host: *Bambusa* sp. Distr.: Brazil (Amazonas). Ref.: Rübsaamen 1907.

Pariana

Adults collected on flowers..

.... *Chauliodontomyia egregia* and *Chauliodontomyia parianae*

Host: *P. stenolemma*. Distr.: Brazil (Pará), Venezuela (Aragua). Ref.: Soderstrom and Calderón 1971.

Paspalum

Imbricated stem gall .... *Cleitodiplosis graminis*

Host: *P. conjugatum*. Distr.: Brazil (Bahia, Rio de Janeiro). Refs.: Tavares 1916 (fig.), 1921 (fig.); Houard 1933 (fig.).

Sorghum

Aborted seed .... *Stenodiplosis sorghicola*

Infestation by this species prevents seed development. It is one of the most serious pests of sorghum. Hosts: *Sorghum* spp. Distr.: Argentina, Brazil, Colombia, Curaçao, Dominican Republic, Puerto Rico, St. Croix, St. Vincent, Trinidad, U.S. Virgin Is., Venezuela; originally Africa, now worldwide, wherever sorghum is grown. Refs.: Harris 1964, 1976; Cermeño et al. 1985, Solinas 1986.

Polygalaceae

Monnina

1a   Unopened, aborted flower............................... cecidomyiid

Larvae are gregarious among flower parts. Host: *M. dictyocarpa*. Distr.: Argentina (Córdoba). Ref.: Tavares 1915a.

1b   Leaf roll .... cecidomyiid

Host: *M. dictyocarpa*. Distr.: Argentina (Córdoba). Ref.: Tavares 1915a.

Securidaca

Unopened, aborted flower. . . . . cecidomyiid

> The species responsible for the damage is an undescribed species of Cecido-myiidi. An inquiline, *Domolasioptera securidacae*, was reared from the galls. Host: *S. sylvestris*. Distr.: El Salvador. Ref.: Möhn 1975.

Polygonaceae

Coccoloba

In addition to the items listed below, an inquiline, *Marilasioptera tripartita*, was reared in El Salvador from an unspecified gall on *Coccoloba* sp. made by an un-described cecidomyiid (Möhn 1975), and another, *Meunieriella magdalenae*, was reared from an unspecified leaf gall on *Coccoloba candolleana* (Wünsch 1979).

1a   Enlarged, deformed flower . . . . cecidomyiid

> Galls are woody, ovoid, 5–7 mm long, reddish-brown, and crowned by a few flower parts. Thick walls surround a larval cavity. Host: *C. populifolia*. Distr.: Brazil (Rio de Janeiro). Ref.: Rübsaamen 1907.

1b   Leaf galls 2
2a   Swollen marginal leaf roll . . . . cecidomyiid

> Host: *Coccoloba* sp. Distr.: Brazil (Amazonas). Ref.: Rübsaamen 1907.

2b   Flat to hemispherical blister galls (330) . . . .
. . . . *Ctenodactylomyia* spp. and cecidomyiids

> *Ctenodactylomyia cocolobae*. Host: *C. uvifera*. Distr.: Cuba. Ref.: Cook 1909.

> *Ctenodactylomyia watsoni*. Host: *C. uvifera*. Distr.: USA (Florida). Refs.: Felt 1915c, Mead 1970 (fig.).

> Cecidomyiids. Host: *Coccoloba* sp. Distr.: Brazil (Amazonas), Honduras, and West Indies. Ref.: Rübsaamen 1905 (described several galls differing in minor details)

Symmeria

1a   Marginal leaf roll . . . . cecidomyiid

Host: *Symmeria* sp. Distr.: Brazil (Amazonas). Ref.: Rübsaamen 1908.

1b   Leaf blister. . . . . cecidomyiid

Host: *Symmeria* sp. Distr.: Brazil (Amazonas). Ref.: Rübsaamen 1908.

Triplaris

Leaf blisters . . . . *Geraldesia polygonarum* and cecidomyiid

Geraldesia polygonarum. An inquiline, *Meunieriella datxeli*, was reared from these galls. Host: *Triplaris/Symmeria/Ruprechtia* sp. Distr.: Colombia (Magdalena). Ref.: Wünsch 1979.

Cecidomyiid. Host: *T. schomburgkiana*. Distr.: Brazil (Amazonas). Ref.: Rübsaamen 1908.

Portulacaceae

Portulaca

1a   Closed, aborted flower. . . . . *Asphondylia portulacae*

Host: *P. oleracea*. Distr.: Colombia (Magdalena), El Salvador, Jamaica, Montserrat, Nevis, St. Kitts, Trinidad. Refs.: Möhn 1959, Wünsch 1979 (fig.). Note: Bennett and Cruttwell (unpub.) record similar damage also from Argentina and Bolivia.

1b   Succulent stem swelling (331) . . . . *Neolasioptera portulacae*

Host: *P.oleracea*. Distr.: Colombia (Magdalena), Cuba, El Salvador, Jamaica, Mexico, Montserrat, Nevis, St. Kitts, St. Vincent, USA (Florida). Refs.: Möhn 1975, Wünsch 1979 (fig.), Gagné 1989 (fig.).

Talinum

Deformed and swollen fruit. . . . . *Asphondylia talini*

Host: *T. paniculatum*. Distr.: El Salvador. Ref.: Möhn 1959.

Ranunculaceae

Clematis

1a   Swollen, enlarged bud.......................... *Asphondylia clematidis*

Hosts: C. *dioica*, C. *drummondii**, and C. *grossa*. Distr.: El Salvador; USA (Texas). Refs.: Felt 1935, Painter 1935 (fig.), Möhn 1960a.

1b   Tapered stem swelling .... *Neolasioptera clematicola*

Host: C. *dioica*. Distr.: El Salvador. Ref.: Möhn 1964a.

Rhamnaceae

Condalia

1a   Enlarged bud with tiny, scattered leaf rudiments .... cecidomyiid

Galls are woody, 5 mm long by 3–4 mm wide, and have two or three separate larval cavities. Host: C. *lineata*. Distr.: Argentina (Córdoba). Refs.: Tavares 1915a (fig.), Houard 1933 (fig.).

1b   Ovoid blister gall .... cecidomyiid

Host: C. *lineata*. Distr.: Argentina (Córdoba). Refs.: Tavares 1915a (fig.), Houard 1933 (fig.).

Rosaceae

Hirtella

Apical rosette bud gall.. cecidomyiid

Host: H. *americana*. Distr.: Brazil (Amazonas). Ref.: Rübsaamen 1907.

Prunus

Irregularly tapered stem swelling............................... cecidomyiid

Host: P. *sphaerocarpa*. Distr.: Brazil (Minas Gerais). Refs.: Tavares 1922, Houard 1933 (fig.).

Rubus

Leaf roll. . . . . cecidomyiid

Larvae are gregarious inside the roll. Host: *Rubus* sp. Distr.: Brazil (Minas Gerais). Refs.: Tavares 1918a, Houard 1933 (fig.).

Spiraea

Ovoid outgrowth of stem, simulating swollen bud. . . . . cecidomyiid

Host: *Spiraea* sp. Distr.: Brazil (Rio Grande do Sul). Ref.: Tavares 1909 (fig.).

Rubiaceae

In addition to the gall midges listed below under the various genera, three species of gall midges were reared from undetermined Rubiaceae. They are *Asphondylia bahiensis* from a spherical, fuzzy, flower bud gall on a plant with the common name corredeira in Brazil (Bahia) (Tavares 1917b [fig.], Houard 1933 [fig.]); *Asphondylia parva* from swollen deformed flowers on carqueja in Brazil (Rio de Janeiro) (Tavares 1917b); and *Gnesiodiplosis itaparicae* from an elongate bud gall, also on carqueja in Brazil (Bahia) (Tavares 1917b [fig.], Houard 1933 [fig.]).

Amaioua

1a   Greatly swollen ovary . . . . cecidomyiid

The infested ovary has several, elongate cavities and may be up to 26 mm long. Host: *A. brasiliana*. Distr.: Brazil (Minas Gerais). Refs.: Tavares 1925 (fig.), Houard 1933 (fig.).

1b   Swollen terminal bud or stem 2
2a   Swollen terminal bud . . . . cecidomyiid

Host: *A. brasiliana*. Distr.: Brazil (Minas Gerais). Ref.: Tavares 1925.

2b   Swollen stem . . . . cecidomyiid

Swellings may be twice as wide as the normal stem and contain numerous ellipsoidal larval cells. Host: *A. brasiliana*. Distr.: Brazil (Minas Gerais). Ref.: Tavares 1925.

Borreria

1a   Swollen, aborted flower . . . . *Asphondylia borreriae*

Host: *Borreria* sp. and *Diodea* sp. Distr.: Brazil (Rio de Janeiro). Ref.: Rübsaamen 1905.

1b   Tapered stem swelling . . . . *Neolasioptera borreriae*

Host: *B. verticillata*. Distr.: El Salvador. Ref.: Möhn 1964a. Note: Rübsaamen 1907 recorded a similar gall on *Borreria* sp. in Brazil.

Coffea (coffee)

Larva from fruit. . . . . *Clinodiplosis coffeae*

Host: *C. liberica*. Distr.: St. Vincent. Ref.: Felt 1911d.

Diodia

1a   Tapered stem swelling . . . . cecidomyiid

Host: *D. hyssopifolia*. Distr.: Brazil (Amazonas). Ref.: Rübsaamen 1907.

1b   Swollen, aborted flower. . . . . *Asphondylia borreriae*

Host: *Diodia* sp. and *Borreria* sp. Distr.: Brazil (Rio de Janeiro). Ref.: Rübsaamen 1907.

Gonzalagunia

Swollen, deformed fruit. . . . . cecidomyiid

Host: *G. hirsuta* var. *dicoccum*. Distr.: Brazil (Rio de Janeiro). Refs.: Tavares 1922 (fig.), Houard 1933 (fig.).

Palicourea

1a   Conical, woody leaf gall (332) . . . . cecidomyiid

The undescribed gall maker belongs to the Oligotrophini. Host: *Palicourea* sp. Distr.: Costa Rica.

1b   Tapered stem swelling . . . . cecidomyiid

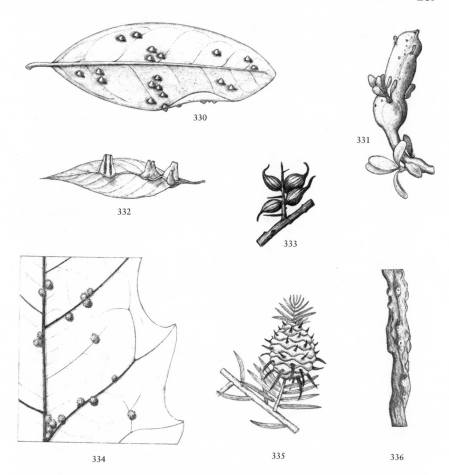

Figures 330–336. Galls of Polygonaceae, Portulacaceae, Rubiaceae, Sapindaceae, Solanaceae, and Taxodiaceae. **330,** Leaf galls, *Coccoloba uvifera* (Florida, USA): *Ctenodactylomyia watsoni.* **331,** Stem gall, *Portulaca oleraceae* (Florida, USA): *Neolasioptera portulacae.* **332,** Leaf galls, *Palicourea* sp. (Costa Rica): cecidomyiid. **333,** Bud galls, *Lycium chilensis* (redrawn from Jörgensen 1917): "*Oligotrophus*" *lyciicola.* **334,** Leaf gall, *Solanum* sp. (Minas Gerais, Brazil): cecidomyiid. **335,** Leaf gall, *Taxodium distichum* (Louisiana, USA): *Taxodiomyia cupressiananassa.* **336,** Stem gall, *Urvillea* sp. (Minas Gerais, Brazil): *Neolasioptera urvilleae.*

The undescribed gall maker belongs to the Lasiopterini. Host: *Palicourea* sp. Distr.: Costa Rica.

Psychotria

1a   Succulent, spheroid, leaf galls . . . .
. . . . *Apodiplosis praecox, Bruggmannia psychotriae,* and cecidomyiids

Apodiplosis praecox. Host: *Psychotria* sp. Distr.: Brazil (Minas Gerais). Refs.: Tavares 1922 (fig.), Houard 1933 (fig.).

Bruggmannia psychotriae. Host: *P. carthaginensis*. Distr.: El Salvador. Ref.: Möhn 1960b.

Cecidomyiids. Host: *Psychotria* sp. Distr.: Brazil (Rio de Janeiro, Rio Grande do Sul). Refs.: Rübsaamen 1908, Tavares 1909 (fig.).

1b   Inflorescence on stem galls. 2
2a   Tapered stem swelling . . . . cecidomyiid

Host: *Psychotria* sp. Distr.: Brazil (Rio de Janeiro). Ref.: Tavares 1922.

2b   Inflorescence galls 3
3a   Swollen flower pedicel . . . . *Bruggmannia psychotriae*

Host: *P. carthaginensis*. Distr.: El Salvador. Ref.: Möhn 1960b.

3b   Closed, swollen flower. . . . . *Asphondylia psychotriae*

Host: *P. pubescens*. Distr.: El Salvador. Ref.: Möhn 1959. Note: Atrophied flowers on *Psychotria* sp. in Brazil (Rio de Janeiro) reported by Tavares (1922) may possibly be made by this or another *Asphondylia*.

Randia

1a   Swollen, deformed fruit . . . . *Asphondylia randiae*

Host: *R. aculeata*. Distr.: El Salvador. Ref.: Möhn 1959.

1b   Spheroid leaf swelling . . . . *Bruggmannia randiae*

An inquiline, *Meunieriella randiae*, has been reared from these galls (Möhn 1975). Host: *R. spinosa*. Distr.: El Salvador. Ref.: Möhn 1960b.

Rondeletia

Swollen flower. . . . . *Asphondylia rondeletiae*

Host: *R. strigosa*. Distr.: El Salvador. Ref.: Möhn 1959.

Rubia

Swollen bud. . . . . *Clinodiplosis rubiae*

Host: *Rubia* sp. Distr.: Brazil (Rio de Janeiro). Ref.: Tavares 1918b.

Rutaceae

Citrus

1a   Larva in flower bud. . . . . *Prodiplosis longifila*

This species feeds on many plant families. Larvae are gregarious and kill the buds. Hosts: *C. aurantifolia*; also alfalfa, castorbean, green beans, potato, tomato, wild cotton, and wormseed. Distr.: Colombia, Ecuador, Peru (Lambayeque), USA (Florida). Refs.: Diaz B. 1981, Gagné 1986, Peña et al. 1989.

1b   Larva under bark. . . . . *Asynapta citrinae*

Host: *Citrus* sp. (grapefruit). Distr.: Puerto Rico. Ref.: Felt 1932.

Sapindaceae

Paullinia

1a   Spheroid gall growing from stem . . . . cecidomyiid

Galls are ovoid, 6 mm long, 4 mm wide, and 4 mm high, and inserted in the bark. Host: *Paullinia* sp. Distr.: Brazil (Amazonas). Ref.: Rübsaamen 1908.

1b   Rolled young leaf . . . . cecidomyiid

Host: *P. weinmanniaefolia*. Distr.: Brazil (Rio de Janeiro).

Sapindus

Rosette bud gall . . . . cecidomyiid

The gall is twice as long as a normal flower, its inner bracts enclose a larva. Host: *Sapindus* sp. Distr.: Brazil (Rio de Janeiro). Ref.: Rübsaamen 1899.

Serjania

1a   Peduncled, cylindrical leaf gall . . . . *Haplopalpus serjaneae*

Host: *Serjania* spp. Distr.: Brazil (Amazonas). Refs.: Rübsaamen 1908, 1916a (fig.); Houard 1933 (fig.).

1b   Stem or flower galls. 2

2a   Tapered stem swelling . . . . *Neolasioptera serjaneae*

       Host: *S. racemosa*. Distr.: El Salvador. Ref.: Möhn 1964a.

2b   Closed, swollen, deformed flowers. . . . .
       . . . . *Asphondylia serjaneae, Schizomyia serjaneae*, and cecidomyiid

       Asphondylia serjaneae. Host: *S. goniocarpa*. Distr.: El Salvador. Ref.: Möhn 1959.

       Schizomyia serjaneae. Host: *S. goniocarpa*. Distr.: El Salvador. Ref.: Möhn 1960a.

       Cecidomyiid. Host: *S. leptocarpa*. Distr.: Brazil (Amazonas). Ref.: Rübsaamen 1908.

Urvillea

Irregularly swollen stem (336) . . . . *Neolasioptera urvilleae*

       Host: *U. uniloba*. Distr.: Brazil (Minas Gerais, Rio Grande do Sul). Refs.: Tavares 1909 (fig.), Houard 1933 (fig.).

Sapotaceae

Chrysophyllum

Tapered stem swelling below bud . . . . cecidomyiid

       Host: *Chrysophyllum* sp. Distr.: Brazil (Bahia). Refs.: Tavares 1921 (fig.), Houard 1933 (fig.).

Manilkara

Spheroid leaf gall . . . . cecidomyiid

       The gall maker is a cecidomyiidi, but does not fit into any known genus. An inquiline, *Trotteria* sp., has been reared from these galls. Host: *M. subseriacea*. Distr.: Brazil (Rio de Janeiro).

Pouteria

Conical, hemispherical, and blister leaf galls . . . . cecidomyiids

Galls are variously shaped and may be caused by several cecidomyiids. An inquiline, *Trotteria* sp., has been reared from these galls. Hosts: *P. laurifolia* and *Pouteria* sp. Distr.: Brazil (Rio de Janeiro). Ref.: Rübsaamen 1908.

Saxifragaceae

Hydrangea

Spherical, smooth leaf gall............................. *Angeiomyia spinulosa*

The galls, more apparent on the lower than upper leaf surface, are 7–8 mm diameter, have hard walls, and may occur isolated or in groups. Host: *H. scandens*. Distr.: Chile. Ref.: Kieffer and Herbst 1909.

Scrophulariaceae

Capraria

Fruit gall... *Asphondylia caprariae*

Host: *C. biflora*. Distr.: Colombia (Magdalena). Ref.: Wünsch 1979.

Scoparia

Fruit gall... *Asphondylia scopariae*

Host: *S. dulcis*. Distr.: El Salvador. Ref.: Möhn 1959.

Smilacaceae

Smilax

1a   Swelling on side of stem................................. cecidomyiid

Swellings are irregular, spongy, and multichambered. Host: *Smilax* sp. Distr.: Brazil (Rio Grande do Sul). Ref.: Tavares 1909 (fig.).

1b   Leaf galls 2
2a   Fusiform or ellipsoidal leaf gall .... *Asphondylia sulphurea*

Galls are 6 mm long, glabrous, longitudinally striate, thick-skinned, inserted obliquely, and show equally on both sides of the leaf. *Compsodiplosis luteo-albida* is apparently an inquiline in the galls. Host: possibly *Smilax* sp. Distr.: Brazil (Rio Grande do Sul). Ref.: Tavares 1909 (fig.).

2b   Blister leaf gall . . . . *Smilasioptera candelariae*

Host: *S. mexicana*. Distr.: El Salvador. Ref.: Möhn 1975. Note: This gall midge may also be responsible for a blister gall on *Smilax* sp. from Brazil (Rio Grande do Sul) reported by Tavares (1909) (fig.).

Solanaceae

Acnistus

Tapered, rough-textured stem swelling . . . . cecidomyiid

Host: *Acnistus* sp. Distr.: Brazil (Rio Grande do Sul). Refs.: Tavares 1909 (fig.), Houard 1933 (fig.).

Capsicum

1a   Larvae in or on fruit, not forming a gall . . . . *Contarinia* sp.

Host: *Capsicum* sp. Distr.: Dominican Republic, Puerto Rico.

1b   Larvae in galls on other plant parts 2
2a   Thorn- or conelike gall on various plant parts . . . . cecidomyiid

Galls occur on branches, leaves, buds, and peduncles of fruit, are up to 4 mm long and 5 mm diameter at base. Larval chambers are numerous and parallel to the axis of the gall. Host: *C. microcarpum*. Distr.: Argentina (Córdoba). Refs.: Tavares 1915a, Houard 1933 (fig.).

2b   Stem swelling. . . . . *Neolasioptera capsici*

Host: *C. baccatum*. Distr.: El Salvador. Ref.: Möhn 1964a.

Cestrum

1a   Fruit or bud galls. 2
1b   Leaf or stem galls 4
2a   Leafy rosette gall. . . . . cecidomyiid

The leaves surround a single larval cell. Host: *Cestrum* sp. Distr.: Brazil (Rio de Janeiro). Ref.: Tavares 1925.

2b   Swollen bud or fruit 3
3a   Enlarged soft, fuzzy, spheroid bud swelling . . . . cecidomyiid

The gall contains several, irregular larval cavities. Host: *Cestrum* sp. Distr.: Brazil (Rio Grande do Sul). Refs.: Tavares 1909 (fig.), Houard 1933 (fig.).

3b   Slightly deformed fruit . . . . *Asphondylia cestri*

Host: *C. nocturnum.* Distr.: El Salvador. Ref.: Möhn 1959.

4a   Marginal leaf roll . . . . cecidomyiid

Host: *Cestrum* sp. Distr.: Brazil (Rio de Janeiro). Ref.: Tavares 1925.

4b   Tapered stem swellings . . . . *Neolasioptera cestri* and cecidomyiid

Neolasioptera cestri. Hosts: *C. aurantiacum* and *C. lanatum.* Distr.: El Salvador. Ref.: Möhn 1964a.

Cecidomyiid. Tavares (1918a) illustrated a larva from this gall that had six lateral papillae on each side of the spatula, so it could not be a *Neolasioptera.* Host: *Cestrum* sp. Distr.: Brazil (Rio de Janeiro). Ref.: Houard 1933 (fig.).

Grabowskia

Leaf blister . . . . *Cystodiplosis longipennis*

Host: *G. obtusa.* Distr.: Argentina (Mendoza). Refs.: Kieffer and Jörgensen 1910, Jörgensen 1917 (fig.).

Lycium

1a   Tapered stem swelling . . . . *Centrodiplosis crassipes*

Host: *L. chilense.* Distr.: Argentina (Mendoza). Refs.: Kieffer and Jörgensen 1910, Jörgensen 1917 (fig.), Houard 1933 (fig.).

1b   Bud galls. 2
2a   Spheroid, long-haired bud gall . . . . *Rhopalomyia bedeguaris*

Galls growing in aggregate may collectively reach 60 mm diameter. Host: *L. chilense.* Distr.: Argentina (Mendoza). Refs.: Kieffer and Jörgensen 1910, Jörgensen 1917 (fig.), Houard 1933 (fig.).

2b   Conical or amorphous, but smooth bud galls. 3
3a   Amorphous, spheroid bud swelling . . . .
 . . . . *Jorgensenia falcigera* and *Lyciomyia gracilis*

The two species were reared together from the same kind of gall. Host: *L. gracile.* Distr.: Argentina (Mendoza). Refs.: Kieffer and Jörgensen 1910, Jörgensen 1917 (fig.), Houard 1933 (fig.).

3b   Conical, tapered bud gall (333) . . . . "*Oligotrophus*" *lyciicola*

Hosts: *L. chilense* and *L. gracile.* Distr.: Argentina (Córdoba, Mendoza). Refs.: Kieffer and Jörgensen 1910, Tavares 1915a (fig.), Jörgensen 1917 (fig.), Houard 1933 (fig.).

Lycopersicon

1a   Larva from flowers . . . . *Contarinia lycopersici*

Host: *L. esculentum.* Distr.: Barbados, Belize, Dominica, Grenada, Guyana, St. Lucia, St. Vincent, Trinidad. Refs.: Felt 1911d, Barnes 1932, Callan 1940, 1941.

1b   Larva from flower buds . . . . *Prodiplosis longifila*

This species feeds on many plant families. Larvae are gregarious and kill the buds. Hosts: *L. esculentum* (tomato); also alfalfa, castorbean, citrus, green beans, potato, wild cotton, and wormseed. Distr.: Colombia, Ecuador, Peru (Lambayeque), USA (Florida). Refs.: Diaz B. 1981, Gagné 1986, Peña et al. 1989.

Physalis

Tapered stem swelling. . . . . *Neolasioptera argentata*

Hosts: *P. viscosa* and *Physalis* spp. Distr.: Argentina (Buenos Aires, Córdoba). Refs.: Brèthes 1917a, Jörgensen 1917 (fig.), Houard 1933 (fig.).

Solanum

1a   Galls on stems or leaves 2
1b   Galls on flowers, fruit, or buds 4
2a   Spherical, fuzzy gall on stem or leaf (334). . . . . cecidomyiid

Host: *Solanum* sp. (bobo). Distr.: Brazil (Bahia). Ref.: Tavares 1918a (fig.).

2b   Tapered stem or midvein swellings 3
3a   Tapered midvein swelling. . . . . cecidomyiid

Host: *S. argenteum*. Distr.: Brazil (Rio de Janeiro). Ref.: Rübsaamen 1908.

3b   Tapered stem swelling . . . . *Neolasioptera exigua*

Host: *S. umbellatum*. Distr.: El Salvador. Ref.: Möhn 1964a.

4a   Larva in fruit. . . . . *Lasioptera kallstroemia*

Hosts: *Solanum* spp. and *Kallstroemia* spp. Distr.: Guatemala, Mexico, Nicaragua; USA (Texas). Refs.: Felt 1935, Gagné 1989.

4b   Larva in bud or flower. . . . . *Prodiplosis longifila* and cecidomyiids

Prodiplosis longifila. This species feeds on many plant families. Larvae are gregarious and kill the buds. Hosts.: *S. tuberosum* (potato); also alfalfa, castorbean, citrus, green beans, tomato, wild cotton, and wormseed. Distr.: Colombia, Ecuador, Peru (Lambayeque), USA (Florida). Refs.: Diaz B. 1981, Gagné 1986, Peña et al. 1989.

Cecidomyiids. It is possible that the following records refer to damage by *Prodiplosis longifila*. (1) An aborted, malformed flower on *Solanum* sp. (rompe-gibão) from Brazil (Bahia) (Tavares 1918a). (2) Swollen, aborted buds on *Solanum* sp. (caissa or cassatinga) from Brazil (Bahia) (Tavares 1918a).

Staphyleaceae

Turpinia

Swollen node. . . . . cecidomyiid

Larvae live in individual, ovoid chambers. Several chambers may occur in one gall. Host: *T. occidentalis*. Distr.: Costa Rica.

Sterculiaceae

Ayenia

Enlarged, multichambered fruit . . . . *Asphondylia ayeniae*

Two inquilines, *Domolasioptera ayeniae* (Möhn 1975) and *Trotteria salvadorensis* (Möhn 1963b), have been reared from these galls. Host: *A. pusilla*. Distr.: El Salvador. Ref.: Möhn 1960a.

Guazuma

Leaf blister . . . . cecidomyiid

> Host: *Guazuma* sp. Distr.: Brazil (Amazonas). Ref.: Rübsaamen 1907.

Helicteres

Aborted, closed flower. . . . . *Asphondylia helicteris*

> Host: *H. mexicana*. Distr.: El Salvador. Ref.: Möhn 1959.

Melochia

Swollen, aborted flower . . . . *Asphondylia yukawai*

> Hosts: *M. caracasana*, *M. nodiflora*, *M. lupulina*, and *Arenaria* sp. (Caryo-
> phyllaceae). Distr.: Colombia (Magdalena). Ref.: Wünsch 1979.

Sterculia

Spheroid, fuzzy, leaf vein gall . . . . *Megaulus sterculiae*

> Rübsaamen (1908) attributed to gall midges several kinds of spheroid *Ster-
> culia* leaf galls, each differing in minor details from *Megaulus sterculiae*, but
> he gave no illustrations. Hosts: *Sterculia* spp. Distr.: Brazil (Amazonas).
> Refs.: Rübsaamen 1908, 1916a (fig.).

Theobroma (cacao)

In addition to the cecidomyiid from leaf galls of *Theobroma*, several gall midges are
associated with flowers of cacao. Among these cecidomyiids are *Mycodiplosis ligulata*
and *Coquillettomyia obliqua* in Costa Rica and Bolivia (Gagné 1984, Young 1985).

Woody, spheroid leaf gall . . . . cecidomyiid

> Host: *T. cacao*. Distr.: Brazil (Amazonas). Ref.: Rübsaamen 1908.

Waltheria

Multicellular, brownish-haired stem swelling. . . . . *Asphondylia waltheriae*

> An inquiline, *Camptoneuromyia waltheriae*, has been reared from these galls
> (Möhn 1975). Host: *W. americana*. Distr.: El Salvador. Ref.: Möhn 1959.

Styracaceae

Styrax

1a   Ovoid, smooth leaf gall . . . . *Styraxdiplosis caetitensis*

An inquiline, *Dialeria styracis*, has been reared from these galls (Tavares 1918b: 80). Host: *Styrax* sp. Distr.: Brazil (Bahia). Refs.: Tavares 1915b, Houard 1933 (fig.).

1b   Spheroid to lenticular, fuzzy leaf and stem galls . . . . *Contodiplosis* spp.

Several galls have been reared from the leaves of *Styrax* in Brazil (Rio de Janeiro), differing only in details. See Houard 1933 for a review of these. Three kinds of galls are formed by the following species:

Contodiplosis friburgensis from fuzzy, spheroid galls on leaves and stems. Refs.: Tavares 1915b (fig.), Houard 1933 (fig.).

Contodiplosis humilis from a lenticular leaf gall. Refs.: Tavares 1915b (fig.), Houard 1933 (fig.).

Contodiplosis tristis, possibly from a fuzzy spheroid gall. Refs.: Tavares 1915b (fig.), Houard 1933 (fig.).

Symplocaceae

Symplocos

Spheroid leaf galls . . . . cecidomyiid

Host: *Symplocos* sp. Distr.: Brazil (Rio de Janeiro). Refs.: Tavares 1922 (fig.), Houard 1933 (fig.).

Taxodiaceae

Taxodium

Pineapple-like twig growths (335). . . . . *Taxodiomyia cupressiananassa*

Host: *T. distichum*. Distr.: USA (Florida; southeastern states). Refs.: Osten Sacken 1878; Chen and Appleby 1984a,b; Gagné 1989.

Tiliaceae

Heliocarpus

Tapered swelling of leaf midvein . . . . *Neolasioptera heliocarpi*

>   Host: *H. donnellsmithii.* Distr.: El Salvador. Ref.: Möhn 1964a.

Luehea

1a   Tapered midvein swelling (339) . . . . cecidomyiid

>   Host: *L. speciosa.* Distr.: Brazil (Minas Gerais, Rio de Janeiro, São Paulo). Ref.: Tavares 1917b (fig.).

1b   Complex leaf galls. 2
2a   Lenticular gall, upper surface deeply concave. . . . . cecidomyiid

>   Host: *L. speciosa.* Distr.: Brazil (Minas Gerais, São Paulo). Ref.: Tavares 1917b (fig.).

2b   Spheroid galls. 3
3a   Rough-surfaced gall without long hairs. cecidomyiid

>   Host: *L. speciosa.* Distr.: Brazil (Rio de Janeiro). Ref.: Tavares 1917b.

3b   Long-haired gall (339) . . . . cecidomyiid

>   Host: *L. divaricata.* Distr.: Brazil (Minas Gerais). Ref.: Fernandes et al. 1988.

Turneraceae

Piriqueta

Stem swelling . . . . *Neolasioptera piriqueta*

>   Host: *P. ovata.* Distr.: Puerto Rico. Ref.: Felt 1917a.

Ulmaceae

Celtis

In addition to the galls keyed below, an unspecified leaf gall made by an undescribed cecidomyiid was reported from *Celtis iguanea* in El Salvador with reference to an inquiline, *Incolasioptera siccida* (Möhn 1975).

1a   Stem swellings.................... *Neolasioptera celtis* and cecidomyiid

        Neolasioptera celtis. Host: *C. iguanea*. Distr.: El Salvador. Ref.: Möhn 1964b.

        Cecidomyiid. Host: *Celtis* sp. Distr.: Brazil (Minas Gerais).

1b   Leaf galls 2
2a   Spheroid leaf gall .... cecidomyiid

        Host: *Celtis* sp. Distr.: Brazil (Rio de Janeiro, Santa Catarina). Refs.: Rübsaamen 1905, Tavares 1925.

2b   Leaf spots or swollen veins 3
3a   Tapered swelling of leaf vein............................. cecidomyiid

        The surface is covered with stiff hairs and the cavity contains a single larva. Host: *Celtis* sp. Distr.: Brazil (Santa Catarina). Ref.: Rübsaamen 1905.

3b   Leaf spot, barely visible................................. cecidomyiid

        Host: *Celtis* sp. Distr.: Argentina (Santa Fé). Ref.: Tavares 1915a.

Urticaceae

Urera

Marginal leaf roll... cecidomyiid

        Host: *U. subpeltata*. Distr.: Brazil (Bahia). Ref.: Tavares 1921.

Verbenaceae

Aegiphila

Tapered petiole swelling .... *Clinodiplosis iheringi*

        Host: *A. arborescens*. Distr.: Brazil (Santa Catarina). Ref.: Tavares 1925.

Avicennia

1a   Tight leaf roll... cecidomyiid

        Host: *A. nitida*. Distr.: Brazil (Bahia). Refs.: Tavares 1918a, Houard 1933 (fig.).

1b   Hemispherical leaf gall (340)................ *"Cecidomyia" avicenniae*

Galls are open and become tubular on the lower leaf surface. It is possible that more than one species is responsible for the gall. Hosts: *A. germinans*, *A. nitida*, and *A. tomentosa*. Distr.: Belize, Brazil, Cuba, USA (Florida). Refs.: Cook 1909, Tavares 1918a, Houard 1933 (fig.).

Citharexylum

Larva in flower .... *Asphondylia pattersoni*

Host: *C. quadrangulare*. Distr.: St. Vincent. Ref.: Felt 1911d.

Clerodendrum

Larva in flower or flower bud........................ *Asphondylia attenuatata*

Host: *C. aculeatum*. Distr.: West Indies. Refs.: Felt 1909, Callan 1940.

Duranta

Enlarged, deformed fruit....................................... cecidomyiid

Host: *D. plumieri*. Distr.: Brazil. Refs.: Tavares 1917a (fig.), 1918a.

Lantana

1a   Leaf galls 2
1b   Galls on stem, bud, or flower 3
2a   Spheroid, fuzzy gall........................ *Schismatodiplosis lantanae*

Galls are largest on the upper leaf surface. Adults exit through a circular hole on the lower surface. An inquiline, *Meunieriella lantanae*, has been reared from the galls (Tavares 1918b). Hosts.: *L. camara*, *L. hispida*, *L. urticifolia*, and *Lantana* spp. Distr.: Brazil (Rio de Janeiro, Santa Catarina), Mexico, Surinam. Refs.: Rübsaamen 1907, Tavares 1909 (fig.), Houard 1933 (fig.).

2b   Tubular gall .... cecidomyiid

Galls are velvety, and the opening at the apex of the tube is filled with dense hairs. Galls occur on branches as well as leaves. Host: *Lantana* sp. Distr.: Brazil (Bahia, Santa Catarina). Refs.: Rübsaamen 1907, Houard 1933 (fig.).

3a   Stem galls.. 4
3b   Bud or flower galls 5

4a Tubular gall perpendicular to stem . . . . cecidomyiid (same as 2b)

4b Tapered stem swellings. . . . .

  . . . . *Neolasioptera camarae* and *Neolasioptera lantanae*

Neolasioptera camarae. Host: *L. camara.* Distr.: possibly Brazil, Colombia (Magdalena), El Salvador. Refs.: Möhn 1964b, Wünsch 1979, Winder 1980.

Neolasioptera lantanae. Host: *Lantana* sp. Distr.: Brazil (Bahia). Refs.: Tavares 1922, C. Houard 1933 (fig.).

5a Swollen, elongated capitulum, the flower aborted, the surrounding bracts larger than normal . . . . *Clinodiplosis pulchra*

Host: *Lantana* sp. Distr.: Brazil (Bahia). Refs.: Tavares 1918a, Houard 1933 (fig.).

5b Swollen, unopened flower. . . . . *Asphondylia camarae*

Hosts: *L. camara* and *L. urticifolia.* Distr.: El Salvador, Mexico (Jalisco), and Panama. Ref.: Möhn 1959.

Lippia

1a Tubular leaf gall . . . . cecidomyiid

Host: *L. turbinata.* Distr.: Argentina (Córdoba). Refs.: Tavares 1915a (fig.), Houard 1933 (fig.).

1b Bud or flower galls 2

2a Closed, swollen flower bud . . . . *Asphondylia lippiae*

Host: *L. cardiostegia.* Distr.: El Salvador. Ref.: Möhn 1959.

2b Complex bud galls 3

3a Ovoid, thin-walled, fuzzy gall . . . . *Rhopalomyia lippiae*

Galls are 3 mm long and 2–3 mm wide. Host: *L. foliolosa.* Distr.: Argentina (Córdoba, Mendoza). Refs.: Kieffer and Jörgensen 1910, Tavares 1915a (fig.), Houard 1933 (fig.).

3b Elongate, woody, smooth, lipped gall (337) . . . . *Pseudomikiola lippiae*

Hosts: *L. turbinata* and *Lippia* sp. Distr.: Argentina (Córdoba, Entre Ríos). Refs.: Brèthes 1917b, Tavares 1915a (fig.), Houard 1933 (fig.).

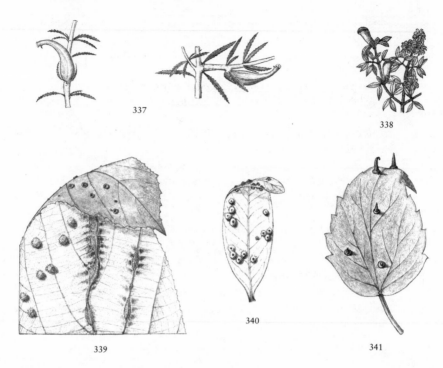

Figures 337–341. Galls of Tiliaceae, Verbenaceae, Vitaceae, and Zygophyllaceae. **337**, Bud galls, *Lippia* sp. (Córdoba, Argentina): *Pseudomikiola lippiae*. **338**, Leaf and bud galls, *Larrea divaricata* (Mendoza): cecidomyiids. **339**, Leaf galls, *Luehea* sp. (Minas Gerais, Brazil): midvein swellings from lower leaf surface, cecidomyiid; spheroid galls from upper and lower leaf surfaces: cecidomyiid. **340**, Leaf gall, *Avicennia germinans* (Florida, USA): "*Cecidomyia*" *avicenniae*. **341**, Leaf galls, *Vitis* sp. (Costa Rica): *Schizomyia* sp.

Stachytarpheta

Larvae from seeds. . . . *Asphondylia stachytarpheta* and *Schizomyia stachytarphetae*

Exterior signs of damage made by these two species were not specified. Host: *S. cayennensis*. Distr.: Trinidad. Ref.: Barnes 1932.

Verbena

Spheroid, fuzzy bud galls . . . . . *Rhopalomyia oreiplana* and *Rhopalomyia verbenae*

Rhopalomyia oreiplana. Host: *V. seriphioides*. Distr.: Argentina (Mendoza). Refs.: Kieffer and Jörgensen 1910, Jörgensen 1917.

Rhopalomyia verbenae. Host: *V. aspera*. Distr.: Argentina (Mendoza). Refs.: Kieffer and Jörgensen 1910, Jörgensen 1917 (fig.), Houard 1933 (fig.).

Vitaceae

Cissus

1a   Irregular, biconvex leaf blister............................ cecidomyiid

Host: *Cissus* sp. Distr.: Brazil (Rio de Janeiro). Refs.: Tavares 1918a, Houard 1933 (fig.).

1b   Stem, bud, or flower galls.. 2
2a   Irregular stem swelling........... *Astrodiplosis speciosa* and cecidomyiid

The larvae attributed to *Astrodiplosis speciosa* belong to an alycauline gall midge, presumably the true gall maker. Galls are woody, 2–7 cm long, and multicellular. Host: *Cissus* sp. Distr.: Guatemala. Refs.: Felt 1918 (fig.), Houard 1933 (fig.).

2b   Bud or flower galls 3
3a   Irregularly deformed lateral bud......................... cecidomyiid

Galls are up to 15 mm long, spheroid, and covered with bracts. Larvae live in separate cells, oriented perpendicular to the gall axis. Host: *Cissus* sp. Distr.: Brazil (Rio de Janeiro). Ref.: Rübsaamen 1905.

3b   Enlarged, swollen, closed flower .... *Asphondylia cissi*

Host: *C. sicyoides*. Distr.: El Salvador. Ref.: Möhn 1959.

Vitis (grape)

1a   Conical leaf gall (341)................................ *Schizomyia* sp.

Galls are similar to those made by *Schizomyia viticola* in North America. Host: *Vitis* sp. Distr.: Costa Rica.

1b   Deformed fruit .... *Asphondylia* sp.

Host: *Vitis* sp. Distr.: Mexico.

Zygophyllaceae

Kallstroemia

Swollen stem or inflorescence........................ *Lasioptera kallstroemia*'

Hosts: *K. parviflora*, *Kallstroemia* sp., and *Solanum* spp. Distr.: Mexico, Nicaragua; USA (Texas). Refs.: Felt 1935, Painter 1935.

Larrea

1a   Larva in seed. . . . . *Dasineura* sp.

Host: *L. divaricata*. Distr.: Argentina.

1b   Larvae in leaf or bud gall. 2

2a   Pair of appressed leaves, much longer than normal, biconvex, and flared at apex (338) . . . . cecidomyiids

Two kinds of larvae, both unidentified to genus, have been found in these galls. One kind was found also in the rosette gall in Figure 338. Host: *Larrea divaricata*. Distr.: Argentina (Mendoza, Neuquen). Ref.: Gagné and Boldt 1991.

2b   Rosette gall caused by extreme foreshortening of stem (338). . . . cecidomyiid

Larvae, unidentifiable to genus at present, have the hind pair of spiracles greatly enlarged and directed caudally. Host: *L. divaricata*. Distr.: Argentina (Mendoza, San Luis, Tucumán). Ref.: Gagné and Boldt 1991.

Tribulus

1a   Stem swelling. . . . . *Neolasioptera tribuli*

Host: *T. cistoides*. Distr.: Colombia (Magdalena). Ref.: Wünsch 1979 (fig.).

1b   Closed flower. . . . . cecidomyiid

Host: *T. cistoides*. Distr.: Colombia (Magdalena). Ref.: Wünsch 1979 (fig.).

Glossary

Anatomical terms defined here are mainly those peculiar to Cecidomyiidae or their galls. Terms that pertain to Diptera in general may be found in McAlpine et al. 1981 or in a general entomology textbook. Botanical terms not defined here can be found in a general botany textbook or a dictionary.

apomorphy derived state of an anatomical character.

bifilar condition of the antenna in which each flagellomere is girdled by two separate circumfila.

binodal condition of antenna with flagellomeres divided into two separate nodes by a narrow, cylindrical internode.

blister as in "leaf blister"; a convex, ovoid to circular gall.

circumfilum (pl.: *circumfila*) threadlike, looped, sensory processes of adult antennae, derived from forked nerve cells.

diapause a state of dormancy during the life cycle.

empodium (pl.: *empodia*) the large pad situated between two tarsal claws.

flagellum (pl.: *flagella*) the adult antenna exclusive of the basal two segments and formed of 7–63 *flagellomeres.*

fold an inpocketing of a leaf blade between veins (leaf fold) or along a vein (vein fold).

gall a predictable and consistent plant deformation that occurs in response to feeding or other stimulus by foreign organisms; *complex gall:* the result of tissue that is completely reorganized into a fundamentally different structure not normally found on the plant; *simple gall:* a gall that does not differ from normal plant tissue except that the tissue is swollen, rolled, or otherwise marked or distorted.

inquiline an organism that may regularly or occasionally be found in a particular gall but does not cause the gall.

instar the larva between molts; gall midges have three larval instars.

monophagous feeding on one species of host.

monophyletic containing species derived from a single ancestor.

307

multivoltine having three or more generations per year.

occipital process narrow vertical extension of head directly behind eyes.

oligophagous feeding on a few closely related species of hosts.

papilla (pl.: *papillae*) dermal sense organ of a larva.

paraphyletic not containing all descendants of a single ancestor.

pedogenetic (also elsewhere as *paedogenetic*) ability of immature stages to reproduce.

pheromone an intraspecific chemical signal.

plesiomorphy primitive state of an anatomical character.

polyphagous feeding on many species and genera of hosts.

polyphyletic containing species derived from unrelated ancestors.

polythalamous condition of a complex gall with two or more separate larval cells.

pouch gall a large vein or leaf fold.

puparium (pl.: *puparia*) of gall midges, the second-instar larval skin when it becomes the cocoon for the third-instar larva and pupa.

roll as in "leaf roll"; a leaf rolled into a cylinder; *marginal leaf roll:* only the leaf margin is rolled into a cylinder.

sensoria sensory receptors.

setiform condition of a larval papilla that bears a seta or hair.

spatula (pl.: *spatulae*) a sclerotized epidermal structure found on the venter of the prothorax of most third instars and occasionally on second instars.

spot as in "leaf spot"; a flat ovoid to circular area with one or more larvae in an open concavity or inside leaf tissue.

symbiont (adj.: *symbiotic*) an organism living in close association with another.

synapomorphy derived state of an anatomical character shared between two or more species.

trifilar condition of the antenna in which each flagellomere is girdled by three looped circumfila.

univoltine having one generation per year.

Literature Cited

Agassiz, L., and H. Loew. 1846. Nomina systematica generum dipterorum. [Pt 4] 42 pp. *In* Agassiz, L. Nomenclator Zoologicus. Fasc. 9/10: Titulum et praefationem operis, etc. Soloduri, Switzerland.

Ahlberg, O. 1939. *Diarthronomyia chrysanthemi* nom. nov. (= *hypogaea* Felt nec Löw). Entomologisk Tidskrift 60: 274–278.

Alexander, C. P. 1936. The maximum number of antennal segments in the order Diptera, with the description of a new genus of Cecidomyiidae. Bulletin of the Brooklyn Entomological Society 31: 12–14.

———. 1937. Change of name in Diptera. Bulletin of the Brooklyn Entomological Society 32: 60.

Anonymous. 1844. [Review/synopsis of] Memoria per servire alla Ditterologia italiana di C. Rondani. Parma I. 1840. 8. 16. t. 1. II. 1840. 28. t. 1. III. 1841. 29. Isis (Oken) 1844: 449–452.

Arduin, M., J. E. Kraus, and M. Venturelli. 1991. Estudo morphológico de galha achatada em folha de *Struthanthus vulgaris* Mart. (Loranthaceae). Revista Brasileira de Botanica 14: 147–156.

Baer, G.-A. 1907. Notes biologiques sur les mouches piqueuses de Goyaz [Dipt.]. Bulletin de la Société Entomologique de France 1907: 140–143.

Barnes, H. F. 1928. Cecidomyiinae (gall midges). Insects of Samoa 6 (2): 103–108.

———. 1930. A new thrips-eating gall midge, *Thripsobremia liothripis*, gen. et sp. n. (Cecidomyidae). Bulletin of Entomological Research 21: 331–332.

———. 1932. Notes on Cecidomyidae. Annals and Magazine of Natural History (10) 9: 475–484.

———. 1939. Gall midges (Cecidomyidae) associated with coffee. Revue de Zoologie et de Botanique Africaines 32: 324–336.

———. 1946. Gall Midges of Economic Importance. Vol. 1: Gall Midges of Root and Vegetable Crops. Crosby Lockwood & Son, London. 104 pp., 10 pls.

———. 1948. Gall Midges of Economic Importance. Vol. 4: Gall Midges of Ornamental Plants and Shrubs. Crosby Lockwood & Son, London. 165 pp., 10 pls.

———. 1949. Gall Midges of Economic Importance. Vol. 6: Gall Midges of Miscellaneous Crops. Crosby Lockwood & Son, London. 229 pp., 14 pls.

———. 1954. Gall-midge larvae as endoparasites, including the description of a species parasitising aphids in Trinidad, B. W. I. Bulletin of Entomological Research 45: 769–775.

——. 1956. Gall Midges of Economic Importance. Vol. 7: Gall Midges of Cereal Crops. Crosby Lockwood & Son, London. 261 pp., 16 pls.

Baylac, M. 1986. Observations sur la biologie et l'écologie de Lestodiplosis sp. (Dipt.: Cecidomyiidae), prédateur de la cochenille du hêtre Cryptococcus fagi (Hom.: Coccoidea). Annales de la Société Entomologique de France 22: 375–386.

——. 1987. Description d'un Lestodiplosis nouveau, prédateur d'Elaeidobius subvittatus (Col. Curculionidae), pollinisateur du palmier à huile Elaeis guineensis en Colombie. Bulletin de la Société Entomologique de France 92: 125–132.

Bennett, F. D., and R. E. Cruttwell. Unpublished. Investigations on the insects attacking Portulaca oleracea L. in the Neotropics. I. Insects recorded from P. oleracea and species encountered in preliminary surveys. Report, West Indian Station, Curepe, Trinidad. Commonwealth Institute of Biological Control 1–12.

Bishop, G. W. 1954. Life history and habits of a new seed midge, Dasyneura gentneri Pritchard. Journal of Economic Entomology 47: 141–147.

Blahutiak, A., and R. Alayo S. 1981. Diadiplosis cocci Felton (Diptera: Cecidomyiidae), un depredador importante de Saissetia hemisphaerica Targioni (Homoptera: Coccoidea) en Cuba. Poeyana 222: 1–11.

——. 1982. Función de Pyemotes boylei (Krczal) (Acarina: Pyemotidae) en la dinámica poblacional de Diadiplosis cocci Felt (Diptera: Cecidomyiidae) en Cuba. Academia de Ciencias de Cuba, Informe Científico-Técnico No. 194: 1–10.

Blanchard, E. 1852. Orden IX. Dipteros, pp. 327–468. In C. Gay, ed. Historia física y política de Chile. Zoologia 7, 471 pp. Paris.

Blanchard, E. E. 1938. Descripciones y anotationes de Dípteros Argentinos. Anales de la Sociedad Cíentifica Argentina 126: 345–386.

——. 1939. Descripción del cecidomyiido productor de la agalla del quebracho blanco. Revista Chilena de Historia Natural 42 (1938): 173–176.

——. 1958. Contarinia palposa sp. nov., parásita del sorgo granífero. Revista de Investigationes Agrícolas 12: 423–425.

Borgmeier, T. 1931. Eine neue zoophage Itonididengattung aus S. Paulo (Diptera, Itonididae). Revista de Entomologia 1: 184–191.

Borkent, A. 1989. A review of the wheat blossom midge, Sitodiplosis mosellana (Géhin) (Diptera: Cecidomyiidae) in Canada. Agriculture Canada Research Branch Technical Bulletin 1989-5E: 2–18, 6 pls.

Brèthes, J. 1914. Description de six Cécidomyiidae (Dipt.) de Buenos Aires. Anales del Museo Nacional de Historia Natural de Buenos Aires 26: 151–156.

——. 1917a. Sur une cécidie de Physalis viscosa: description de la cécidie et de la cécidomyie. Physis 3: 239–240.

——. 1917b. Description d'une cécidie et de sa cécidomyie d'une Lippia d'Entre Rios. Physis 3: 411–413.

——. 1918. Description de la galle et de sa cécidomyie d'Aeschynomene montevidensis. Physis 4: 312–313.

——. 1922. Himenópteros y Dípteros de varias procedencias. Anales de la Sociedad Científica Argentina 93: 119–146.

Bromley, S. W. 1944. Ephraim Porter Felt—1868–1943. Journal of the New York Entomological Society 52: 223–236.

Callan, E. M. 1940. The gall midges (Diptera, Cecidomyidae) of the West Indies. Revista de Entomologia 11: 730–758.

——. 1941. The gall midges (Diptera, Cecidomyidae) of economic importance in the West Indies. Tropical Agriculture 18(6): 117–127.

Cermeño L., G. F., R. Galvan V., and V. Lobatón G. 1985. Ciclo de vida y fluctuacion poblacional de la mosquita del ovario *Contarinia sorghicola* (Coquillett) (Diptera: Cecidomyiidae). Revista Colombiana de Entomologia 10 (1984): 15–19.

Chen, C., and J. E. Appleby. 1984a. Biology of the cypress twig gall midge, *Taxodiomyia cupressiananassa* (Diptera: Cecidomyiidae), in Central Illinois. Annals of the Entomological Society of America 77: 203–207.

——. 1984b. Chemical control of the cypress twig gall midge, *Taxodiomyia cupressiananassa* in Central Illinois. Journal of Environmental Horticulture 2: 38–40.

Chiang, H. C. 1968. Ecology of insect swarms, V. Movement of individual midges of *Anarete pritchardi* Kim within a swarm. Annals of the Entomological Society of America 61: 584–587.

Chiang, H. C., and A. Okubo. 1977. Quantitative analysis and mathematical model of horizontal orientation of midges, *Anarete pritchardi* Kim, in a swarm (Diptera: Cecidomyiidae). University of Minnesota Agricultural Experiment Station Technical Bulletin 310: 27–30.

Cockerell, T. D. A. 1892. A cecid bred from Coccidae. Entomologist 25: 180–182.

——. 1900. Notes and observations. Entomologist 33: 201.

——. 1907. A gall-gnat of the prickly-pear cactus. Canadian Entomologist 39: 324.

Collin, J. E. 1922. Description of a new genus and two new species of Cecidomyidae, and six new species of Acalyptrate Muscidae (Ephydridae and Milichidae). Transactions of the Entomological Society of London 1922: 504–517, pls. XIV–XVII.

Cook, M. T. 1906. Algunas agallas de Cuba producidas por insectos. Primer Informe Anual de la Estación Central Agronómica de Cuba 1: 247–252, pls. 47–49.

——. 1909. Some insect galls of Cuba. Second Report Estación Central Agronómica de Cuba 2: 143–146, pls. 37–44.

Coquillett, D. W. 1899. A cecidomyiid injurious to the seeds of sorghum. U.S. Department of Agriculture, Division of Entomology, Bulletin 18 (N.S.): 81–82.

——. 1905. A new cecidomyiid on cotton. Canadian Entomologist 37: 200.

——. 1910. The type-species of the North American genera of Diptera. Proceedings of the U.S. National Museum 37: 499–647.

Coutin, R. 1959. Quelques particularités du cycle évolutif des cécidomyies. La diapause prolongée des larves et l'apparition diférées des imagos. Annales de Epiphyties 4: 491–500.

——. 1980. *Pseudasphondylia rauwolfiae*, nov. sp. cécidomyie des fleurs de *Rauwolfia schumanniana* (Schl.) Boiteau, en Nouvelle-Calédonie. Annales de la Société Entomologique de France 16: 501–508.

Cruttwell, R. E. Unpublished. Insects and mites attacking *Eupatorium odoratum* in the Neotropics. 4. An annotated list of the insects and mites recorded from *Eupatorium odoratum* L., with a key to the types of damage found in Trinidad.

Dallas, E. D. 1928. Dr. Juan Brèthes. Bio-Bibliographia. Revista de la Sociedad Entomológica Argentina 2: 103–112.

DeClerck, R. A., and T. A. Steeves. 1988. Oviposition of the gall midge *Cystiphora sonchi* (Bremi) (Diptera: Cecidomyiidae) via the stomata of perennial sowthistle (*Sonchus arvensis* L.). Canadian Entomologist 120: 189–193.

Diatloff, G., and W. A. Palmer. 1987. The host specificity of *Neolasioptera lathami* Gagné (Diptera: Cecidomyiidae) with notes on its biology and phenology. Proceedings of the Entomological Society of Washington 89: 122–125.

Diaz B., W. 1981. Prodiplosis n. sp. (Diptera: Cecidomyiidae) plaga de la alfalfa y otros cultivos. Revista Peruana de Entomologia 24: 95–97.

Dreger-Jauffret, F., and J. D. Shorthouse. 1992. Diversity of gall-inducing insects and their galls, pp. 8–33. In J. D. Shorthouse and O. Rohfritsch, eds. Biology of Insect-Induced Galls. Oxford University Press, New York, Oxford.

Eberhard, W. G. 1991. *Chrosiothes tonala* (Araneae, Theridiidae): a web-building spider specializing on termites. Psyche 98: 7–19.

Edwards, F. W. 1929. p. 2 (footnote). In A. L. Tonnoir. Diptera of Patagonia and South Chile. Part 2. Fascicle 1—Psychodidae. British Museum, London. 32 pp., pls. I–IV.

——. 1930. Bibionidae, Scatopsidae, Cecidomyiidae, Culicidae, Thaumaleidae (Orphnephilidae), Anisopodidae (Rhyphidae). Diptera of Patagonia and South Chile, Part 2. Fascicle 3: 77–119, pls. IX–XI.

——. 1938. On the British Lestremiinae, with notes on exotic species. I. (Diptera, Cecidomyiidae). Proceedings of the Royal Entomological Society of London. Series B. Taxonomy. 7: 18–32.

Enderlein, G. 1940. Die Dipterenfauna der Juan-Fernandez-Inseln und der Oster-Insel, pp. 643–680. In C. Skottsberg, ed. The Natural History of Juan Fernandez and Easter Island, vol. 3. Zoology. Uppsala, Almquist & Wiksells.

Esben-Petersen, J. 1938. Peter (Pedro) Jørgensen. Entomologiske Meddelelser 20: 105–106.

Farquharson, C. O. 1922. A. The habits of two myrmecophilous Cecidomyiidae, pp. 437–443. In E. B. Poulton, ed. Transactions of the Entomological Society of London 1921: 319–531, pls. IX–XIX.

Feil, J. P., and S. S. Renner. 1991. The pollination of *Siparuna* (Monimiaceae) by gall-midges (Cecidomyiidae): another likely ancient association. American Journal of Botany 78: 186.

Felt, E. P. 1907a. New Species of Cecidomyiidae. Albany, N.Y. 53 pp.

——. 1907b. New Species of Cecidomyiidae. New York State Museum Bulletin 110: 97–165.

——. 1907c. New species of Cecidomyiidae II. Albany, N.Y. 26 pp.

——. 1908a. Contarinia gossypii n. sp. Entomological News 19: 210–211.

——. 1908b. Appendix D. New York State Museum Bulletin 124: 286–422, pls. 33–44.

——. 1909. New species of West Indian Cecidomyiidae. Entomological News 20: 299–302.

——. 1910a. Schizomyia ipomoeae n. sp. Entomological News 21: 160–161.

——. 1910b. West Indian Cecidomyiidae. Entomological News 21: 268–270.

——. 1910c. Gall midges of *Aster, Carya, Quercus,* and *Salix.* Journal of Economic Entomology 3: 347–356.

——. 1911a. A new Lestodiplosis. Entomological News. 22: 10–11.

——. 1911b. Two new gall midges (Dipt.). Entomological News 22: 109–111.

——. 1911c. Endaphis Kieff. in the Americas (Dipt.). Entomological News 22: 128–129.

——. 1911d. Four new gall midges (Dipt.). Entomological News 22: 301–305.

——. 1911e. Hosts and galls of American gall midges. Journal of Economic Entomology 4: 451–475.

——. 1911f. Two new gall midges. Canadian Entomologist 43: 194–196.

——. 1911g. New species of gall midges. Journal of Economic Entomology 4: 546–560.

——. 1911h. A generic synopsis of the Itonidae. Journal of the New York Entomological Society 19: 31–62.

——. 1911i. Three new gall midges (Dipt.). Journal of the New York Entomological Society 19: 190–193.

——. 1911j. A new species of *Lasioptera* with observations on certain homologies. Psyche 18: 84–86.

——. 1911k. 26th Report of the State Entomologist on injurious and other insects of the State of New York. 1910. New York State Museum Bulletin 147: 5–180, pls. 1–35.

——. 1912a. New West Indian gall midges (Dipt.). Entomological News 23: 173–177.

——. 1912b. Observations on Uleela Rubs. (Dipt.). Entomological News 23: 353–354.

——. 1912c. New Itonididae (Dipt.). Journal of the New York Entomological Society 20: 102–107.

——. 1912d. New gall midges or Itonidae (Dipt.). Journal of the New York Entomological Society 20: 146–156.

——. 1912e. Lasiopteryx manihot, n. sp. (Diptera). Canadian Entomologist 43: 144.

——. 1913a. Three new gall midges (Diptera). Canadian Entomologist 45: 304–308.

——. 1913b. Descriptions of gall midges (Diptera). Journal of the New York Entomological Society 21: 213–219.

——. 1913c. *Cystodiplosis eugeniae* n. sp. (Dipt.). Entomological News 24: 175–176.

——. 1914a. New gall midges (Itonididae). Insecutor Inscitiae Menstruus 2: 117–123.

——. 1914b. *Arthrocnodax constricta* n. sp. Journal of Economic Entomology 7: 481.

——. 1915a. New South American gall midges. Psyche 22: 152–157.

——. 1915b. New North American gall midges. Canadian Entomologist 47: 226–232.

——. 1915c. New genera and species of gall midges. Proceedings of the United States Museum 48: 195–211.

——. 1916. Appendix. A study of gall midges IV. New York State Museum Bulletin 186: 101–172, pls. 14–18.

——. 1917a. New gall midges. Journal of the New York Entomological Society 25: 193–196.

——. 1917b. *Asphondylia websteri* n. sp. Journal of Economic Entomology 10: 562.

——. 1918. Key to American insect galls. New York State Museum Bulletin 200: 5–310, pls. 5–16.

——. 1921a. Three new sub-tropical gall midges (Itonididae, Dipt.). Entomological News 32: 141–143.

——. 1921b. A new *Diadiplosis*. Zoologica (Scientific Contributions of the New York Zoological Society) 3: 225–226.

——. 1921c. Appendix. A study of gall midges, VII. New York State Museum Bulletin 231–232: 81–288, pls. 8–20.

——. 1922a. A new cecidomyiid parasite of the white fly. Proceedings of the United States National Museum 61 (23): 1–2.

——. 1922b. A new and remarkable fig midge. Florida Entomologist 6: 5–6.

——. 1925. Key to gall midges (a resumé of Studies I–VII, Itonididae). New York State Museum Bulletin 257: 3–239, pls. 1–8.

——. 1932. A new citrus cambium miner from Puerto Rico. Journal of the Department of Agriculture of Puerto Rico 16: 117–118.

——. 1933. A new enemy of the pineapple mealybug and a list of gall midge enemies of mealybugs. Journal of the New York Entomological Society 41: 87–89.

——. 1934. A new gall midge on fig (Diptera: Itonididae). Entomological News 45: 131–133.

——. 1935. New species of gall midges from Texas. Journal of the Kansas Entomological Society 8: 1–8.

——. 1938. A new species of gall midge predacious on mealybugs. Proceedings of the Hawaiian Entomological Society 10: 43.

——. 1940. Plant Galls and Gall Makers. Comstock Publishing Co., Ithaca, N.Y. viii and 364 pp.

Fernandes, G. W. 1990. Hypersensitivity: a neglected plant resistance mechanism against insect herbivores. Environmental Entomology 19: 1173–1182.

Fernandes, G. W., R. P. Martins, and E. T. Neto. 1987. Food web relationships involving *Anadiplosis* sp. galls (Diptera: Cecidomyiidae) on *Machaerium aculeatum* (Leguminosae). Revista Brasileira de Botanica 10: 117–123.

Fernandes, G. W., E. T. Neto, and R. P. Martins. 1988. Ocorrência e caracterização de galhas entomogenas na vegetação do campus Pampulha da Universidade Federal de Minas Gerais. Revista Brasileira de Zoologia 5: 11–29.

Fleur, E. 1929. Monsieur l'Abbé Kieffer professeur au Collège de Bitche (Moselle). Bulletin de la Société d'Histoire Naturelle de la Moselle (3) 8: 7–29, 2 pls. (photographs).

Frauenfeld, G. 1858. Mein Aufenthalt in Rio Janeiro. Verhandlungen der Zoologisch-Botanischen Gesellschaft in Wien 8: 253–262.

Freeman, B. E., and A. Geoghagen. 1987. Size and fecundity in the Jamaican gall-midge *Asphondylia boerhaaviae*. Ecological Entomology 12: 239–249.

——. 1989. A population study in Jamaica on the gall-midge *Asphondylia boerhaaviae*: a contribution to spatial dynamics. Journal of Animal Ecology 58: 367–382.

Freeman, P. 1953. Los insectos de las Islas Juan Fernandez. 13. Mycetophilidae, Sciaridae, Cecidomyiidae and Scatopsidae (Diptera). Revista Chilena de Entomologia 3: 23–40.

Fyles, T. W. 1883. Description of a dipterous parasite of *Phylloxera vastatrix*. Canadian Entomologist 14: 237–239.

Gagné, R. J. 1968a. [Fascicle] 23. Family Cecidomyiidae, pp. 1–62. *In* N. Papavero, ed. A Catalogue of the Diptera of the Americas South of the United States. Departamento de Zoologia, Secretaria da Agricultura, São Paulo, Brazil.

——. 1968b. *Chrybaneura harrisoni*, a new genus and species from Central America. Proceedings of the Entomological Society of Washington 70: 33–34.

——. 1968c. A taxonomic revision of the genus *Asteromyia* (Diptera: Cecidomyiidae). Miscellaneous Publications of the Entomological Society of America 6: 1–40.

———. 1969a. Revision of the gall midges of bald cypress. Entomological News 79: 269–274.

———. 1969b. A tribal and generic revision of the Nearctic Lasiopteridi (Diptera: Cecidomyiidae). Annals of the Entomological Society of America 62: 1348–1364.

———. 1969c. A new genus and two new species of Cecidomyiidae associated with *Pariana* species (Gramineae) in South America (Diptera). Proceedings of the Entomological Society of Washington 71: 108–111.

———. 1971. Two new species of North American *Neolasioptera* from *Baccharis* (Diptera: Cecidomyiidae—Compositae). Proceedings of the Entomological Society of Washington 73: 153–157.

———. 1973a. Cecidomyiidae from Mexican Tertiary amber. Proceedings of the Entomological Society of Washington 75: 169–171.

———. 1973b. A generic synopsis of the Nearctic Cecidomyiidi (Diptera: Cecidomyiidae). Annals of the Entomological Society of America 66: 857–889.

———. 1974. A new species of Cecidomyiidae (Diptera) from *Nothofagus* in Chile. Studia Entomologica 16 (1973): 447–448.

———. 1975a. A revision of the Nearctic Stomatosematidi (Diptera: Cecidomyiidae: Cecidomyiinae). Annals of the Entomological Society of America 68: 86–90.

———. 1975b. A redefinition of *Diarthronomyia* Felt as a subgenus of *Rhopalomyia* Rübsaamen (Diptera: Cecidomyiidae: Oligotrophidi). Annals of the Entomological Society of America 68: 482–484.

———. 1975c. The gall midges of ragweed, *Ambrosia*, with descriptions of two new species (Diptera: Cecidomyiidae). Proceedings of the Entomological Society of Washington 77: 50–55.

———. 1976. New Nearctic records and taxonomic changes in the Cecidomyiidae (Diptera). Annals of the Entomological Society of America 69: 26–28.

———. 1977a. Cecidomyiidae (Diptera) from Canadian amber. Proceedings of the Entomological Society of Washington 79: 57–62.

———. 1977b. The Cecidomyiidae (Diptera) associated with *Chromolaena odorata* (L.) K. and R. (Compositae) in the Neotropical Region. Brenesia 12/13: 113–131.

———. 1978a. A new species of *Asphondylia* (Diptera: Cecidomyiidae) from Costa Rica with taxonomic notes on related species. Proceedings of the Entomological Society of Washington 80: 514–516.

———. 1978b. New synonymy and a review of *Haplusia* (Diptera: Cecidomyiidae). Proceedings of the Entomological Society of Washington 80: 517–519.

———. 1978c. A systematic analysis of the pine pitch midges, *Cecidomyia* spp. (Diptera: Cecidomyiidae). U.S. Department of Agriculture Technical Bulletin 1575: 1–18.

———. 1981. Cecidomyiidae, pp. 257–292. *In* J. F. McAlpine, B. V. Peterson, G. E. Shewell, H. J. Teskey, J. R. Vockeroth, and D. M. Wood, eds. Manual of Nearctic Diptera, vol. 1. Monograph 27. Research Branch, Agriculture Canada, Hull, Quebec. vi and 674 pp.

———. 1984. Five new species of Neotropical Cecidomyiidae (Diptera) associated with cacao flowers, killing the buds of Clusiaceae, or preying on mites. Brenesia 22: 123–138.

———. 1985. Descriptions of new Nearctic Cecidomyiidae (Diptera) that live in xylem vessels of fresh-cut wood, and a review of *Ledomyia* (s. str.). Proceedings of the Entomological Society of Washington 87: 116–134.

———. 1986. Revision of *Prodiplosis* (Diptera: Cecidomyiidae) with descriptions of three new species. Annals of the Entomological Society of America 79: 235–245.

———. 1989. The Plant-Feeding Gall Midges of North America. Cornell University Press, Ithaca, N.Y. xi and 356 pp., 4 pls.

Gagné, R. J., and P. E. Boldt. 1989. A new species of *Neolasioptera* (Diptera: Cecidomyiidae) from Baccharis (Asteraceae) in southern United States and the Dominican Republic. Proceedings of the Entomological Society of Washington 91: 169–174.

———. 1991. *Asphondylia* (Diptera: Cecidomyiidae) does not reflect the disjunct distribution of *Larrea* (Zygophyllaceae). Proceedings of the Entomological Society of Washington 93: 509–510.

Gagné, R. J., and G. W. Byers. 1985. A remarkable new Neotropical species of *Contarinia* (Diptera: Cecidomyiidae). Journal of the Kansas Entomological Society 57: 736–738.

Gagné, R. J., and J. H. Hatchett. 1989. Instars of the Hessian fly (Diptera: Cecidomyiidae). Annals of the Entomological Society of America 82: 73–79.

Gagné, R. J., and M. L. A. Rios de Saluso. 1987. Una nueva especie de *Mycodiplosis* (Diptera: Cecidomyiidae) asociada a la roya amarilla de *Avena sativa* (Poaceae) en Argentina. INTA Secretaria de Estado de Agricultura, Ganaderia y Pesca de la Nación. Estación Experimental Agropecuaria Paraná Entre Rios. Serie Técnica 54: 1–12.

Gagné, R. J., and C. E. Stegmaier, Jr. 1971. A new species of *Chilophaga* on *Aristida* (Gramineae) in Florida (Diptera: Cecidomyiidae). Florida Entomologist 54: 335–338.

Gagné, R. J., and K. Valley. 1984. Two new species of Cecidomyiidae (Diptera) from honeylocust, *Gleditsea triacanthos* L. (Fabaceae), in eastern United States. Proceedings of the Entomological Society of Washington 86: 543–549.

Gagné, R. J., and W. M. Woods. 1988. Native American plant hosts of *Asphondylia websteri* (Diptera: Cecidomyiidae). Annals of the Entomological Society of America 81: 447–448.

Gagné, R. J., and A. L. Wuensche. 1986. Identity of the *Asphondylia* (Diptera: Cecidomyiidae) on guar, *Cyamopsis tetragonoloba* (Fabaceae), in the southwestern United States. Annals of the Entomological Society of America 79: 246–250.

Haliday, A. H. 1833. Catalogue of Diptera occurring about Holywood on Downshire. Entomological Magazine (London) 1: 147–180.

Hallberg, E., and I. Åhman. 1987. Sensillar types of the ovipositor of *Dasineura brassicae*: structure and relation to oviposition behaviour. Physiological Entomology 12: 51–58.

Haridass, E.T. 1987. Midge-fungus interactions in a cucurbit stem gall. Phytophaga. 1: 57–74.

Harris, K. M. 1964. The sorghum midge complex (Diptera, Cecidomyiidae). Bulletin of Entomological Research 55: 233–247.

———. 1966. Gall midge genera of economic importance (Diptera: Cecidomyiidae). Part I: Introduction and subfamily Cecidomyiinae; supertribe Cecidomyiidi. Transactions of the Royal Entomological Society of London 118: 313–358.

———. 1968. A systematic revision and biological review of the cecidomyiid predators (Diptera: Cecidomyiidae) on world Coccoidea (Hemiptera-Homoptera). Transactions of the Royal Entomological Society of London 119: 401–494.

——. 1973. Aphidophagous Cecidomyiidae (Diptera): taxonomy, biology and assessments of field populations. Bulletin of Entomological Research 63: 305–325.

——. 1976. The sorghum midge. Annals of Applied Biology 84: 114–118.

——. 1981a. *Bremia legrandi*, sp.n. [Diptera, Cecidomyiidae], a predator on eggs of a dragonfly, *Malgassophlebia aequatoris* Legrand [Odonata, Libellulidae] in Gabon. Revue Française d'Entomologie (N.S.) 3: 27–30.

——. 1981b. *Dicrodiplosis manihoti*, sp. n. (Diptera: Cecidomyiidae), a predator on cassava mealybug, *Phenacoccus manihoti* Matile-Ferraro (Homoptera: Coccoidea: Pseudococcidae) in Africa. Annales de la Société Entomologique de France 17: 337–344.

Hatchett, J. H., G. L. Kreitner, and R. L. Elzinga. 1990. Larval mouthparts and feeding mechanism of the Hessian fly (Diptera: Cecidomyiidae). Annals of the Entomological Society of America. 83: 1137–1147.

Henriksen, K. L. 1936. Pp. 382–385 in Oversigt over Dansk Entomologis Historie. Entomologiske Meddelelser 15: 289–578.

Holz, B. 1970. Revision in Mitteleuropa vorkommender mycophager Gallmücken der *Mycodiplosis*-Gruppe (Diptera, Cecidomyiidae) unter Berücksichtigung ihrer Wirtsspezifität. Paul Jllg Photo-Offsetdruck, Stuttgart. 238 pp.

Houard, C. 1908. Les Zoocécidies des Plantes d'Europe et du Bassin de la Méditerranée, vol. 1. Librairie Scientifique A. Hermann et Fils, Paris. 569 pp.

——. 1909. Les Zoocécidies des Plantes d'Europe et du Bassin de la Méditerranée, vol. 2. Librairie Scientifique A. Hermann et Fils, Paris. 775 pp. [573–1247].

——. 1922. Les Zoocécidies des Plantes d'Afrique, d'Asie et d'Océanie, vol. 1. Librairie Scientifique Jules Hermann, Paris. 496 pp.

——. 1923. Les Zoocécidies des Plantes d'Afrique, d'Asie et d'Océanie, vol. 2. Librairie Scientifique Jules Hermann, Paris. 554 pp. [503–1056].

——. 1924. Les collections cécidologiques du laboratoire d'entomologie du Muséum d'Histoire Naturelle de Paris: galles de la Guyane Française. Marcellia 21: 97–128.

——. 1932. Joaquim da Silva Tavares, cécidologue portugais (1866–1931). Marcellia 27: 107–119.

——. 1933. Les Zoocécidies des Plantes de l'Amérique du Sud et de l'Amérique Central. Hermann et Cie, Paris. 519 pp.

——. 1940. Les Zoocécidies des Plantes de l'Amérique du Nord: Galles des Chênes. Hermann et Cie, Paris. 549 pp.

Houard, J. 1948. C. A. V. Houard: sa vie, son oeuvre, pp. 254–268. *In* A. Trotter, Prof. C. L. Houard. Marcellia 30: 252–268.

International Commission on Zoological Nomenclature. 1958. Opinion 526, pp. 291–300. *In* F. Hemming, ed. Opinions and declarations rendered by the International Commission on Zoological Nomenclature 19: 436 pp. London.

——. 1963. Opinion 678. The suppression under the Plenary Powers of the pamphlet published by Meigen, 1800. Bulletin of Zoological Nomenclature 20: 339–342.

——. 1970. Opinion 929. *Lasioptera* Meigen, 1818 (Insecta, Diptera). Preservation under the Plenary Powers in its accustomed meaning. Bulletin of Zoological Nomenclature 27: 95–96.

Isidoro, N. 1987. *Allocontarinia* (*Contarinia*) *sorghicola* (Coq.) (Diptera, Cecidomyiidae): morfologia, biologia, danni, controllo. Entomologica 22: 35–73.

Isidoro, N., and A. Lucchi. 1989. Eggshell fine morphology of *Allocontarinia sorghicola* (Coq.) (Diptera: Cecidomyiidae). Entomologica 24: 127–138.

Jones, R. G., R. J. Gagné, and W. F. Barr. 1983. A systematic and biological study of the gall midges (Cecidomyiidae) of *Artemisia tridentata* Nuttall (Compositae) in Idaho. Contributions of the American Entomological Institute 81: ii, 1–79.

Jörgensen, P. 1916. Zoocecidios argentinos. Physis (Revista de la Sociedad Argentina de Ciencias Naturales) 2: 349–365.

——. 1917. Zoocecidios argentinos. Physis (Revista de la Sociedad Argentina de Ciencias Naturales) 3: 1–29.

Karsch, F. A. F. 1877. Revision der Gallmücken. E. C. Brunn, Münster i. W. 58 pp., 1 pl.

——. 1880. Neue Zoocecidien und Cecidozöen. Zeitschrift für Naturwissenschaften 53: 288–309.

Kieffer, J.-J. 1889. Neue Beiträge zur Kenntniss der Gallmücken. Entomologische Nachrichten 15: 149–156, 171–176, 183–194, 202–212.

——. 1894a. [Untitled]. Bulletin Bimensuel de la Société Entomologique de France 63: xxviii–xxix.

——. 1894b. Description de quelques larves de Cécidomyes. Feuille des Jeunes Naturalistes 24: 83–88, 119–121, 147–152, 185–189.

——. 1894c. Sur le groupe *Epidosis* de la famille des Cecidomyidae. Annales de la Société Entomologique de France 63: 311–350, 2 pls.

——. 1894d. Ueber die Heteropezinae. Wiener Entomologische Zeitung 13: 200–211, pl. I.

——. 1894e. [Untitled]. Bulletin Bimensuel de la Société Entomologique de France 63 (1894): clxxv–clxxvi.

——. 1895a. [Untitled]. Bulletin Bimensuel de la Société Entomologique de France 63 (1894): cclxxx.

——. 1895b. Ueber moosbewohnende Gallmückenlarven. Entomologische Nachrichten 21: 113–123.

——. 1895c. Nouvelles observations sur le groupe des *Diplosis* et description de cinq genres nouveaux (Dipt.). Bulletin de la Société Entomologique de France 64: cxcii–cxciv.

——. 1895d. Essai sur le groupe *Campylomyza*. Miscellanea Entomologica 3: 46–47, 57–65, 73–79, 91–97, 109–113, 129–133, pls. I–II.

——. 1895e. Changement de nom. Bulletin de la Société Entomologique de France 64: cccxx.

——. 1896. Neue Mittheilungen über Gallmücken. Wiener Entomologische Zeitung 15: 85–105.

——. 1897. Meine Antwort an den Herrn Zeichenlehrer Rübsaamen und an den Herrn Docenten Dr. H. Karsch nebst Beschreibung neuer Gallmücken. Trier. 21 pp.

——. 1898. Synopse des Cécidomyies d'Europe et d'Algérie décrites jusqu'à ce jour. Bulletin de la Société d'Histoire Naturelle de Metz (2) 8: 1–64.

——. 1900. Monographie des cécidomyides d'Europe et d'Algérie. Annales de la Société Entomologique de France 69: 181–472, pls. 15–44.

——. 1901. Synopsis des zoocécidies d'Europe. Annales de la Société Entomologique de France 70: 233–578.

——. 1902. Notice critique sur le catalogue des zoocécidies de MM Darboux, Houard et Giard. Bulletin de la Société d'Histoire Naturelle de Metz 22: 79–88.

——. 1903. Descriptions de cécidomyies nouvelles du Chili. Revista Chilena de Historia Natural 7: 226–228.

Literature Cited **319**

——. 1904. Nouvelles cécidomyies xylophiles. Annales de la Société Scientifique de Bruxelles 28 (1903): 367–410, 1 pl.

——. 1908. Description de quelques galles et d'insectes gallicoles de Colombie. Marcellia 7: 140–142.

——. 1909. Contributions à la connaissance des insectes gallicoles. Bulletin de la Société d'Histoire Naturelle de Metz 28: 1–35.

——. 1910a. Cécidomyies parasites de *Diaspis* sur le mûrier. Portici 4: 128–133.

——. 1910b. Zusatz. P. 442, *in* J. J. Kieffer and P. Jörgensen. Gallen und Gallentiere aus Argentinien. Centralblatt für Bakteriologie und Parasitenkunde. (2) 27: 362–444.

——. 1912a. Neue Gallmücken Gattungen. Bitsche. 2 pp. Reprinted in Marcellia 11: x–xi.

——. 1912b. Description de quatre nouveaux insectes exotiques. Portici 6: 171–175.

——. 1912c. Nouvelle contribution à la connaissance des cécidomyies. Marcellia 11: 219–235.

——. 1913a. Diptera. Fam. Cecidomyidae. Genera Insectorum. Fasc. 152: 1–346 pp., 15 pls.

——. 1913b. Glanures Diptérologiques. Bulletin de la Société d'Histoire Naturelle de Metz 28: 45–55.

Kieffer, J.-J., and P. Herbst. 1905. Über Gallen und Gallenerzeuger aus Chile. Zeitschrift für wissenschaftliche Insekten-Biologie. 1: 63–66.

——. 1906. Description de galles et d'insectes gallicoles du Chili. Annales de la Société Scientifique de Bruxelles 30: 223–236, 1 pl.

——. 1909. Ueber einige neue Gallen und Gallenerzeuger aus Chile. Centralblatt für Bakteriologie und Parisitenkunde (2) 23: 119–126.

——. 1911. Über Gallen und Gallentiere aus Chile. Centralblatt für Bakteriologie und Parisitenkunde (2) 29: 696–704.

Kieffer, J.-J., and P. Jörgensen. 1910. Gallen und Gallentiere aus Argentinien. Centralblatt für Bakteriologie und Parisitenkunde. (2) 27: 362–444.

Kim, K. C. 1967. The North American species of the genus *Anarete* (Diptera: Cecidomyiidae). Annals of the Entomological Society of America 60: 521–530.

Kirkpatrick, T. W. 1954. Notes on *Pseudendaphis maculans* Barnes, a cecidomyiid endoparasite of aphids of Trinidad, B. W. I. Bulletin of Entomological Research 45: 777–781.

Kleesattel, W. 1979. Beiträge zu einer Revision der Lestremiinae (Diptera, Cecidomyiidae) unter besonderer Berücksichtigung ihrer Phylogenie. Privately published. 203 pp.

Köhler, P. 1932. Una diptero-cecidia nueva de la Argentina. Physis 11: 127–129.

Korytkowski G., C., and L. Llontop B. 1967. Dos moscas Cecidomyiidae dañinas a la Sandia. Revista Peruana de Entomologia 10: 21–27.

Korytkowski G., C., and A. Sarmiento P. 1967. *Hyperdiplosis* sp. (Dipt.: Cecidomyiidae), un insecto formador de agallas en las hojas de la yuca. Revista Peruana de Entomologia 10: 44–50.

Küster, E. 1903. Pathologische Pflanzenanatomie in ihren Grundzügen. Jena, 312 pp.

Lagrange, E. B. 1980. Notas sobre cecidomidos neotropicales (Insecta: Diptera). Neotropica 26: 95–98.

Larew, H. G., R. J. Gagné, and A. Y. Rossman. 1987. Fungal gall caused by a new species of *Ledomyia* (Diptera: Cecidomyiidae) on *Xylaria enterogena* (Ascomycetes: Xylariaceae). Annals of the Entomological Society of America 80: 502–507.

LeClerg, E. L. 1953. Dr. Melville Thurston Cook. Annals of the Entomological Society of America 46: 172.

Leite, S. 1931. J. S. Tavares. Broteria, Série Mensal: Fé-Sciências-Letras 13: 273–297.

Loew, H. 1850. Dipterologische Beiträge. Vierter Teil. Oeffentl. Prüfung der Schüler des Königlichen Friedrich-Wilhelms-Gymnasiums zu Posen 1850: 1–40.

Luisier, A. 1932. In Memoriam. Le R. P. J. da Silva Tavares, S. J. Broteria, Série de Ciências Naturais 28: 9–34.

McAlpine, J. F., B. V. Peterson, G. E. Shewell, H. J. Teskey, J. R. Vockeroth, and D. M. Wood, eds. 1981. Manual of Nearctic Diptera, vol. 1. Monograph 27. Research Branch, Agriculture Canada, Hull, Quebec, vi and 674 pp.

McKay, P. A., and J. H. Hatchett. 1984. Mating behavior and evidence of a female sex pheromone in the Hessian fly, *Mayetiola destructor* (Say) (Diptera: Cecidomyiidae). Annals of the Entomological Society of America 77: 616–620.

Macquart, J. 1826. Insectes Diptères du nord de la France. Tipulaires. Recueil des Travaux de la Société d'Amateurs des Sciences, de l'Agriculture et des Arts à Lille 1823/1824: 59–224, 4 pls.

Madrid, F. J. 1974. *Rhopalomyia nothofagi* Gagne, biologia y daño en roble (Diptera: Cecidomyiidae). Boletin de la Sociedad Biologia de Concepción 48: 395–402.

Maes, J.-M. 1990. Catalogo de los Diptera de Nicaragua. 10. Cecidomyiidae (Nematocera). Revista Nicaraguense de Entomologia 14B: 33–41.

Mallea, A. R., G. S. Nacola, J. G. Garcia Saez, and S. J. Lanati. 1983. Presencia en Mendoza del "mosquito de la agalla del crisantemo," *Diarthronomyia chrysanthemi* (Diptera-Cecidomyiidae). Intersectum 15: 1–4.

Mamaev, B. M. 1967. [Gall midges of the USSR. 7. New speices of noncecidogenous gall midges of the tribe Oligotrophini (Diptera, Cecidomyiinae).] Entomologicheskoye Obozreniye 46: 873–883. In Russian, English translation published in Entomological Review 46: 522–528.

——. 1968. [Evolution of Gall-Forming Insects—the Gall Midges.] Akademia Nauk, Leningrad. 235 pp. In Russian, English translation published in 1975. British Library, Boston Spa, United Kingdom. iii and 316 pp.

Mamaev, B. M., and E. P. Krivosheina. 1965. [Larvae of Gall Midges. Diptera, Cecidomyiidae.] Akademia Nauk USSR, Moscow. 278 pp. In Russian, English translation, edited by J. C. Roskam, published in 1992. A. A. Balkema Publishers, Rotterdam, The Netherlands. 304 pp.

Mamaeva, K. P. 1964. New and little-known gall midges in Moscow Region (Diptera, Itonididae). Zoologicheskii Zhurnal 43: 206–213.

Mann, J. 1969. Cactus-feeding insects and mites. United States National Museum Bulletin 256: i–x, 1–156, pls. 1–8.

Mansour, M. H. 1975. The role of plants as a factor affecting oviposition by *Aphidoletes aphidimyza* (Diptera: Cecidomyiidae). Entomologia Experimentalis et Applicata 18: 173–179.

Maresquelle, H. J. 1944. C. A. V. Houard (1873–1943). Bulletin de la Société Botanique de France 91: 14–15, pl. III (portrait).

Matile, L. 1990. Recherches sur la systématique et l'évolution des Keroplatidae (Diptera, Mycetophiloidea). Mémoires du Muséum National d'Histoire Naturelle. Série A, Zoologie 148: 1–682.

Mead, F. W. 1970. *Ctenodactylomyia watsoni* Felt, a gall midge pest of seagrape, *Coccoloba uvifera* L., in Florida. Florida Department of Agriculture and Consumer Services, Division of Plant Industry Entomology Circular 97: 1–2.

Meigen, J. W. 1800. Nouvelle classification des mouches à deux ailes (Diptera L.) d'après un plan tout nouveau. Paris. 40 pp.

———. 1803. Versuch einer neuen Gattungseintheilung der europäischen zweiflügeligen Insekten. Magazin für Insektenkunde 2: 259–281.

———. 1818. Systematische Beschreibung der bekannten europäischen zweiflügeligen Insekten, vol. 1. Aachen. xxxvi and 333 pp., pls. 1–11.

———. 1830. Systematische Beschreibung der bekannten europäischen zweiflügeligen Insekten, vol. 6. Hamm. iv and 401 pp., pls. 55–66.

Meinert, F. 1864. *Miastor metroloas*: Yderligere oplysning om den af Prof. Nic. Wagner nyligt beskrevne insektlarve, som formerer sig ved spiredannelse. Naturhistorisk Tidsskrift (3) 3: 37–43.

Metcalfe, M. M. 1933. The morphology and anatomy of the larva of *Dasyneura leguminicola* Lint. (Diptera). Proceedings of the Zoological Society of London 1933: 119–130, pls. I–IV.

Meunier, F. 1904. Monographie des Cecidomyidae, des Sciaridae, des Mycetophilidae et des Chironomidae de l'ambre de la Baltique. Annales de la Société Scientifique de Bruxelles 28: 12–275 and 16 pls.

Meyer, J. 1952. Cécidogenèse de la galle de *Lasioptera rubi* Heeger et rôle nouricier d'un mycélium symbiotique. Comptes Rendus de l'Académie des Sciences de Paris 234: 2556–2558.

———. 1987. Plant Galls and Gall Inducers. Gebrüder Borntraeger, Berlin and Stuttgart. viii and 291 pp.

Milne, D. L. 1961. The function of the sternal spatula in gall midges. Proceedings of the Royal Entomological Society of London 36: 126–131.

Möhn, E. 1955a. Beiträge zur Systematik der Larven der Itonididae (= Cecidomyiidae, Diptera). 1. Teil: Porricondylinae und Itonidinae Mitteleuropas. Zoologica 105 (1 and 2); 1–247 and 30 pls.

———. 1955b. Neue freilebende Gallmücken-Gattungen. Deutsche Entomologische Zeitschrift 2: 127–151.

———. 1959. Gallmücken (Diptera, Itonididae) aus El Salvador. 1. Teil. Senckenbergiana Biologica 40: 297–368.

———. 1960a. Gallmücken (Diptera, Itonididae) aus El Salvador. 2. Teil. Senckenbergiana Biologica 41: 197–240.

———. 1960b. Gallmücken (Diptera, Itonididae) aus El Salvador. 3. Teil. Senckenbergiana Biologica 41: 333–358.

———. 1961a. Neue Asphondyliidi-Gattungen (Diptera, Itonididae). Stuttgarter Beiträge zur Naturkunde 49: 1–14.

———. 1961b. Gallmücken (Diptera, Itonididae) aus El Salvador. 4. Zur Phylogenie der Asphondyliidi der neotropischen und holarktischen Regionen. Senckenbergiana Biologica 42: 131–330.

———. 1962. Studien über neotropische Gallmücken (Diptera, Itonididae). 1. Teil. Brotéria, Série de Ciências Naturais 31: 211–239.

———. 1963a. Studien über neotropische Gallmücken (Diptera, Itonididae). 1. Teil. (Fortsetzung). Brotéria, Série de Ciências Naturais 32: 3–23.

——. 1963b. Gallmücken (Diptera, Itonididae) aus El Salvador. 5. Teil, Lasiopteridi. Stuttgarter Beiträge zur Naturkunde 116: 1–7.

——. 1964a. Gallmücken (Diptera, Itonididae) aus El Salvador. 6. Teil, Lasiopteridi. Deutsche Entomologische Zeitschrift 11: 47–143.

——. 1964b. Gallmücken (Diptera, Itonididae) aus El Salvador. 7. Teil, Lasiopteridi. Beiträge zur Entomologie 14: 553–600.

——. 1973. Studien über neotropische Gallmücken (Diptera, Itonididae). 2. Teil. Stuttgarter Beiträge zur Naturkunde (A) 257: 1–9.

——. 1975. Gallmücken (Diptera, Itonididae) aus El Salvador. 8. Teil, Lasiopteridi. Stuttgarter Beiträge zur Naturkunde (A) 276: 1–101.

Molliard, M. 1903. La galle du *Cecidomyia cattleyae* n. sp. Marcellia 1 (1902): 165–170, pl. 2.

Morrill, W. L. 1982. Hessian fly: host selection and behavior during oviposition, winter biology, and parasitoids. Journal of the Georgia Entomological Society 17: 156–167.

Moutia, L. A. 1958. Contribution to the study of some phytophagous Acarina and their predators in Mauritius. Bulletin of Entomological Research 49: 59–75.

Muesebeck, C. F. W., and C. W. Collins. 1944. Ephraim Porter Felt 1868–1943. Proceedings of the Entomological Society of Washington 46: 27–29.

Needham, J. G. 1941. Insects from seed pods of the primrose willow, *Jussiaea angustifolia*. Proceedings of the Entomological Society of Washington 43: 2–6.

Nijveldt, W. C. 1967. *Neolasioptera martelli* sp. n., a gall midge injuring *Agave* in Mexico. Entomologische Berichte 27: 125–127.

——. 1968. Two gall midges from Surinam, pp. 61–66. *In* Studies on the Fauna of Suriname and Other Guyanas, vol. 10. Martinus Nijhoff, The Hague.

Nominé, M. H. 1926. A la Mémoire de M. l'Abbé J.-J. Kieffer Professeur au Collège Saint-Augustin à Bitche (Moselle) 1857–1925. Imprimerie Lorraine, Metz. 49 pp.

——. 1929. L'oeuvre de J.-J. Kieffer (1857–1925). Bibliographie complète. Bulletin de la Société d'Histoire Naturelle de Moselle (3) 8: 31–59.

Osgood, E. A., and R. J. Gagné. 1978. Biology and taxonomy of two gall midges (Diptera: Cecidomyiidae) found in galls on balsam fir needles with description of a new species of *Paradiplosis*. Annals of the Entomological Society of America 71: 85–91.

Osten Sacken, R. 1878. Catalogue of the described Diptera of North America. Smithsonian Miscellaneous Collections 16 (2): 1–276.

Painter, R. H. 1935. The biology of some dipterous gall-makers from Texas. Journal of the Kansas Entomological Society 8: 81–97.

Panelius, S. 1965. A revision of the European gall midges of the subfamily Porricondylinae. Acta Zoologica Fennica 113: 1–157.

Parnell, J. R. 1964. Investigations on the biology and larval morphology of the insects associated with the galls of *Asphondylia sarothamni* H. Loew (Diptera: Cecidomyiidae) on broom (*Sarothamnus scoparius* (L.) Wimmer). Transactions of the Royal Entomological Society of London 116: 255–273.

——. 1966. Observations on the larval morphology and life history of *Olesicoccus costalimai* Borgmeier (Diptera: Cecidomyidae) in Jamaica. Proceedings of the Royal Entomological Society of London (A) 41: 51–54.

——. 1969. The biology and morphology of all stages of *Toxomyia fungicola* Felt (Diptera: Cecidomyidae) in Jamaica. Proceedings of the Royal Entomological Society of London (A) 44: 113–122.

——. 1971. A revision of the Nearctic Porricondylinae (Diptera: Cecidomyiidae) based largely on an examination of the Felt types. Miscellaneous Publications of the Entomological Society of America 7: 275–348.

Parnell, J. R., and G. P. Chapman. 1966. Observations on a gall midge (Diptera: Cecidomyidae) infesting pimento (*Pimenta dioica* (L.) Merrill) in Jamaica. Marcellia (1965) 32: 237–245.

Peña, J. E., R. J. Gagné, and R. Duncan. 1989. Biology and characterization of *Prodiplosis longifila* (Diptera: Cecidomyiidae) on lime in Florida. Florida Entomologist 72: 444–450.

Pettersson, J. 1976. Ethology of *Dasineura brassicae* Winn. (Dipt., Cecidomyidae). 1. Laboratory studies of olfactory reactions to the host-plant. Symposia Biologica Hungarica 16: 203–208.

Philippi, R. A. 1865. Aufzählung der chilenischen Dipteren. Verhandlungen der Zoologisch-Botanischen Gesellschaft in Wien 15: 595–782, pls. 23–29.

——. 1873. Chilenische Insekten. Stettiner Entomologische Zeitung 34: 296–316, pls. 1–2.

Pitcher, R. S. 1957. The abrasion of the sternal spatula of the larva of *Dasyneura tetensi* (Rübs.) (Diptera: Cecidomyidae) during the postfeeding phase. Proceedings of the Royal Entomological Society of London (A) 32: 83–88.

Prasad, S. N. 1971. The Mango Midge Pests. Cecidological Society of India, Allahabad. vi and 172 pp.

Pritchard, A. E. 1947. The North American gall midges of the tribe Micromyini; Itonididae (Cecidomyiidae); Diptera. Entomologica Americana 27 (N.S.): 1–44, 45–87.

——. 1951. The North American gall midges of the tribe Lestremiini; Itonididae (Cecidomyiidae); Diptera. University of California Publications in Entomology 8: 239–275.

Readshaw, J. L. 1966. The ecology of the swede midge, *Contarinia nasturtii* (Kieff.) (Diptera, Cecidomyiidae). I. Life history and influence of temperature and moisture on development. Bulletin of Entomological Research 56: 685–700.

Reuter, E. 1895. Zwei neue Cecidomyinen. Acta Societas pro Fauna et Flora Fennica 11 (8): 1–15, pls. I–II.

Riley, C. V. 1870. Cypress-gall. American Entomologist and Botanist 2: 244.

Rock, E. A., and D. Jackson. 1985. The biology of xylophilic Cecidomyiidae (Diptera). Proceedings of the Entomological Society of Washington 87: 135–141.

Rohfritsch, O. 1980. Relations hôte-parasite au début de la cécidogenèse du *Hartigiola annulipes* Hartig sur le hêtre. Bulletin de la Société Botanique de France, Actualités Botaniques 127: 199–207.

——. 1990. Two gall midges (Diptera: Cecidomyiidae) on *Tilia*: two different patterns of interaction with their host tree. Phytophaga 3: 13–21.

——. 1992a. Patterns in gall development, pp. 60–86. *In* J. D. Shorthouse and O. Rohfritsch, eds. Biology of Insect-Induced Galls. Oxford University Press, New York.

——. 1992b. A fungus associated gall midge, *Lasioptera arundinis* (Schiner), on *Phragmites communis* (Cav.) Trin. Bulletin de la Société Bontanique de France 139, Lettres Botaniques: 45–59.

Rondani, C. 1840. Memoria per servir alla ditterologia italiana. No. 2: Sopra alcuni nuovi generi di insetti ditteri. Parma. 28 pp., 1 pl.

——. 1846. Compendio della seconda memoria ditterologica (publicata 1840) con algune aggiunte et correzioni. Nuovi Annali delle Scienze Naturali di Bologna 6: 363–376, 1 pl.

——. 1847. Osservazioni sopra parechiae specie di esapodi afidicidi e sui loro nemeci. Nuovi Annali delle Scienze Naturali di Bologna 8: 337–351, 432–448.

——. 1856. Dipterologiae Italicae prodromus. Vol. 1: Genera Italica ordinis dipterorum ordinatum disposita et distincta et in familias et stires aggregata. Parma. 228 pp.

——. 1860. Stirpis cecidomyarum. Genera revisa. Nota undecima, pro dipterologia italica. Atti della Societá Italiana di Scienze Naturali (1859–1860) 2: 286–294, 1 pl.

Roskam, J. C. 1977. Biosystematics of insects living in female birch catkins. I. Gall midges of the genus *Semudobia* Kieffer (Diptera, Cecidomyiidae). Tijdschrift voor Entomologie 120: 153–197.

——. 1979. Biosystematics of insects living in female birch catkins. II. Inquiline and predaceous gall midges belonging to various genera. Netherlands Journal of Zoology 29: 283–351.

Roskam, J. C., and H. Nadel. 1990. Redescription and immature stages of *Ficiomyia perarticulata* (Diptera: Cecidomyiidae), a gall midge inhabiting syconia of *Ficus citrifolia*. Proceedings of the Entomological Society of America 92: 778–792.

Rossi, A. M., and D. R. Strong. 1990. A new species of *Asphondylia* (Diptera: Cecidomyiidae) on *Borrichia* (Asteraceae) from Florida. Proceedings of the Entomological Society of Washington 92: 732–735.

Rübsaamen, E. H. 1892. Die Gallmücken des Königl. Museums für Naturkunde zu Berlin. Berliner Entomologische Zeitschrift 37: 319–411, pls. VII–XVIII.

——. 1895a. Ueber Cecidomyiden. Wiener Entomologische Zeitung 14: 181–193, pl. I.

——. 1895b. Cecidomyidenstudien. Entomologische Nachrichten 21: 177–194.

——. 1896. Zurückweisung der Angriffe in J. J. Kieffer's Abhandlung: Die Unterscheidungsmerkmale der Gallmücken. Entomologische Nachrichten 1896: 119–127, 154–158, 181–187, 202–211.

——. 1899. Mitteilung über neue und bekannte Gallen aus Europa, Asien, Afrika, und Amerika. Entomologische Nachrichten 25: 225–282, pls. I–II.

——. 1905. Beiträge zur Kenntnis aussereuropäischer Zoocecidien. II. Beitrag: Gallen aus Brasilien und Peru. Marcellia 4: 65–85, 115–138.

——. 1907. Beiträge zur Kenntnis aussereuropäischer Zoocecidien. III. Beitrag. Gallen aus Brasilien und Peru. Marcellia 6: 110–173.

——. 1908. Beiträge zur Kenntnis aussereuropäischer Zoocecidien. III. Beitrag [cont.]: Gallen aus Brasilien und Peru. Marcellia 7: 15–79.

——. 1910. Ueber deutsche Gallmücken und Gallen. Zeitschrift für wissenschaftliche Insektenbiologie 6: 283–289.

——. 1916a. Beitrag zur Kenntnis aussereuropäischer Gallmücken. Sitzungsberichte der Gesellschaft Naturforschender Freunde zu Berlin 1915: 431–481.

——. 1916b. Cecidomyidenstudien V. Revision der deutschen Asphondylarien. Sitzungsberichte der Gesellschaft Naturforschender Freunde zu Berlin 1916: 1–12.

——. 1929. Gallbildungen, pp. 219–248. *In* C. W. M. Schröder, ed. Handbuch der Entomologie, vol. 2. Jena.

Rübsaamen, E. H., and H. Hedicke. 1926–1939. Die Zoocecidien, durch Tiere erzeugte Pflanzengallen Deutschlands und ihre Bewohner. Band II: Die Cecidomyiden (Gallmücken) und ihre Cecidien. Zoologica 77: 1–350, 42 pls.

Schaffnitt, E. 1928. Professor Ewald Rübsaamen. Zeitschrift für Angewandte Entomologie 13: 210–217.

Schlee, D., and H.-G. Dietrich. 1970. Insektenführender Bernstein aus der Unterkreide des Libanon. Neues Jahrbuch für Geologie und Paläontologie, Monatshefte 1970: 40–50.

Seitner, M. 1906. *Resseliella piceae*, die Tannensamen-Gallmücke. Verhandlungen der zoologisch-botanischen Gesellschaft in Wien 56: 174–186.

Shorthouse, J. D., and O. Rohfritsch. 1992. Biology of Insect-Induced Galls. Oxford University Press, New York. x and 285 pp.

Shorthouse, J. D., and R. J. West. 1986. Role of the inquiline, *Dasineura balsamicola* (Diptera: Cecidomyiidae), in the balsam fir needle gall. Proceedings of the Entomological Society of Ontario 117: 1–7.

Silvestri, F. 1901. Descrizione di nuovi termitofili e relazioni di essi con gli ospiti. Bollettino dei Musei di Zoologia ed Anatomia Comparata della Università di Torino 16 (395): 1–6.

——. 1903. Contribuzione alla conoscenza dei termitidi e termitofili dell'America meridionale. Redia 1: 1–234, pls. I–V.

Sivinski, J., and M. Stowe. 1981. A kleptoparasitic cecidomyiid and other flies associated with spiders. Psyche 87: 337–348.

Skuhravá, M. 1986. Family Cecidomyiidae, pp. 72–297. *In* A. Soós. Catalogue of Palaearctic Diptera. 4. Sciaridae—Anisopodidae. Akademiae Kiado, Budapest. 441 pp.

——. 1989. Taxonomic changes and records in Palaearctic Cecidomyiidae (Diptera). Acta Entomologica Bohemoslovaca 86: 202–233.

Slifer, E. H., and S. S. Sekhon. 1971. Circumfila and other sense organs on the antenna of the sorghum midge (Diptera, Cecidomyiidae). Journal of Morphology 133: 281–301.

Soderstrom, T. R., and C. E. Calderón. 1971. Insect pollination in tropical rain forest grasses. Biotropica 3: 1–16.

Solinas, M. 1965. Studi sui Ditteri Cecidomyiidi. I. *Contarinia medicaginis* Kieffer. Bollettino dell'Istituto di Entomologia della Università di Bologna 27: 249–300.

——. 1968. Morfologia, anatomia e organizzazione funzionale del capo della larva matura de *Phaenobremia aphidimyza* (Rondani). Entomologica 4: 7–44.

——. 1971. Adattamenti morfologici nel capo delle larve di ditteri cecidomiidi a regime dietetico specializzato, pp. 70–92. *In* Atti dell'VIII Congresso Nazionale Italiano di Entomologia, Firenze 1969. Tipografia Compositori, Bologna.

——. 1986. *Allocontarinia* n. g., *A. sorghicola* (Coq.) n. comb. (Diptera, Cecidomyiidae). Entomologica 21: 23–35.

Solinas, M., and M. Bucci. 1982. An ecological investigation into the flower galls on *Diplotaxis muralis* D. C. caused by the gall midge *Paragephyraulus diplotaxis* Solinas. Entomologica 17: 5–22.

Solinas, M., and N. Isidoro. 1991. Identification of the female sex pheromone gland in the sorghum midge, *Allocontarinia sorghicola* (Coq.) Solinas (Diptera, Cecidomyiidae). Redia 74: 441–466.

Solinas, M., G. Nuzzaci, and N. Isidoro. 1987. Antennal sensory structures and their ecological-behavioural meaning in Cecidomyiidae (Diptera) larvae. Entomologica 22: 165–184.

Stiling, P., A. M. Rossi, D. R. Strong, and D. M. Johnson. 1992. Life history and parasites of *Asphondylia borrichiae* (Diptera: Cecidomyiidae), a gall maker on *Borrichia frutescens*. Florida Entomologist 75: 130–137.

Stuart, J. J., and J. H. Hatchett. 1987. Morphogenesis and cytology of the salivary gland of the Hessian fly, *Mayetiola destructor* (Diptera: Cecidomyiidae). Annals of the Entomological Society of America 80: 475–482.

Sunose, T. 1978. Studies on extended diapause in *Hasegawaia sasacola* Monzen (Diptera, Cecidomyiidae) and its parasites. Kontyû 46: 400–415.

Sylvén, E. 1975. Study on relationships between habits and external structures in Oligotrophidi larvae (Diptera, Cecidomyiidae). Zoologica Scripta 4: 55–92.

———. 1979. Gall midges (Diptera, Cecidomyiidae) as plant taxonomists. Symbolae Botanicae Upsalienses 22: 62–69.

Sylvén, E., and U. Carlbäcker. 1981. Allometric relationships concerning antenna and wing in Oligotrophini (Diptera: Cecidomyiidae). Acta Entomologica Fennica 42: 90–102.

Tastás-Duque, R., and E. Sylvén. 1989. Sensilla and cuticular appendages on the female abdomen of *Lasioptera rubi* (Schrank) (Diptera, Cecidomyiidae). Acta Zoologica 70: 163–174.

Tavares, J. S. 1906. Descripção de uma Cecidomyia nova do Brazil, pertencente a um genero novo. Brotéria 5: 81–84.

———. 1908. Contributio prima ad cognitionem cecidologiae Regionis Zambeziae (Moçambique, Africa Orientalis). Brotéria, Série Zoológica 7: 133–171, pls. VII–XV.

———. 1909. Contributio prima ad cognitionem cecidologiae Braziliae. Brotéria, Série Zoológica 8: 5–28, pls. I–VIII.

———. 1915a. Cécidologie Argentine. Brotéria, Série Zoológica 13: 88–126, pls. II–V.

———. 1915b. As cecidias das plantas do genero *Styrax* no Brazil. Brotéria, Série Zoológica 13: 145–160, pl. VI.

———. 1916. Cecidomyias novas do Brazil. Brotéria, Série Zoológica 14: 36–57.

———. 1917a. As cecidias do Brazil que se criam nas plantas da familia das Melastomataceae. Brotéria, Série Zoológica 15: 18–49, pls. I–V.

———. 1917b. Cecidias do Brazileiras que se criam em plantas das Compositae, Rubiaceae, Tiliaceae, Lythraceae e Artocarpaceae. Brotéria, Série Zoológica 15: 113–181, pls. VI–XI.

———. 1918a. Cecidologia Brazileira. Cecidias que se criam nas plantas das familias das Verbenaceae, Euphorbiaceae, Malvaceae, Anacardiaceae, Labiatae, Rosaceae, Anonaceae, Ampelidaceae, Bignoniaceae, Aristolochiaceae e Solanaceae. Brotéria, Série Zoológica 16: 21–68, pls. I–II.

———. 1918b. Cecidomyias novas do Brazil. Segunda série. Brotéria, Série Zoológica 16: 69–84, pls. III–IV.

———. 1920a. O género Bruggmanniella Tav. com a descripção de uma espécie nova e a clave dichotómica dos géneros das Asphondyliariae. Brotéria, Série Zoológica 18: 33–42.

——. 1920b. Cecidologia brazileira. Cecidias que se criam em plantas das familias das Leguminosae, Sapotaceae, Lauraceae, Myrtaceae, Punicaceae, Aurantiaceae, Malpighiaceae, Sapindaceae, Umbelliferae, Loranthaceae, Apocynaceae, Urticaceae, Salicaceae e Gramineae. Brotéria, Série Zoológica 18: 82–125, pl. III.

——. 1920c. Anadiplosariae, nouvelle tribu de Cecidomyinae (Dipt.). Marcellia 17: 57–72, pl. I.

——. 1921. Cecidologia brazileira. Cecidias que se criam em plantas das familias das Leguminosae, Sapotaceae, Lauraceae, Myrtaceae, Punicaceae, Aurantiaceae, Malpighiaceae, Sapindaceae, Umbelliferae, Loranthaceae, Apocynaceae, Urticaceae, Salicaceae e Gramineae. Brotéria, Série Zoológica 19: 76–112, pl. I.

——. 1922. Cecidologia brazileira. As restantes familias. Brotéria, Série Zoológica 20: 5–48c, pls. XI–XIX.

——. 1925. Nova contribuïção para o conhecimento da cecidologia brazileira. Brotéria, Série Zoológica 22: 5–55, pls. I–V.

——. 1926. Dr. J. J. Kieffer (1857–1925). Broteria 23: 126–148.

Teetes, G. L. 1985. Insect resistant sorghums in pest management. Insect Science and Its Application 6: 443–451.

Walker, F. 1856. Insecta Saundersiana, vol. 1. London. 474 pp., 8 pls.

Williston, S. W. 1896. On the Diptera of St. Vincent (West Indies). Transactions of the Entomological Society of London 1896: 253–446, pls. VIII–XIV.

Wilson, L. F., and G. C. Heaton. 1987. Life history, damage, and gall development of the gall midge, *Neolasioptera brevis* (Diptera: Cecidomyiidae), injurious to honeylocust in Michigan. Great Lakes Entomologist 20: 111–118.

Winder, J. A. 1980. Cecidomyiid leaf galls of *Lantana* spp. (Verbenaceae) and their associated Hymenoptera. Dusenia 12: 33–36.

Wolcott, G. N. 1936. "Insectae Borinquenses." A revised annotated check-list of the insects of Puerto Rico. Journal of Agriculture of the University of Puerto Rico 20: 1–600.

Wood, D. M., and A. Borkent. 1989. Phylogeny and Classification of the Nematocera, pp. 1333–1370. *In* J. F. McAlpine, and D. M. Wood, eds. Manual of Nearctic Diptera, vol. 3. Monograph 32. Research Branch, Agriculture Canada, Hull, Quebec. vi and pp. 1333–1581.

Wünsch, A. 1979. Gallenerzeugende Insekten Nordkolumbiens, speziell Asphondyliidi und Lasiopteridi (Diptera, Cecidomyiidae) aus dem Küstenbereich um Santa Marta. A. Wünsch, Waiblingen, West Germany. 238 pp.

Wyatt, I. J. 1959. A new genus and species of Cecidomyiidae (Diptera) infesting mushrooms. Proceedings of the Royal Entomological Society of London (B) 28: 175–179.

——. 1967. Pupal paedogenesis in the Cecidomyiidae (Diptera). 3—A reclassification of the Heteropezini. Transactions of the Royal Entomological Society of London 119: 71–98.

Young, A. M. 1985. Studies of cecidomyiid midges (Diptera: Cecidomyiidae) as cocoa pollinators (*Theobroma cacao* L.) in Central America. Proceedings of the Entomological Society of Washington 87: 49–79.

Yukawa, J. 1974. Descriptions of new Japanese gall midges (Diptera, Cecidomyiidae, Asphondyliidi) causing leaf galls on Lauraceae. Kontyû 42: 293–304.

Yukawa, J., and K. Miyamoto. 1979. Redescription of *Asphondylia sphaera* Monzen (Diptera, Cecidomyiidae), with notes on its bionomics. Memoirs of the Faculty of Agriculture Kagoshima University 15: 99–106.

Yukawa, J., K. Takahashi, and N. Ohsaki. 1976. Population behaviour of the *Neolitsea* leaf gall midge, *Pseudasphondylia neolitseae* Yukawa (Diptera, Cecidomyiidae). Kontyû 44: 358–365.

Index

Cecidomyiid names in italic are junior synonyms or emendations. Numbers in boldface indicate pages for the main mention of Neotropical gall midge taxa and plant host genera and families. Numbers in italic refer to pages with illustrations. Unidentified gall midge species in generic and higher category entries are separated by semicolons. Author names follow only cecidomyiid species names and identical generic names.